Remote Sensing: Technology and Applied Principles

Remote Sensing: Technology and Applied Principles

Editor: Matt Weilberg

www.callistoreference.com

Callisto Reference,
118-35 Queens Blvd., Suite 400,
Forest Hills, NY 11375, USA

Visit us on the World Wide Web at:
www.callistoreference.com

ISBN: 978-1-63239-891-8 (Hardback)

Cataloging-in-Publication Data

Remote sensing : technology and applied principles / edited by Matt Weilberg.
    p. cm.
Includes bibliographical references and index.
ISBN 978-1-63239-891-8
1. Remote sensing. 2. Remote-sensing images. I. Weilberg, Matt.
G70.4 .R46 2017
621.367 8--dc23

Table of Contents

Preface.. VII

Chapter 1 **Fusion of Airborne Discrete-Return LiDAR and Hyperspectral Data for Land Cover Classification**.. 1
Shezhou Luo, Cheng Wang, Xiaohuan Xi, Hongcheng Zeng, Dong Li, Shaobo Xia and Pinghua Wang

Chapter 2 **Quantifying Multi-Decadal Change of Planted Forest Cover using Airborne LiDAR and Landsat Imagery**... 20
Xiaoyi Wang, Huabing Huang, Peng Gong, Gregory S. Biging, Qinchuan Xin, Yanlei Chen, Jun Yang and Caixia Liu

Chapter 3 **Application of Synthetic NDVI Time Series Blended from Landsat and MODIS Data for Grassland Biomass Estimation**.................................. 36
Binghua Zhang, Li Zhang, Dong Xie, Xiaoli Yin, Chunjing Liu and Guang Liu

Chapter 4 **Remote Sensing Based Simple Models of GPP in both Disturbed and Undisturbed Piñon-Juniper Woodlands in the Southwestern U.S**... 57
Dan J. Krofcheck, Jan U. H. Eitel, Christopher D. Lippitt, Lee A. Vierling, Urs Schulthess and Marcy E. Litvak

Chapter 5 **Scaling up Ecological Measurements of Coral Reefs using Semi-Automated Field Image Collection and Analysis**.............................73
Manuel González-Rivero, Oscar Beijbom, Alberto Rodriguez-Ramirez, Tadzio Holtrop, Yeray González-Marrero, Anjani Ganase, Chris Roelfsema, Stuart Phinn and Ove Hoegh-Guldberg

Chapter 6 **Spatially-Explicit Testing of a General Aboveground Carbon Density Estimation Model in a Western Amazonian Forest using Airborne LiDAR**... 93
Patricio Xavier Molina, Gregory P. Asner, Mercedes Farjas Abadía, Juan Carlos Ojeda Manrique, Luis Alberto Sánchez Diez and Renato Valencia

Chapter 7 **Multispectral Radiometric Analysis of Façades to Detect Pathologies from Active and Passive Remote Sensing**............................. 108
Susana Del Pozo, Jesús Herrero-Pascual, Beatriz Felipe-García, David Hernández-López, Pablo Rodríguez-Gonzálvez and Diego González-Aguilera

Chapter 8   **Modeling and Mapping Agroforestry Aboveground Biomass in the Brazilian Amazon using Airborne Lidar Data**..124
Qi Chen, Dengsheng Lu, Michael Keller, Maiza Nara dos-Santos,
Edson Luis Bolfe, Yunyun Feng and Changwei Wang

Chapter 9   **User Validation of VIIRS Satellite Imagery**...141
Don Hillger, Tom Kopp, Curtis Seaman, Steven Miller, Dan Lindsey,
Eric Stevens, Jeremy Solbrig, William Straka III, Melissa Kreller,
Arunas Kuciauskas and Amanda Terborg

Chapter 10   **Use of SSU/MSU Satellite Observations to Validate Upper Atmospheric Temperature Trends in CMIP5 Simulations**.. 165
Lilong Zhao, Jianjun Xu, Alfred M. Powell, Zhihong Jiang and
Donghai Wang

Chapter 11   **Investigation on the Weighted RANSAC Approaches for Building Roof Plane Segmentation from LiDAR Point Clouds**...............................181
Bo Xu, Wanshou Jiang, Jie Shan, Jing Zhang and Lelin Li

Chapter 12   **Cloud and Snow Discrimination for CCD Images of HJ-1A/B Constellation Based on Spectral Signature and Spatio-Temporal Context**... 204
Jinhu Bian, Ainong Li, Qiannan Liu and Chengquan Huang

Permissions

List of Contributors

Index

Preface

Remote sensing has developed by leaps and bounds in the 20th century. This has helped in developing efficient and reliable navigation technologies, weather predictions that benefit agriculture, aquaculture, disaster warning, etc. The objective of this book is to give a general view of the different areas of remote sensing, and its applications. While understanding the long-term perspectives of the topics, the book makes an effort in highlighting their impact as a modern tool for the growth of the discipline. This book discusses the advanced technologies and applied principles in the field of remote sensing. As this field is emerging at a rapid pace, the contents of this book will help the readers understand the modern concepts and applications of remote sensing.

The information contained in this book is the result of intensive hard work done by researchers in this field. All due efforts have been made to make this book serve as a complete guiding source for students and researchers. The topics in this book have been comprehensively explained to help readers understand the growing trends in the field.

I would like to thank the entire group of writers who made sincere efforts in this book and my family who supported me in my efforts of working on this book. I take this opportunity to thank all those who have been a guiding force throughout my life.

Editor

1

Fusion of Airborne Discrete-Return LiDAR and Hyperspectral Data for Land Cover Classification

Shezhou Luo [1,2], Cheng Wang [1,*], Xiaohuan Xi [1], Hongcheng Zeng [3], Dong Li [1], Shaobo Xia [1] and Pinghua Wang [1]

Academic Editors: Parth Sarathi Roy and Prasad S. Thenkabail

[1] Key Laboratory of Digital Earth Science, Institute of Remote Sensing and Digital Earth, Chinese Academy of Sciences, Beijing 100094, China; luoshezhou@163.com (S.L.); Xixh@radi.ac.cn (X.X.); lidong@radi.ac.cn (D.L.); xiasb@radi.ac.cn (S.X.); wangph@radi.ac.cn (P.W.)

[2] Department of Geography and Program in Planning, University of Toronto, 100St. George St., Room 5047, Toronto, ON M5S 3G3, Canada

[3] Faculty of Forestry, University of Toronto, 33 Willcocks Street, Toronto, ON M5S 3B3, Canada; hongcheng.zeng@utoronto.ca

* Correspondence: wangcheng@radi.ac.cn

Abstract: Accurate land cover classification information is a critical variable for many applications. This study presents a method to classify land cover using the fusion data of airborne discrete return LiDAR (Light Detection and Ranging) and CASI (Compact Airborne Spectrographic Imager) hyperspectral data. Four LiDAR-derived images (DTM, DSM, nDSM, and intensity) and CASI data (48 bands) with 1 m spatial resolution were spatially resampled to 2, 4, 8, 10, 20 and 30 m resolutions using the nearest neighbor resampling method. These data were thereafter fused using the layer stacking and principal components analysis (PCA) methods. Land cover was classified by commonly used supervised classifications in remote sensing images, *i.e.*, the support vector machine (SVM) and maximum likelihood (MLC) classifiers. Each classifier was applied to four types of datasets (at seven different spatial resolutions): (1) the layer stacking fusion data; (2) the PCA fusion data; (3) the LiDAR data alone; and (4) the CASI data alone. In this study, the land cover category was classified into seven classes, *i.e.*, buildings, road, water bodies, forests, grassland, cropland and barren land. A total of 56 classification results were produced, and the classification accuracies were assessed and compared. The results show that the classification accuracies produced from two fused datasets were higher than that of the single LiDAR and CASI data at all seven spatial resolutions. Moreover, we find that the layer stacking method produced higher overall classification accuracies than the PCA fusion method using both the SVM and MLC classifiers. The highest classification accuracy obtained (OA = 97.8%, kappa = 0.964) using the SVM classifier on the layer stacking fusion data at 1 m spatial resolution. Compared with the best classification results of the CASI and LiDAR data alone, the overall classification accuracies improved by 9.1% and 19.6%, respectively. Our findings also demonstrated that the SVM classifier generally performed better than the MLC when classifying multisource data; however, none of the classifiers consistently produced higher accuracies at all spatial resolutions.

Keywords: LiDAR; hyperspectral image; land cover classification; data fusion; support vector machine; maximum likelihood classification

1. Introduction

Land cover information is an essential variable in main environmental problems of importance to the human-environmental sciences [1–3]. Land cover classification has been widely used in the modeling of carbon budgets, forest management, crop yield estimation, land use change identification, and global environmental change research [4,5]. Accurate and up-to-date land cover classification information is fundamentally vital for these applications [1], since it significantly affects the uncertainty of these applications. Therefore, the accurate classification of land cover, notably in areas that are rapidly changing, is essential [6].

Remote sensing techniques provide the advantage of rapid data acquisition of land cover information at a low cost compared with ground survey [7,8], which are an attractive information source for land cover at different spatial and temporal scales [1]. Remotely sensed data are most frequently used for land cover classification [6,9,10]. Many studies focusing on land cover classification using passive optical remotely sensed data have been conducted [11,12]. However, accurate land cover classification using remotely sensed data remains a challenging task. Multiple studies have been performed to improve the land cover classification accuracy when using remotely sensed data, e.g., [13,14]. However, such classifications have largely relied upon passive optical remotely sensed data alone [5]. Therefore, to improve classification accuracies of optical remotely sensed data, some studies have been conducted through combining other data, e.g., [15,16]. A limitation of this approach is that passive optical remote sensing data neglect the three-dimensional characteristics of ground objects and will reduce the land cover classification accuracy [5]. However, optical remote sensing data can provide abundant spectral information of Earth surface and can be easily acquired at a relative low cost.

Light detection and ranging (LiDAR) systems are an active remote sensor which uses laser light as an illumination source [17]. The LiDAR systems consist of a laser scanning system, global positioning system (GPS) and inertial navigation system (INS) or inertial measurement units (IMU) [18]. Airborne LiDAR can provide horizontal and vertical information with high spatial resolutions and vertical accuracies in the format of three-dimensional laser point cloud [19–21]. In recent years, LiDAR has been rapidly developed and widely used in estimating vegetation biomass, height, canopy closure and leaf area index (LAI) [22–26].

Although LiDAR data are increasingly used for land cover classification [27–29], it is still difficult to employ only discrete return LiDAR data [30,31]. LiDAR can provide accurate vertical information, but has limited spatial coverage mostly due to its cost, especially for large-area LiDAR data acquisition [32]. However, the incorporation of additional data sources, such as passive optical remote sensing data, may mitigate the cost of data acquisition, and improve the accuracy and efficiency of LiDAR applications [33]. Previous studies have shown that the combination of LiDAR and passive optical remote sensing data can provide complementary information and optimize the strengths of both data sources [34,35]. Therefore, this process can be useful for improving the accuracy of information extraction and land cover mapping [36–38]. There are fewer land cover classification studies combining airborne LiDAR and spectral remotely sensed data than traditional optical remote sensing classifications. However, this technique is gaining popularity [6]. Several studies have employed the fusion classification method of LiDAR and other data sources; these studies noted improvements in the classification accuracy [38–42]. Reese, *et al.* [34] combined airborne LiDAR and optical satellite (SPOT 5) data to classify alpine vegetation, and the results showed that the fusion classification method generally obtained a higher classification accuracy for alpine and subalpine vegetation. Mesas-Carrascosa, *et al.* [43] combined LiDAR intensity with airborne camera image to classify agricultural land use, and the results showed that the overall classification accuracy improved 30%–40%. Mutlu, *et al.* [44] developed an innovative method to fuse LiDAR data and QuickBird image, and improved the overall accuracy by approximately 13.58% compared with Quickbird data alone in the image classification of surface fuels. Bork and Su [40] found that the integration of the LiDAR and multispectral imagery classification schedules resulted in accuracy improvements of 16%–20%. Previous studies have highlighted the positive contribution and benefit of integrating LiDAR data and passive optical remote sensed data when classifying objects. Moreover, other application fields have also reported in previous studies using a fusion method combining LiDAR and other data sources [45–47].

The main goal of our study is to explore the potential of fused LiDAR and passive optical data for improving land cover classification accuracy. To achieve this goal, three specific objectives of this paper are established: (1) fuse LiDAR and passive optical remotely sensed data with different fusion methods; (2) classify land cover using different classifiers and input data with seven spatial resolutions; and (3) assess the accuracies of the land cover classification and derive the best fusion and classification method for the study area. Previous studies have shown that the spatial scale of the input data has a substantial influence on the classification accuracy [48,49]. Consequently, all input data with a range of spatial resolutions were classified to obtain the optimal spatial scale for the land cover classification. The land cover classification map was then produced based on the optimal fusion method, classifier and spatial resolution.

2. Study Areas and Data

2.1. Study Areas

The study was conducted in Zhangye City in Gansu Province, China (see Figure 1). This study was carried out based on the HiWATER project, and the detailed scientific objectives can be found in Li, *et al.* [50]. Historically, Zhangye was a famous commercial port on the Silk Road. The study area is characterized as a dry and temperate continental climate with approximately 129 mm of precipitation per year and a mean annual potential evapotranspiration of 2047 mm. The mean annual air temperature is approximately 7.3 °C. The land use status of Zhangye in 2011 year was derived from the website of Zhangye people's government. Zhangye has a total area of 42,000 km², including arable land (6.44%), woodland (9.47%), grassland (51.28%), urban land (1.05%), water area (2.47%), and other lands (29.29%).

Figure 1. Location of the study area in Zhangye City in Gansu Province, China. The image is a true color composite of the Compact Airborne Spectrographic Imager (CASI) hyperspectral image with 1 m resolution, which will be used in this study.

2.2. Field Measurements

Fieldwork was performed from the 10 to 17 of July 2012 across the study area to collect training samples for the supervised classification. Altogether 2086 points for the training and accuracy assessment data of supervised classification. The selection principle of the sampled points is that the identical object is required within a radius of 3 m around the sample points. The sample point coordinates were recorded by a high-accuracy Real Time Kinematic differential Global Position System (RTK dGPS) with a 1-cm survey accuracy. However, for the training samples of buildings and water bodies, we directly selected the training data from remotely sensed images because we did not determine the locations of the plot centers of buildings and water bodies with a GPS. In total, 240 buildings and 66 water bodies were selected based on a CASI image. Therefore, a total of 2392 samples were used for the land cover classification.

2.3. Remotely Sensed Data Acquisition and Processing

2.3.1. Hyperspectral Data

Airborne hyperspectral data were obtained using the Compact Airborne Spectrographic Imager (CASI-1500). The CASI-1500 sensor was on board a Harbin Y-12 aircraft on 29 June 2012. The CASI-1500 is developed by Itres Research of Canada and is a two-dimensional charge coupled device (CCD) pushbroom imager designed to acquire the visible to near infrared hyperspectral images. The CASI data have a 1 m spatial resolution covering the wavelength range from 380 nm to 1050 nm. The specific characteristics of the CASI dataset used in this study are shown in Table 1. To reduce atmospheric effects related to particulate and molecular scattering, CASI images were atmospherically corrected using the FLAASH (Fast-Line-of-sight Atmospheric Analysis of Spectral Hypercube) [45], an atmospheric correction module in the commercial software ENVI. Subsequently the hyperspectral data were orthorectified using a digital terrain model (DTM) derived from LiDAR data. To retain the original pixel values, the nearest neighbor resampling method was used [39,51]. To obtain the entire study area, CASI images were mosaicked using the mosaicking function in the ENVI software.

Table 1. Specification of the CASI dataset used.

Parameter	Specification
Flight height	2000 m
Swath width	1500 m
Number of spectral bands	48
Spatial resolution	1.0 m
Spectral resolution	7.2 nm
Field of view	40°
Wavelength range	380–1050 nm

2.3.2. LiDAR Data

LiDAR point clouds were collected in July 2012 using the laser scanner of a Leica ALS70 system. The LiDAR data used were provide by the HiWATER [52]. In this study, the laser wavelength was 1064 nm with an average footprint size of 22.5 cm. The scan angle was $\pm18°$ with a 60% flight line side overlap. The average flying altitude was 1300 m. To acquire consistent point density data across the study area, we removed LiDAR point clouds in overlapping strips. After the removal of overlapping points, the average point density was 2.9 points/m^2 with an average point spacing of 0.59 m, and the average point density of first echoes was 2.1 points/m^2. Raw LiDAR point clouds were processed using the TerraScan software (TerraSolid, Ltd., Finland), and vegetation and ground returns were separated using the progressive triangulated irregular network (TIN) densification method proposed in the TerraScan software.

The digital surface model (DSM) (a grid size of 1.0 m) was created using the maximum height to each grid cell (Figure 2a). Empty cells were assigned elevation values by a nearest neighbor interpolation from neighboring pixels to avoid introducing new elevation values into the generated DTM [53]. The DTM was produced by computing the mean elevation value of the ground returns within each 1.0 m × 1.0 m grid cell (Figure 2b). The normalized DSM (nDSM) was then computed by subtracting the DTM from the DSM to obtain relative object height above the ground level (Figure 2c).

LiDAR intensity values are the amount of energy backscattered from features to the LiDAR sensor; these values are increasingly used [54–57]. The LiDAR intensity is closely related to laser power, incidence angle, object reflectivity and range of the LiDAR sensor to the object [43]. All these factors might cause LiDAR intensity differences for the identical type of object. Therefore, the LiDAR intensity must be corrected before they are applied to obtain a better comparison among different strips, flights and regions. Previous studies showed that normalized LiDAR intensity values can improve the accuracy of classification [58]. Consequently, to improve the classification accuracy of the land cover in our study, LiDAR intensity values were normalized by Equation (1) [59].

$$I_{normalized} = I \frac{R^2}{R_s \cos \alpha} \tag{1}$$

where $I_{normalized}$ is normalized intensity, I is raw intensity, R is sensor to object distance, R_s is reference distance or average flying altitude (in this study R_s was 1300 m) and a is incidence angle.

After normalization of the LiDAR intensity, the LiDAR intensity image with a 1.0 m spatial resolution was created using the intensity field of the LiDAR points (Figure 2d). To alleviate the variation in intensity values for the identical feature, only the first echoes from the LiDAR data were used to create the intensity image [55]. When more than one laser point fell within the same pixel, the mean value of the LiDAR intensity was assigned to that pixel [60].

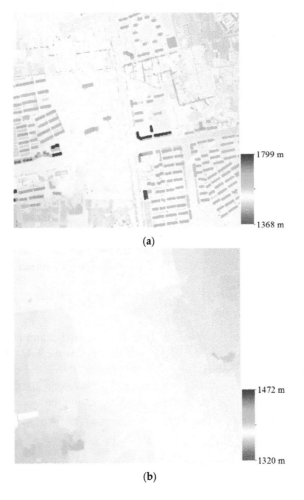

(a)

(b)

Figure 2. *Cont.*

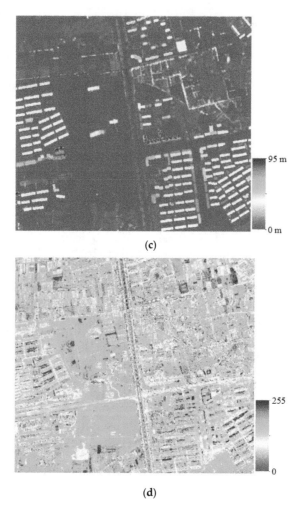

(c)

(d)

Figure 2. Four LiDAR-derived raster images with 1 m spatial resolution: (**a**) digital surface model (DSM); (**b**) digital terrain model (DTM); (**c**) normalized digital surface model (nDSM); and (**d**) normalized LiDAR intensity image.

3. Methodology

The flowchart for the data processing and land cover classification is presented in Figure 3. This figure summarizes the steps of the land cover classification using LiDAR and CASI data. Four main steps were performed including (1) pre-processing of the LiDAR and CASI data; (2) the fusion of the LiDAR and CASI data; (3) classification of the LiDAR data alone, the CASI data alone and the fused data, and (4) analyzing the classification results and assessing the accuracy. First of all, the data pre-processing was performed (see Section 2.3).

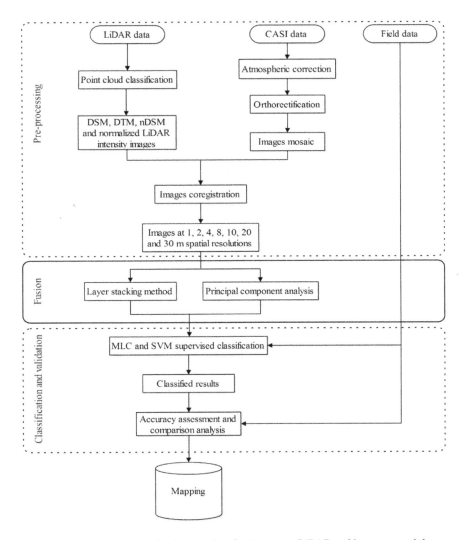

Figure 3. Flowchart for the land cover classification using LiDAR and hyperspectral data.

3.1. Fusion of the LiDAR and CASI Data

Before the data were fused, the cell values of the four LiDAR-derived raster images were linearly stretched to the 10 bit data range from 0 to 1023. In this study, the CASI data and LiDAR data were collected at different times, along with differences between the data types, flight conditions and paths produced inconsistencies in registration between the images [61]. As depicted in Figure 3a prerequisite for all fusion approaches is the accurate geometric alignment of two different data sources. Therefore, image-to-image registration of two data sources was necessary. To produce the best possible overlap between the hyperspectral CASI and LiDAR data, we coregistered the hyperspectral to the LiDAR data. The manual coregistration of these datasets was performed using the ENVI software by selecting tie points. The coregistration accuracy resulted in a root mean square error (RMSE) of less than 1 pixel (1 m).

To investigate the effect of the spatial resolution on classification accuracy, we fused LiDAR-derived images and CASI data at a range of spatial scales. Therefore, LiDAR-derived and CASI data with 1 m spatial resolution were spatially resampled to 2, 4, 8, 10, 20 and 30 m resolutions using a nearest neighbor resampling method before data fusion.

Data fusion is the integration of two or more different datasets to form a new data utilizing a certain algorithm. It can be implemented at different levels: at the decision level, feature level, and pixel level [62]. The goal of data fusion is to integrate complementary data to acquire high quality data with more abundant information. This process has been extensively used in the field of remote sensing [63]. Data fusion of LiDAR and other remote sensing data is a critical part of this study. The selection of data fusion methods between LiDAR and CASI data could have an influence on the land cover classification results. Therefore, to obtain the best fusion method for different data sources and applications, a trial and error method is necessary. In this study, to obtain the optimal fusion method of LiDAR and CASI data for land classification, both layer stacking and principal component analysis (PCA) fusion methods were tested. PCA is an orthogonal linear transformation, which transforms the data to a new coordinate system to reduce the multidimensional datasets into lower dimensions [64]. PCA transformations will produce some uncorrelated variables called principal components, and only the first few principal components may account for meaningful amounts of variance in the original data [65]. We used the layer stacking method to combine four LiDAR-derived images (DTM, DSM, nDSM, and intensity) and CASI data (48 bands) into one multiband image at the pixel level. This new multiband image includes a total of 52 bands. Figure 4a shows the 3 band image (near infrared image of CASI, the intensity image and nDSM) of the LiDAR-CASI stacked image with 52 bands. For the stacked image, we used two methods to process the image. The first method was directly classifying the stacked image with 52 bands. The second method was transforming the stacked image using a PCA algorithm, and then the first few principal components derived from the PCA were used as the data for the land cover classification. The PCA transformation could reduce data redundancy and retain the uncorrelated variables (*i.e.*, independent principal components). In this study, we found that the first five principal components account for 98.93% of the total variance in the whole LiDAR and CASI data. Therefore, the first five principal components were used to replace the stacked image (52 bands), and the redundancy, noise, and size of the dataset were significantly reduced [44]. Figure 4b illustrates the image of the first three principal components image in the principal component analysis of the LiDAR-CASI stacked image with 52 bands.

(a)   (b)

Figure 4. LiDAR-CASI fusion images with 1 m spatial resolution: (**a**) LiDAR-CASI stacked image (near infrared image of CASI, intensity image and nDSM); and (**b**) principal component analysis (first three principal components).

3.2. Classification Methods

Generally, remote sensing classification methods are divided into two broad types: parametric and non-parametric methods [66]. Parametric methods are based on the assumption that the data of each class are normally distributed [7,49,67]; however, non-parametric techniques do not assume specific data class distributions [11,41]. Supervised classification is one of the most commonly used techniques for the classification of remotely sensed data. In this study, two supervised classification methods were performed: (1) maximum likelihood classification (MLC) method; (2) support vector machine (SVM) classification.

The MLC method is a simple, yet robust, classifier [39,51]. The MLC is a parametric classifier based on the Bayesian decision theory and is the most popular conventional supervised classification technique [68,69]. However, for the MLC method, it is difficult to satisfy the assumption of a normally distributed dataset. Another disadvantage of conventional parametric classification approaches is that they are limited in their ability to classify multisource and high dimensional data [67]. However, the SVM classifier is a non-parametric classifier for supervised classification [45,70], which is able to handle non-normal distributed datasets [11,30]. The primary advantage of the SVM method is that it requires no assumptions in terms of the distribution of the data and can achieve a good result with relatively limited training sample data [5,67]. In addition, the SVM method is not sensitive to high data dimensionality [11,41,71], and is especially suited to classify data with high dimensionality and multiple sources [9,67]. In general, previous studies have shown that the SVM is more accurate than the MLC method, e.g., [66,72]. To validate the better performance of the SVM classifier compared with other parametric approaches in our study, the MLC was performed and the results were compared with those produced from the SVM classifier.

In this study, we performed a land cover classification at seven different spatial scales (1, 2, 4, 8, 10, 20 and 30 m) to determine an optimal spatial scale for the classification using different datasets. In addition, to compare the classification accuracies of the fusion data with those of a single data source, single source LiDAR and CASI data were also classified. Therefore, four different remotely sensed datasets were trained: (a) LiDAR-derived images alone; (b) the CASI images alone (48 bands); (c) the layer stacking image with 52 bands (CASI: 48 bands, LiDAR: four images); and (d) the first five principal components derived from the PCA transformation. Finally, twenty-eight datasets were classified using the MLC and SVM supervised classifiers.

The land cover category was classified into seven classes, *i.e.*, buildings, road, water bodies, forests, grassland, cropland and barren land. For each land cover category, about two-thirds of the samples were randomly selected as training data, with the remaining samples used to assess the classification accuracies. Table 2 lists the per-class numbers for the classification training and validation data (a total of 2392 samples).

Table 2. The number of classification training and validation data per class.

Class	Number of Training Sample (Points)	Number of Validation Sample (Points)
Buildings	160	80
Road	225	113
Water bodies	44	22
Forests	397	198
Grassland	278	139
Cropland	307	154
Barren land	183	92
	1594	798

3.3. Accuracy Assessment

The number of validation samples per class based on separate test data is shown in Table 2. A total of 798 samples were applied to assess the classification accuracies. Classification accuracies were assessed based on confusion matrices (also called the error matrixes), which is the most standard method for remote sensing classification accuracy assessments [72]. The accuracy metrics include the overall accuracy (OA), the producer's accuracy (PA), the error of omission (*i.e.*, 1-PA), the user's accuracy (UA), the error of commission (*i.e.*, 1-UA) and the kappa coefficient (K). These accuracy metrics can be acquired from the confusion matrixes; these matrixes are widely used in classification accuracy assessment of remote sensing [13]. We assessed and compared the classification accuracies of 56 classification maps produced by the MLC and SVM classifiers.

4. Results and Discussion

Supervised classifications were performed on four datasets with seven different spatial resolutions using the MLC and SVM classifiers. Each classifier produced 28 classification results, and a total of 56 classification results were obtained. The classification accuracies were evaluated using one-third of the samples collected which were not used in the supervised classification. The classification accuracies of the 56 classification maps produced by the MLC and SVM classifiers are shown in Table 3.

Table 3. Classification accuracies of the maximum likelihood classification (MLC) and support vector machine (SVM) classifiers using four types of datasets with different spatial resolutions. The accuracy metrics included the overall accuracy (OA) and kappa coefficient (K).

| Resolution (meters) | Accuracy Metrics | LiDAR and CASI Data Alone | | | | Fused Data of LiDAR and CASI | | | |
| | | LiDAR Data | | CASI Data | | PCA | | Layer stacking | |
		MLC	SVM	MLC	SVM	MLC	SVM	MLC	SVM
1	OA (%)	25	75.6	84.7	88.7	91.9	95.3	92.9	97.8
	K	0.181	0.582	0.758	0.836	0.868	0.923	0.888	0.964
2	OA (%)	27.8	78.2	82.8	87.1	90.7	96.5	92.1	97.7
	K	0.203	0.654	0.743	0.825	0.862	0.943	0.884	0.963
4	OA (%)	34.1	74.5	80.2	85.6	89.9	94.5	91.2	96.3
	K	0.262	0.626	0.726	0.803	0.86	0.922	0.878	0.948
8	OA (%)	37.8	69.5	76.6	81.1	87.1	88.9	89	92.8
	K	0.292	0.579	0.694	0.757	0.829	0.849	0.855	0.904
10	OA (%)	39.3	68.2	75.6	80.5	85.9	87.9	87.7	91.2
	K	0.302	0.561	0.692	0.746	0.814	0.836	0.84	0.883
20	OA (%)	53.5	62.6	74.8	77.3	81.5	81.2	84.5	86.5
	K	0.401	0.445	0.679	0.692	0.741	0.723	0.783	0.805
30	OA (%)	48.7	60.3	73.1	71.2	81.4	79.7	83.3	82
	K	0.298	0.401	0.619	0.589	0.696	0.644	0.727	0.686
Mean	OA (%)	38	69.8	78.3	81.6	86.9	89.1	88.7	92
	K	0.277	0.55	0.702	0.75	0.81	0.834	0.836	0.879

4.1. Comparison of Classification Results with Different Datasets

Data fusion has been extensively used in the remote sensing field. Data fusion can implement data compression, image enhancement and complementary information for multiple data sources. To determine the performance of different fused methods for land cover classification, the PCA and the layer stacking were applied to fuse LiDAR and CASI data. For the four types of datasets (*i.e.*, LiDAR data, CASI data, PCA data and layer stacking images), the LiDAR data alone produced the worst classification result; the lowest overall classification accuracy (OA = 25.0%, K = 0.181) was produced using the MLC classifier at 1 m resolution, and the highest accuracy was only 78.2% with a kappa of 0.654 at 2 m spatial resolution. This poor classification accuracy could be explained by the lack spectral information in LiDAR data, which could provide useful information related to land cover. Therefore, it was difficult to obtain accurate land cover classification using only LiDAR data. The layer stacking fusion data obtained the best classification result, and the highest accuracy was 97.8% with a kappa of 0.964 at 1 m spatial resolution. The highest overall classification accuracies for the PCA fused data and CASI data alone were 96.5% and 88.7%, respectively. The results showed that the overall classification accuracies of LiDAR and CASI fusion data outperformed the LiDAR and CASI data alone (Table 3). This result was consistent with findings reported by Singh, *et al.* [39]. The increased accuracies could be attributed to complementary vertical and spectral information from the LiDAR and CASI data.

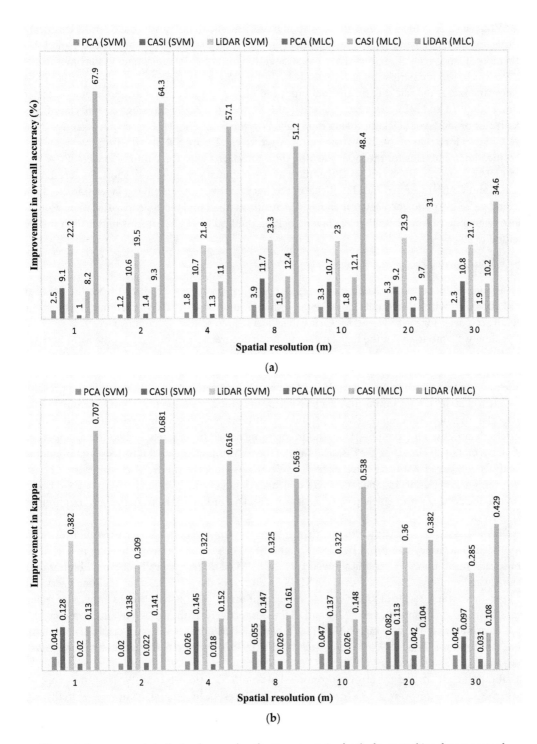

Figure 5. Improvements in the land cover classification accuracies for the layer stacking data compared with three other datasets (PCA, CASI and LiDAR) using the SVM and MLC classifiers at seven spatial resolutions; (a) overall accuracy and (b) kappa coefficient.

For the fused data, all overall classification accuracies for the seven spatial resolutions were greater than or equal to 79.7%. Moreover, the overall accuracies of fusion datasets exceeded 94% using the SVM classifier at the 1 m, 2 m and 4 m spatial resolutions. In our study, the overall classification accuracies of the CASI and LiDAR fusion data were 9.1% and 19.6% higher than the CASI and LiDAR data alone, respectively. Previous study has shown that different data fusion methods have an effect on the land cover classification accuracies [44]. In this study, we obtained similar results. Table 3 demonstrates that the stacking fusion data produced higher overall accuracies and kappa coefficients using both SVM and MLC classifiers than that of PCA fusion data. The average overall classification accuracies of the layer stacking fusion data based on the SVM and MLC classifier were 2.90% and 1.76% higher than that of the PCA fusion data, respectively. Figure 5 shows the improvements in the classification accuracies for the layer stacking data compared with three other datasets (PCA, CASI and LiDAR) using the SVM and MLC classifiers at seven spatial resolutions. In short, the fused data of LiDAR and CASI improved the classification accuracy, and we found that the layer stacking fusion method generally performed better than the PCA fusion method for land cover classification. However, these results do not indicate that the layer stacking method always performs better than the PCA method in all cases.

4.2. Classification Performance of the MLC and SVM Classifiers

Figure 6 illustrates the comparison of the classification results produced from the MLC and SVM classifiers with all four datasets at seven spatial resolutions. For the stacked image, the average overall accuracies for the SVM and MLC classifiers at all seven spatial resolutions were 92% and 88.7%, respectively. Comparison of the highest classification accuracies produced from the SVM and MLC classifier showed that the SVM classifier obtained a slightly higher accuracy, improving the overall accuracy of approximately 4.9%. The result indicated that the distribution-free SVM classifier generally performed better than the MLC classifier. This result was consistent with the findings presented by García et al. [67], who found that the SVM classifier could improve the overall accuracy of land cover classification. Therefore, the SVM is more suitable for a high dimensionality and multiple sources dataset than the parametric methods commonly used in remotely sensed image supervised classification [9,41]. However, the disadvantage of the SVM classifier is that it increases computational demands compared with the MLC classifier [73]. Consequently, the SVM classifier may be time consuming when classifying the land cover of a large data volume. We must acquire a good balance between the accuracy and efficiency of the classification according to the purposes and requirements of the work.

Table 4 shows the accuracy of each class from the confusion matrices of the highest overall classification accuracy produced by the SVM and MLC classifiers for layer stacking fusion data at 1 m spatial resolution. The results showed that the SVM classifier generally obtained higher user's accuracy and producer's accuracy than the MLC classifier. For the SVM classifier, the user's accuracies and producer's accuracies of all seven classes (i.e., buildings, road, water bodies, forests, grassland, cropland and barren land) exceed 80.1% and 84.5%, respectively. Although the overall accuracy of the MLC classifier was relatively high (OA = 92.9%, kappa = 0.888), the two largest errors of commission were observed for forests (63.67%) and grassland (43.99%), and the two largest errors of omission were road (29.85%) and forests (18.05%). Correspondingly, low errors were obtained using the SVM classification method, and the errors of commission were 1.93% (forests) and 19.87% (grassland), and the errors of omission were 8.85% (road) and 15.47% (forests). In general, compared with the MLC classifier, the SVM improved the user's accuracies of forests and grassland and producer's accuracies of road and forests in this study. Therefore, we produced a land cover classification map of the study area using the SVM classifier, obtaining the highest overall accuracy using the layer stacking fusion data at a 1 m spatial resolution. Following the supervised classification, a post-classification process based on a majority filter was used to remove isolated pixels [68]. Figure 7 is the subset of the study area classification map with an overall accuracy of 97.8% (kappa = 0.964).

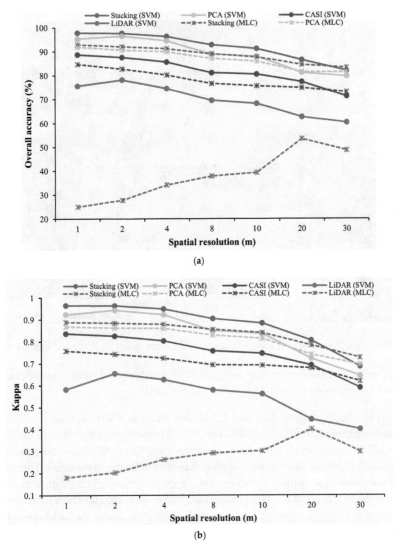

Figure 6. Comparison of the classification accuracies using different classifiers, datasets and spatial resolutions; (**a**) overall accuracy and (**b**) kappa coefficient.

Table 4. Per-class accuracies obtained with the highest overall accuracy produced by the support vector machine (SVM) and maximum likelihood (MLC) classifiers for layer stacking fusion data at a 1 m spatial resolution. The accuracies included the producer's accuracy (PA), user's accuracy (UA), error of commission (EC) and error of omission (EO).

Class Name	SVM Classifier				MLC Classifier			
	PA (%)	UA(%)	EO (%)	EC (%)	PA (%)	UA (%)	EO (%)	EC (%)
Buildings	94.78	99.58	5.22	0.42	94.73	97.85	5.27	2.15
Road	91.15	94.58	8.85	5.42	70.15	88.42	29.85	11.58
Water bodies	99.99	99.92	0.01	0.08	96.05	100	3.95	0
Forests	84.53	98.07	15.47	1.93	81.95	36.33	18.05	63.67
Grassland	93.44	80.13	6.56	19.87	89.55	56.01	10.45	43.99
Cropland	99.8	95.42	0.2	4.58	99.53	94.2	0.47	5.8
Barren land	94.7	94.48	5.3	5.52	89.76	99.4	10.24	0.6
OA (%)		97.8				92.9		
K		0.964				0.888		

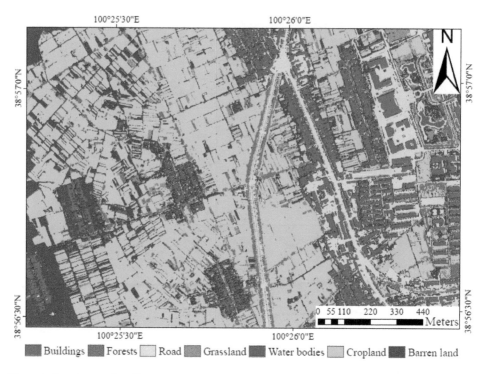

Figure 7. Land cover classification map using the SVM classifier based on the layer stacking fusion method at a 1 m spatial resolution (OA = 97.8%, kappa = 0.964).

For the layer stacking fusion data and CASI data alone at a 30 m spatial resolution, the MLC classifier obtained better results (OA = 83.3% and 73.1%, respectively) than that of the SVM classifier (OA = 82% and 71.2%, respectively). For the PCA fusion data, the overall classification accuracies of the MLC classifier both at 20 m and 30 m were higher than that of the SVM classifier. Therefore, the SVM classifier was not always the optimal classifier at all seven resolutions. This result was consistent with the results observed by Ghosh, *et al.* [48], who claimed that none of the classifiers produced consistently higher accuracies at all spatial resolutions. For the LiDAR data alone, the SVM classifier displayed higher accuracies than the MLC classifiers at all seven spatial resolutions. In general, our results demonstrated that the classification accuracies were affected by the dataset types, spatial resolutions and classifiers.

4.3. Influence of the Spatial Resolution on the Classification Accuracy

Spatial resolution of remotely sensed data is one of the most critical factors that affects the classification accuracies [74]. In this study, therefore, the land cover was classified at seven spatial scales (1, 2, 4, 8, 10, 20 and 30 m) to obtain an optimal spatial resolution for the classification with four types of datasets. We analyzed the optimal resolutions for each datasets and classifier. Different optimal resolutions were found for the four datasets and two classification methods, and similar results were also presented by Ke, *et al.* [74]. For the PCA and LiDAR data using the SVM classifier, the highest classification accuracies were obtained at a 2 m spatial resolution (Figure 6). After that, the overall classification accuracy started to decrease. However, for the LiDAR data using the MLC classifier, the classification accuracies increased as spatial resolutions became coarser (from 1 m to 20 m), and the overall accuracy was the highest at a 20 m spatial resolution. Except for the above-mentioned results, the classification results showed that classification accuracy generally decreased as spatial resolution of the input datasets became coarser, and the highest classification accuracies were found at a 1 m spatial resolution (Figure 6). Therefore, the higher classification accuracy does not always occur for the higher spatial resolutions in the different datasets. In addition, although the classification accuracy is high for select datasets with a fine resolution, a substantial amount of data covers large areas for land cover classification. Thus, data management and processing will become a problem, and the classification performance will decrease. In conclusion, when classifying land cover, we should select the optimal spatial resolution based on the application and research purposes.

5. Conclusions

This study explored the potential for fusing airborne LiDAR and hyperspectral CASI data to classify land cover. We fused LiDAR and CASI data using layer stacking and PCA fusion methods, and successfully performed supervised image classification using the SVM and MLC classifiers. Overall, the use of fused data between LiDAR and CASI for land cover classification outperformed the classification accuracies produced from LiDAR and CASI data alone (Table 3). The highest overall accuracy (OA = 97.8% with a kappa of 0.964) was generated by the non-parametric SVM classifier when using layer stacking fusion data at 1 m spatial resolution. Compared with the best classification results of the CASI and LiDAR data alone, the overall classification accuracies improved by 9.1% and 19.6%, respectively. Moreover, we found that the classification accuracies based on the layer stacking data were higher than that based on a PCA for all spatial resolutions. In this study, therefore, the layer stacking fusion method was more suitable for the land cover classification compared to the PCA fusion method. Using the SVM classifier to classify the fused LiDAR and CASI data produced a higher accuracy than that of the MLC classifier. This result showed that the SVM classifier possesses a higher potential than the MLC for land cover classification using multisource fusion data from LiDAR and CASI data.

LiDAR and multi/hyperspectral remote sensing fusion data have provided complementary information that LiDAR and multi/hyperspectral data alone have not. Therefore, the fusion of LiDAR and multi/hyperspectral data has a great potential for highly accurate information extraction of objects and land cover mapping. Our future work will focus on fusing LiDAR and passive optical remotely sensed data to classify different vegetation types and land covers to estimate the vegetation biomass and to detect land cover changes.

Acknowledgments: This work was supported by the National Natural Science Foundation of China (Nos. 41371350 and 41271428); the International Postdoctoral Exchange Fellowship Program 2014 by the Office of China Postdoctoral Council (20140042); Beijing Natural Science Foundation (4144074). We are grateful to the four anonymous reviewers and associate editor Danielle Yao for their valuable comments and suggestions on the manuscript.

Author Contributions: Shezhou Luo and Cheng Wang contributed the main idea, designed the study method, and finished the manuscript. Xiaohuan Xi and Hongcheng Zeng processed LiDAR data. Dong Li, Shaobo Xia and Pinghua Wang processed CASI data. The manuscript was improved by the contributions of all of the co-authors.

Conflicts of Interest: The authors declare no conflict of interest.

References

1. Foody, G.M. Assessing the accuracy of land cover change with imperfect ground reference data. *Remote Sens. Environ.* **2010**, *114*, 2271–2285. [CrossRef]
2. Turner, B.L.; Lambin, E.F.; Reenberg, A. The emergence of land change science for global environmental change and sustainability. *Proc. Natl. Acad. Sci. USA* **2007**, *104*, 20666–20671. [CrossRef] [PubMed]
3. Gong, P.; Wang, J.; Yu, L.; Zhao, Y.; Zhao, Y.; Liang, L.; Niu, Z.; Huang, X.; Fu, H.; Liu, S.; *et al.* Finer resolution observation and monitoring of global land cover: First mapping results with Landsat TM and ETM+ data. *Int. J. Remote Sens.* **2012**, *34*, 2607–2654. [CrossRef]
4. Liu, Y.; Zhang, B.; Wang, L.-M.; Wang, N. A self-trained semisupervised svm approach to the remote sensing land cover classification. *Comput. Geosci.* **2013**, *59*, 98–107. [CrossRef]
5. Fieber, K.D.; Davenport, I.J.; Ferryman, J.M.; Gurney, R.J.; Walker, J.P.; Hacker, J.M. Analysis of full-waveform LiDAR data for classification of an orange orchard scene. *ISPRS J. Photogramm. Remote Sens.* **2013**, *82*, 63–82. [CrossRef]

6.   Chasmer, L.; Hopkinson, C.; Veness, T.; Quinton, W.; Baltzer, J. A decision-tree classification for low-lying complex land cover types within the zone of discontinuous permafrost. *Remote Sens. Environ.* **2014**, *143*, 73–84. [CrossRef]

7.   Pal, M.; Mather, P.M. Assessment of the effectiveness of support vector machines for hyperspectral data. *Future Gener. Comput. Syst.* **2004**, *20*, 1215–1225. [CrossRef]

8.   Szuster, B.W.; Chen, Q.; Borger, M. A comparison of classification techniques to support land cover and land use analysis in tropical coastal zones. *Appl. Geogr.* **2011**, *31*, 525–532. [CrossRef]

9.   Waske, B.; Benediktsson, J.A. Fusion of support vector machines for classification of multisensor data. *IEEE Trans. Geosci. Remote Sens.* **2007**, *45*, 3858–3866. [CrossRef]

10. Berger, C.; Voltersen, M.; Hese, O.; Walde, I.; Schmullius, C. Robust extraction of urban land cover information from HSR multi-spectral and LiDAR data. *IEEE J. Sel. Topics Appl. Earth Obs. Remote Sens.* **2013**, *6*, 2196–2211. [CrossRef]

11. Klein, I.; Gessner, U.; Kuenzer, C. Regional land cover mapping and change detection in central asia using MODIS time-series. *Appl. Geogr.* **2012**, *35*, 219–234. [CrossRef]

12. Adam, E.; Mutanga, O.; Odindi, J.; Abdel-Rahman, E.M. Land-use/cover classification in a heterogeneous coastal landscape using Rapideye imagery: Evaluating the performance of random forest and support vector machines classifiers. *Int. J. Remote Sens.* **2014**, 3440–3458. [CrossRef]

13. Puertas, O.L.; Brenning, A.; Meza, F.J. Balancing misclassification errors of land cover classification maps using support vector machines and Landsat imagery in the Maipo River Basin (Central Chile, 1975–2010). *Remote Sens. Environ.* **2013**, *137*, 112–123. [CrossRef]

14. Lu, D.; Chen, Q.; Wang, G.; Moran, E.; Batistella, M.; Zhang, M.; Vaglio Laurin, G.; Saah, D. Aboveground forest biomass estimation with Landsat and LiDAR data and uncertainty analysis of the estimates. *Int. J. For. Res.* **2012**, *2012*, 1–16. [CrossRef]

15. Edenius, L.; Vencatasawmy, C.P.; Sandström, P.; Dahlberg, U. Combining satellite imagery and ancillary data to map snowbed vegetation important to Reindeer rangifer tarandus. *Arct. Antarct. Alp. Res.* **2003**, *35*, 150–157. [CrossRef]

16. Amarsaikhan, D.; Blotevogel, H.H.; van Genderen, J.L.; Ganzorig, M.; Gantuya, R.; Nergui, B. Fusing high-resolution sar and optical imagery for improved urban land cover study and classification. *Int. J. Image Data Fusion* **2010**, *1*, 83–97. [CrossRef]

17. Dong, P. LiDAR data for characterizing linear and planar geomorphic markers in tectonic geomorphology. *J. Geophys. Remote Sens.* **2015**. [CrossRef]

18. Richardson, J.J.; Moskal, L.M.; Kim, S.-H. Modeling approaches to estimate effective leaf area index from aerial discrete-return LiDAR. *Agric. For. Meteorol.* **2009**, *149*, 1152–1160. [CrossRef]

19. Lee, H.; Slatton, K.C.; Roth, B.E.; Cropper, W.P. Prediction of forest canopy light interception using three-dimensional airborne LiDAR data. *Int. J. Remote Sens.* **2009**, *30*, 189–207. [CrossRef]

20. Brennan, R.; Webster, T.L. Object-oriented land cover classification of LiDAR-derived surfaces. *Can. J. Remote Sens.* **2006**, *32*, 162–172. [CrossRef]

21. Ko, C.; Sohn, G.; Remmel, T.; Miller, J. Hybrid ensemble classification of tree genera using airborne LiDAR data. *Remote Sens.* **2014**, *6*, 11225–11243. [CrossRef]

22. Luo, S.; Wang, C.; Pan, F.; Xi, X.; Li, G.; Nie, S.; Xia, S. Estimation of wetland vegetation height and leaf area index using airborne laser scanning data. *Ecol. Indicators* **2015**, *48*, 550–559. [CrossRef]

23. Tang, H.; Brolly, M.; Zhao, F.; Strahler, A.H.; Schaaf, C.L.; Ganguly, S.; Zhang, G.; Dubayah, R. Deriving and validating leaf area index (LAI) at multiple spatial scales through LiDAR remote sensing: A case study in Sierra National Forest. *Remote Sens. Environ.* **2014**, *143*, 131–141. [CrossRef]

24. Hopkinson, C.; Chasmer, L. Testing LiDAR models of fractional cover across multiple forest EcoZones. *Remote Sens. Environ.* **2009**, *113*, 275–288. [CrossRef]

25. Glenn, N.F.; Spaete, L.P.; Sankey, T.T.; Derryberry, D.R.; Hardegree, S.P.; Mitchell, J.J. Errors in LiDAR-derived shrub height and crown area on sloped terrain. *J. Arid Environ.* **2011**, *75*, 377–382. [CrossRef]

26. Zhao, K.; Popescu, S.; Nelson, R. LiDAR remote sensing of forest biomass: A scale-invariant estimation approach using airborne lasers. *Remote Sens. Environ.* **2009**, *113*, 182–196. [CrossRef]

27. Antonarakis, A.S.; Richards, K.S.; Brasington, J. Object-based land cover classification using airborne LiDAR. *Remote Sens. Environ.* **2008**, *112*, 2988–2998. [CrossRef]

28. Qin, Y.; Li, S.; Vu, T.-T.; Niu, Z.; Ban, Y. Synergistic application of geometric and radiometric features of LiDAR data for urban land cover mapping. *Opt. Express* **2015**, *23*, 13761–13775. [CrossRef] [PubMed]

29. Sherba, J.; Blesius, L.; Davis, J. Object-based classification of abandoned logging roads under heavy canopy using LiDAR. *Remote Sens.* **2014**, *6*, 4043–4060. [CrossRef]

30. Mallet, C.; Bretar, F.; Roux, M.; Soergel, U.; Heipke, C. Relevance assessment of full-waveform LiDAR data for urban area classification. *ISPRS J. Photogramm. Remote Sens.* **2011**, *66*, S71–S84. [CrossRef]

31. Hellesen, T.; Matikainen, L. An object-based approach for mapping shrub and tree cover on grassland habitats by use of LiDAR and CIR orthoimages. *Remote Sens.* **2013**, *5*, 558–583. [CrossRef]

32. Stojanova, D.; Panov, P.; Gjorgjioski, V.; Kobler, A.; Džeroski, S. Estimating vegetation height and canopy cover from remotely sensed data with machine learning. *Ecol. Inf.* **2010**, *5*, 256–266. [CrossRef]

33. Debes, C.; Merentitis, A.; Heremans, R.; Hahn, J.; Frangiadakis, N.; van Kasteren, T.; Wenzhi, L.; Bellens, R.; Pizurica, A.; Gautama, S.; *et al.* Hyperspectral and LiDAR data fusion: Outcome of the 2013 GRSS data fusion contest. *IEEE J. Sel. Topics Appl. Earth Obs. Remote Sens.* **2014**, *7*, 2405–2418. [CrossRef]

34. Reese, H.; Nyström, M.; Nordkvist, K.; Olsson, H. Combining airborne laser scanning data and optical satellite data for classification of alpine vegetation. *Int. J. Appl. Earth Obs. Geoinf.* **2014**, *27*, 81–90. [CrossRef]

35. Wulder, M.A.; Han, T.; White, J.C.; Sweda, T.; Tsuzuki, H. Integrating profiling LiDAR with Landsat data for regional boreal forest canopy attribute estimation and change characterization. *Remote Sens. Environ.* **2007**, *110*, 123–137. [CrossRef]

36. Buddenbaum, H.; Seeling, S.; Hill, J. Fusion of full-waveform LiDAR and imaging spectroscopy remote sensing data for the characterization of forest stands. *Int. J. Remote Sens.* **2013**, *34*, 4511–4524. [CrossRef]

37. Kim, Y. Improved classification accuracy based on the output-level fusion of high-resolution satellite images and airborne LiDAR data in urban area. *IEEE Geosci. Remote Sens. Lett.* **2014**, *11*, 636–640.

38. Hartfield, K.A.; Landau, K.I.; Leeuwen, W.J.D. Fusion of high resolution aerial multispectral and LiDAR data: Land cover in the context of urban mosquito habitat. *Remote Sens.* **2011**, *3*, 2364–2383. [CrossRef]

39. Singh, K.K.; Vogler, J.B.; Shoemaker, D.A.; Meentemeyer, R.K. LiDAR-Landsat data fusion for large-area assessment of urban land cover: Balancing spatial resolution, data volume and mapping accuracy. *ISPRS J. Photogramm. Remote Sens.* **2012**, *74*, 110–121. [CrossRef]

40. Bork, E.W.; Su, J.G. Integrating LiDAR data and multispectral imagery for enhanced classification of rangeland vegetation: A meta analysis. *Remote Sens. Environ.* **2007**, *111*, 11–24. [CrossRef]

41. Koetz, B.; Morsdorf, F.; Linden, S.; Curt, T.; Allgöwer, B. Multi-source land cover classification for forest fire management based on imaging spectrometry and LiDAR data. *For. Ecol. Manag.* **2008**, *256*, 263–271. [CrossRef]

42. Liu, X.; Bo, Y. Object-based crop species classification based on the combination of airborne hyperspectral images and LiDAR data. *Remote Sens.* **2015**, *7*, 922–950. [CrossRef]

43. Mesas-Carrascosa, F.J.; Castillejo-González, I.L.; de la Orden, M.S.; Porras, A.G.-F. Combining LiDAR intensity with aerial camera data to discriminate agricultural land uses. *Comput. Electron. Agric.* **2012**, *84*, 36–46. [CrossRef]

44. Mutlu, M.; Popescu, S.; Stripling, C.; Spencer, T. Mapping surface fuel models using LiDAR and multispectral data fusion for fire behavior. *Remote Sens. Environ.* **2008**, *112*, 274–285. [CrossRef]

45. Jones, T.G.; Coops, N.C.; Sharma, T. Assessing the utility of airborne hyperspectral and LiDAR data for species distribution mapping in the coastal Pacific Northwest, Canada. *Remote Sens. Environ.* **2010**, *114*, 2841–2852. [CrossRef]

46. Holmgren, J.; Persson, Å.; Söderman, U. Species identification of individual trees by combining high resolution LiDAR data with multi-spectral images. *Int. J. Remote Sens.* **2008**, *29*, 1537–1552. [CrossRef]

47. Vaglio Laurin, G.; del Frate, F.; Pasolli, L.; Notarnicola, C.; Guerriero, L.; Valentini, R. Discrimination of vegetation types in Alpine sites with alos palsar-, radarsat-2-, and LiDAR-derived information. *Int. J. Remote Sens.* **2013**, *34*, 6898–6913. [CrossRef]

48. Ghosh, A.; Fassnacht, F.E.; Joshi, P.K.; Koch, B. A framework for mapping tree species combining hyperspectral and LiDAR data: Role of selected classifiers and sensor across three spatial scales. *Int. J. Appl. Earth Obs. Geoinf.* **2014**, *26*, 49–63. [CrossRef]

49. Dalponte, M.; Orka, H.O.; Gobakken, T.; Gianelle, D.; Naesset, E. Tree species classification in boreal forests with hyperspectral data. *IEEE Trans. Geosci. Remote Sens.* **2013**, *51*, 2632–2645. [CrossRef]

50. Li, X.; Cheng, G.; Liu, S.; Xiao, Q.; Ma, M.; Jin, R.; Che, T.; Liu, Q.; Wang, W.; Qi, Y.; *et al.* Heihe watershed allied telemetry experimental research (HiWATER): Scientific objectives and experimental design. *B. Am. Meteorol. Soc.* **2013**, *94*, 1145–1160. [CrossRef]

51. Geerling, G.W.; Labrador-Garcia, M.; Clevers, J.G.P.W.; Ragas, A.M.J.; Smits, A.J.M. Classification of floodplain vegetation by data fusion of spectral (CASI) and LiDAR data. *Int. J. Remote Sens.* **2007**, *28*, 4263–4284. [CrossRef]

52. Xiao, Q.; Wen, J. HiWATER: Airborne LiDAR raw data in the middle reaches of the Heihe River Basin. *Inst. Remote Sens. Digi. Earth Chin. Aca. Sci.* **2014**. [CrossRef]

53. Salah, M.; Trinder, J.C.; Shaker, A. Performance evaluation of classification trees for building detection from aerial images and LiDAR data: A comparison of classification trees models. *Int. J. Remote Sens.* **2011**, *32*, 5757–5783. [CrossRef]

54. Luo, S.; Wang, C.; Xi, X.; Pan, F. Estimating fpar of maize canopy using airborne discrete-return LiDAR data. *Opt. Express* **2014**, *22*, 5106–5117. [CrossRef]

55. Donoghue, D.; Watt, P.; Cox, N.; Wilson, J. Remote sensing of species mixtures in conifer plantations using LiDAR height and intensity data. *Remote Sens. Environ.* **2007**, *110*, 509–522. [CrossRef]

56. Kwak, D.-A.; Cui, G.; Lee, W.-K.; Cho, H.-K.; Jeon, S.W.; Lee, S.-H. Estimating plot volume using LiDAR height and intensity distributional parameters. *Int. J. Remote Sens.* **2014**, *35*, 4601–4629. [CrossRef]

57. Wang, C.; Glenn, N.F. Integrating LiDAR intensity and elevation data for terrain characterization in a forested area. *IEEE Geosci. Remote Sens. Lett.* **2009**, *6*, 463–466. [CrossRef]

58. Yan, W.Y.; Shaker, A.; Habib, A.; Kersting, A.P. Improving classification accuracy of airborne LiDAR intensity data by geometric calibration and radiometric correction. *ISPRS J. Photogramm. Remote Sens.* **2012**, *67*, 35–44. [CrossRef]

59. Höfle, B.; Pfeifer, N. Correction of laser scanning intensity data: Data and model-driven approaches. *ISPRS J. Photogramm. Remote Sens.* **2007**, *62*, 415–433. [CrossRef]

60. Chust, G.; Galparsoro, I.; Borja, Á.; Franco, J.; Uriarte, A. Coastal and estuarine habitat mapping, using LiDAR height and intensity and multi-spectral imagery. *Estuar. Coast. Shelf Sci.* **2008**, *78*, 633–643. [CrossRef]

61. Lucas, R.M.; Lee, A.C.; Bunting, P.J. Retrieving forest biomass through integration of CASI and LiDAR data. *Int. J. Remote Sens.* **2008**, *29*, 1553–1577. [CrossRef]

62. Pohl, C.; Van Genderen, J.L. Review article multisensor image fusion in remote sensing: Concepts, methods and applications. *Int. J. Remote Sens.* **1998**, *19*, 823–854. [CrossRef]

63. Zhang, J. Multi-source remote sensing data fusion: Status and trends. *Int. J. Image Data Fusion* **2010**, *1*, 5–24. [CrossRef]

64. Dópido, I.; Villa, A.; Plaza, A.; Gamba, P. A quantitative and comparative assessment of unmixing-based feature extraction techniques for hyperspectral image classification. *IEEE J. Sel. Topics Appl. Earth Obs. Remote Sens.* **2012**, *5*, 421–435. [CrossRef]

65. Huang, B.; Zhang, H.; Yu, L. Improving Landsat ETM+ urban area mapping via spatial and angular fusion with misr multi-angle observations. *IEEE J. Sel. Top. Appl. Earth Obs. Remote Sens.* **2012**, *5*, 101–109. [CrossRef]

66. Srivastava, P.K.; Han, D.; Rico-Ramirez, M.A.; Bray, M.; Islam, T. Selection of classification techniques for land use/land cover change investigation. *Adv. Space Res.* **2012**, *50*, 1250–1265. [CrossRef]

67. García, M.; Riaño, D.; Chuvieco, E.; Salas, J.; Danson, F.M. Multispectral and LiDAR data fusion for fuel type mapping using support vector machine and decision rules. *Remote Sens. Environ.* **2011**, *115*, 1369–1379. [CrossRef]

68. Hladik, C.; Schalles, J.; Alber, M. Salt marsh elevation and habitat mapping using hyperspectral and LiDAR data. *Remote Sens. Environ.* **2013**, *139*, 318–330. [CrossRef]

69. Oommen, T.; Misra, D.; Twarakavi, N.K.C.; Prakash, A.; Sahoo, B.; Bandopadhyay, S. An objective analysis of support vector machine based classification for remote sensing. *Math. Geosci.* **2008**, *40*, 409–424. [CrossRef]

70. Huang, C.; Davis, L.S.; Townshend, J.R.G. An assessment of support vector machines for land cover classification. *Int. J. Remote Sens.* **2002**, *23*, 725–749. [CrossRef]

71. Pal, M.; Mather, P.M. Some issues in the classification of dais hyperspectral data. *Int. J. Remote Sens.* **2006**, *27*, 2895–2916. [CrossRef]

72. Paneque-Gálvez, J.; Mas, J.-F.; Moré, G.; Cristóbal, J.; Orta-Martínez, M.; Luz, A.C.; Guèze, M.; Macía, M.J.; Reyes-García, V. Enhanced land use/cover classification of heterogeneous tropical landscapes using support vector machines and textural homogeneity. *Int. J. Appl. Earth Obs. Geoinf.* **2013**, *23*, 372–383. [CrossRef]

73. Mountrakis, G.; Im, J.; Ogole, C. Support vector machines in remote sensing: A review. *ISPRS J. Photogramm. Remote Sens.* **2011**, *66*, 247–259. [CrossRef]

74. Ke, Y.; Quackenbush, L.J.; Im, J. Synergistic use of quickbird multispectral imagery and LiDAR data for object-based forest species classification. *Remote Sens. Environ.* **2010**, *114*, 1141–1154. [CrossRef]

Quantifying Multi-Decadal Change of Planted Forest Cover Using Airborne LiDAR and Landsat Imagery

Xiaoyi Wang [1,2], Huabing Huang [1,2,*], Peng Gong [1,3,4,*], Gregory S. Biging [2], Qinchuan Xin [5], Yanlei Chen [2], Jun Yang [3] and Caixia Liu [1]

Academic Editors: Sangram Ganguly, Compton Tucker, Nicolas Baghdadi and Prasad S. Thenkabail

[1] State Key Laboratory of Remote Sensing Science, Institute of Remote Sensing and Digital Earth, Chinese Academy of Sciences (CAS), Beijing 100101, China; xiaoyiwangchina@gmail.com (X.W.); liucx@radi.ac.cn (C.L.)

[2] Department of Environmental Sciences, Policy & Management, University of California, Berkeley, CA 94720, USA; biging@berkeley.edu (G.S.B.); yanleichen@berkeley.ed (Y.C.)

[3] Ministry of Education Key Laboratory for Earth System Modeling, Center for Earth System Science, Tsinghua University, Beijing 100084, China; larix@tsinghua.edu.cn

[4] Joint Center for Global Change Studies, Beijing 100875, China

[5] School of Geography and Planning, Sun Yat-sen University, Guangzhou 510275, China; xqcchina@gmail.com

[*] Correspondences: huanghb@radi.ac.cn (H.H.); penggong@tsinghua.edu.cn (P.G.)

Abstract: Continuous monitoring of forest cover condition is key to understanding the carbon dynamics of forest ecosystems. This paper addresses how to integrate single-year airborne LiDAR and time-series Landsat imagery to derive forest cover change information. LiDAR data were used to extract forest cover at the sub-pixel level of Landsat for a single year, and the Landtrendr algorithm was applied to Landsat spectral data to explore the temporal information of forest cover change. Four different approaches were employed to model the relationship between forest cover and Landsat spectral data. The result shows incorporating the historic information using the temporal trajectory fitting process could infuse the model with better prediction power. Random forest modeling performs the best for quantitative forest cover estimation. Temporal trajectory fitting with random forest model shows the best agreement with validation data ($R^2 = 0.82$ and $RMSE = 5.19\%$). We applied our approach to Youyu county in Shanxi province of China, as part of the Three North Shelter Forest Program, to map multi-decadal forest cover dynamics. With the availability of global time-series Landsat imagery and affordable airborne LiDAR data, the approach we developed has the potential to derive large-scale forest cover dynamics.

Keywords: forest inventory; forest monitoring; Three-North Shelter Forest Program; afforestation; time-series

1. Introduction

Terrestrial forest ecosystems play a key role in global carbon and hydrologic cycling, and influence the climate system through biogeochemical processes [1,2]. Variation in forest cover has been recognized as an important indicator to changes in ecosystem services and functions [3]. Continuous monitoring of forest cover is essential for assessing forest growth condition and exploring effective management strategies.

Earth observing satellites, such as the Landsat series, provide frequent and consistent large-scale observations, which allow us to monitor land surface changes and supporting climate change studies [4]. Early studies on change detection of forest cover using satellite data have largely been based on multi-date analysis [5,6]. To derive trend information of forest cover change while minimizing inter-annual noise introduced by phenological differences, atmospheric interference, solar angle variation and imperfection in geometric registration and radiometric calibration, recent studies tried to detect forest cover changes using qualitative time-series analysis [7,8]. The ability to quantitatively evaluate forest cover change over a relatively long time period at Landsat's sub-pixel level is necessary for detecting the time-series tendency of changes and for providing more straightforward and detailed transient and perpetual information on the transition of forest ecosystems. However, one key challenge is that references of forest cover is often missing when using optical satellite sensor alone in large-scale studies [6].

LiDAR has proven suitable for measuring forest structural parameters and deriving forest cover with high accuracy [9–14]. Airborne LiDAR could offer a wall-to-wall solution to mapping forest cover at high spatial resolution. Owing to the high costs, the use of airborne LiDAR data is often limited in both the spatial extent and time scale. Therefore, combining airborne LiDAR and Landsat imagery provides an opportunity to take advantage of both data sources toward better mapping of forest cover change. Recent studies have built models based on LiDAR measurements to derive canopy information spatially; Chen et al. [15] adopted linear spectral mixture analyses at pixel scale. Ahmed et al. [16] investigated the object-based model with Landsat-derived Tasseled Cap Angle and spectral mixture analysis and demonstrated the optimum mean object size of 2.5 hectares. Sexton et al. [17] rescaled the 250-m MOderate-resolution Imaging Spectroradiometer Vegetation Continuous Fields to Landsat-scale and calibrated with LiDAR data. With the free availability of long time-series Landsat imagery, more efforts should be focused on the expansion on the time scale.

The Chinese government initiated the Three North Shelter Forest Program back in the 1970s, which aims to plant trees to prevent desertification. The forest cover has been reported by the government to increase thereafter in related areas, but the exact amount of the increased forest cover remains controversial. The time-series quantification of forest cover would provide detailed monitoring information. Joint use of LiDAR and Landsat time-series data could enable long-time monitoring over large areas.

The objective of this research is to develop a new approach to map time-series sub-pixel forest cover with Landsat and LiDAR data. We (1) integrate Landsat and LiDAR data for continuous sub-pixel forest cover monitoring; (2) compare four statistical modeling algorithms for deriving forest cover; and (3) provide LiDAR-based evidence for documenting forest cover change in the Three North Shelter Forest Program area over the past three decades.

2. Study Area and Data

2.1. Study Area

The research was conducted for Youyu county in northwestern Shanxi province in China (Figure 1). Youyu county is part of the Three North Shelter Forest Program in China. With continuous afforestation and ecological constructions, Youyu has now become the "fortress oasis". The forest cover of Youyu was only 0.3% before the 1950s, and the desertification area caused by soil erosion accounted for 76.2% of its total area. Residents of Youyu have started planting trees since the "Three North Shelter Forest Program" was initiated. After decades of unremitting afforestation efforts, Youyu's forest coverage has increased substantially, and the ecological environment has been improved. Youyu, therefore, represents an ideal case for forest cover studies to evaluate the Three-North Shelter Forest Program.

Youyu county is located in the forefront of Mu Us Desert, and is surrounded by mountains from which flows the Cangtou River. In this region, the elevations are generally higher in the south than in the north. The county of Youyu reaches a north-south extent of 67.7 km and east-west width of 45.7 km, with a total area of 1967 square kilometers.

Youyu has a monsoon-influenced, semi-arid continental climate, with warm humid summers, and cold dry winters. The average elevation is about 1344 m. Due to its high elevation and dry winter climate, monthly average temperature ranges from −14.4 °C to 19.5 °C in January and July, respectively. The annual precipitation is 410 millimeters, three-fourths of which occurs from June through September.

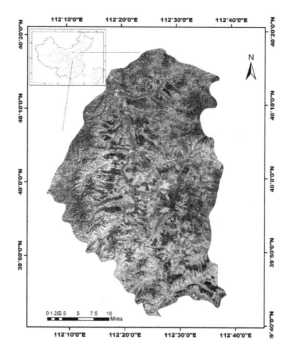

Figure 1. Youyu county in Shanxi Province, China. The imagery represent a false-color composite of one Landsat image acquired on 10 July 2009.

2.2. Data Collection and Preprocessing

2.2.1. Airborne LiDAR Data

Airborne LiDAR data were acquired during 19 to 28 September in 2009 with a Leica ALS60 laser system at an altitude of approximately 1000 m. The system was equipped with Multiple Pulses in Air (MPiA) technique, such that high density data can be obtained when flying at a high altitude.

The lateral overlaps of flight lines were set at least 5% on each side. The LiDAR sensor operated at a wavelength of 1064 nm. The recorded data have a nominal pulse spacing of 1.45 m and a point density of 0.46 pulses per m^2. Up to four returns per pulse were recorded.

2.2.2. Landsat TM and ETM+ Imagery

Our study area is covered by one Landsat path 34 row 32. We downloaded the Landsat time-series stacks between 1986 and 2013 over the study region from the United States Geological Survey (USGS) data archive, as is shown in Table 1. The downloaded data are terrain corrected as level 1T (L1T) products. Radiometric calibration and atmospheric correction were performed and the surface reflectance were acquired with Landsat Ecosystem Disturbance Adaptive Processing System (LEDAPS) [16].

To screen clouds and cloud shadows, we preprocessed the Landsat imagery using the FMASK algorithm [17]. To reduce the influences of vegetation phenology and illumination geometry on surface spectral reflectance, only images acquired during the growing season (defined as early May to late September) were used in analysis. There may still exist some phenology difference within these days, it will be further reduced with Landtrendr algorithm [8], details can be find in Section 3.2.2. There are no cloud-free images available for the years of 1988, 1991, 1992, 1996, and 1997. To ensure a high level of co-register accuracy, LiDAR data was co-registered with Landsat image in 2009. We selected 35 ground control points and performed a second-order polynomial transformation and nearest neighbor resampling. Results showed a root mean square error (RMSE) of 0.63 m.

Table 1. Landsat imagery information.

Year	Julian Day	Day/Month	Year	Julian Day	Day/Month
1986	159	9 June	2003	190	10 July
1987	258	16 September	2004	241	29 August
1989	199	19 July		161	10 June
	215	4 August	2005	251	9 September
1990	138	19 May		187	7 July
	234	23 August	2006	166	16 June
1993	258	16 September	2007	153	3 June
1994	181	1 July	2008	244	1 September
1995	264	22 September		260	17 September
1998	181	1 July	2009	190	10 July
1999	267	25 September	2010	193	13 July
2000	206	25 July	2011	228	17 August
	182	1 July	2012	239	27 August
2001	152	2 June	2013	177	27 June
2002	235	24 August		257	15 September

2.2.3. Field Experiment Data

The field experiment was carried out in 78 plots, each plot with the size of 30 m × 30 m. To reduce the location error, the location of each plot was determined according to the pixel center location of Landsat image in 2003. Within each plot tree stands share similar structure, and differentiate from its neighborhood plots. The sampling is designed to avoid spatial autocorrelation. The coverage of each plot was surveyed during 2003–2004 by line intercepts method [18], with an interval of 1 m, and only counted tree heights greater than 3 m.

3. Method

Our approach mainly includes four steps. The first step involves extracting the referenced sub-pixel forest cover from LiDAR data. By "sub-pixel forest cover", we refer to evaluating the percentage of forest within each 30 m resolution Landsat pixel. The second step builds the trajectory Landsat spectral data with the Landtrendr noise reduction and fitting method. The third step is to build the statistical relationships between sub-pixel forest cover and Landtrendr fitted spectral data, as well as original Landsat spectral data. Lastly, we predict forest cover for other years using the derived models. The process flowchart was represented in Figure 2. Four steps were shown in different background.

3.1. Extracting Reference Forest Cover

Sub-pixel forest cover in 2009 was extracted from LiDAR as the reference for forest cover evaluation over time. Since a high accuracy forest cover dataset could lay a better foundation for subsequent modeling, our strategy is to identify the forest area according to land cover dataset and maximum height, and then calculate the forest cover with a Beer's Law modified model [19].

Non-vegetated areas, including impervious, water, and agriculture areas, were masked out from the raw LiDAR point cloud with a 30 m resolution global land cover dataset named Finer Resolution Observation and Monitoring-Global Land Cover (FROM-GLC) [19]. FROM-GLC that covers our study area was built based on Landsat TM imagery of 2009. The classification results in terms of vegetation and non-vegetation areas were considered satisfactory in our visual inspection of Landsat imagery.

Since this study was designed to detect forest cover change, we excluded areas occupied by short vegetation such as shrub and grass. Following previous studies, forests are defined as area with tree heights greater than 3 meter [20]. To derive the digital elevation model (DEM) and canopy height model (CHM), the raw LiDAR point cloud was processed with RiSCAN PRO 2.0 software package (Riegl GmbH, Horn, Austria), which incorporated a multi-level iterative terrain filter [21]. If the maximum height within a Landsat pixel is less than 3 m, we consider the pixel is dominated by short vegetation and exclude the corresponding Landsat pixel for analysis. DEM data would also assist to further separate the canopy and below-canopy LiDAR point cloud.

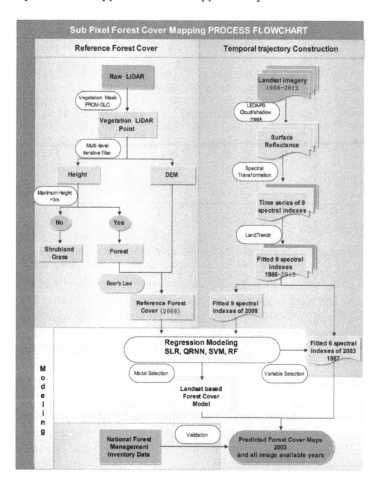

Figure 2. Flowchart for the sub-pixel forest cover process.

For the forested area, forest cover (FC) can be calculated by dividing the number of returns above a height threshold (h) by the total number of returns for each Landsat pixel within a 30-m resolution. In this research, we adopted a derivative model [19] based on Beer's law to extract forest cover reference in 2009. The potential two-way transmission loss of pulse energy was taken into account for intermediate or last returns in this LiDAR-based model. This model is less sensitive to sensor pulse repetition frequency and canopy height and, as a result, the least amount of calibration was required when applied to large areas with varied forest types. The model has shown to provide accurate forest cover as a basis for further analysis. In this model, the forest cover can be expressed as follows:

$$FC = \frac{I_h}{I_{Total}} = 1 - \sqrt{\frac{\sum I_{Ground}}{\sqrt{I_{Total}}}} = 1 - \left(\frac{\left(\frac{\sum I_{GroundSingle}}{\sum I_{Total}}\right) + \sqrt{\frac{\sum I_{GroundLast}}{\sum I_{Total}}}}{\left(\frac{\sum I_{First} + \sum I_{Single}}{\sum I_{Total}}\right) + \sqrt{\frac{\sum I_{Intermediate} + \sum I_{Last}}{\sum I_{Total}}}} \right) \quad (1)$$

In the equation above, $\sum I_{Ground}$ represented the sum of all intensity below canopy, which was determined by the canopy height threshold (see below). I_{Total} was the intensity of all returns including single, first, intermediate, and last returns. The calculated reference forest cover in 2009 was shown in result Section 4.1. In agreement with the global definition of forest, pixels with forest cover exceeding ten percent were reserved [22].

A height threshold was involved in the model to separate the ground and canopy point. This model parameter could vary with forest type and local forest condition. Ahmed and Smith employed 1.50 m and found that the results performed reasonably well [16,23]. In this study, we defined the threshold by the transition zone between the forest over-story and under-story where the least number of LiDAR returns were detected. We counted the number of returns in each vertical layer using voxel statistics with an interval of 0.1 m, and determined the appropriate threshold as 1.4 m, details can be found in supplementary Figure S1.

A stratified sampling [24] was applied to the referenced data to diminish the uncertainty of the training sample and auto-correlation. The dataset was ordered according to descending forest cover values and was split into 20% percentiles. We randomly selected half of the sample within each percentile.

3.2. Temporal Trajectory Construction

3.2.1. Spectral Indexes Extraction for the Time-Series Trajectory

A suite of spectral indexes were extracted from Landsat time-series imagery between 1986 and 2013. Each spectral index was used to build temporal trajectories for each Landsat pixel and construct empirical models with sub-pixel forest cover from LiDAR data.

We used four vegetation indexes: Normalized Difference Water Index (NDWI) [25], Normalized Burn Ratio (NBR), Normalized Difference Vegetation Index (NDVI), and Enhanced Vegetation Index (EVI). NDWI has been interpreted as an indicator of changes in water content of plant canopies, and has been used to detect forest changes [26]. NBR represents the vegetation condition and is sensitive to forest disturbance [27]. NDVI is a widely used measure of live green vegetation, and EVI complements NDVI in vegetation studies by improving the ability of forest change detection [28].

We also used five indexes based on the tasseled-cap transformation [29]: tasseled cap brightness (TCB), greenness (TCG), wetness (TCW), tasseled cap distance (TCD), and tasseled cap angle (TCA). As the names indicated, TCB, TCG, and TCW contain significant information for background reflectance, green vegetation, and wetness. TCD describes vegetation structure and composition. TCA is related to the proportion of vegetation cover within a Landsat pixel [30].

3.2.2. Temporal Fitting of Spectral Indexes Trajectories

The forest cover time-series trajectories mainly depend on its history (plantation, disturbance, and recovery). So, for each spectral index described above, we took temporal factors into account and processed the time-series trajectories with LandTrendr, a temporal segmentation and smoothing algorithm developed by Kennedy [27]. By reducing the inter-annual noise introduced by atmospheric condition, solar angle, phenology, and imperfection in geometric registration and radiometric calibration, LandTrendr was able to capture the short-term changes and maintain the long-term trends related to disturbance and recovery. As a result, the algorithm outperformed the single-date evaluation [31] or two-date change assessment [30,32] in previous studies, and could improve forest cover estimation at any given point along the temporal trajectory and enhance the forest cover change detection capability.

The spectral trajectory for each forest pixel was identified as a series of sequential straight-line segments. LandTrendr algorithm removed singular value according to the consistency before and after the singular value year, and extracted up to six straight-line segments to simplify the temporal trajectory of the Landsat time series at each pixel location. Then a fitting algorithm was applied to each segment, not only to minimize the noise coming from those aforementioned factors, but also capture the changes during all the time periods.

Briefly, the LandTrendr algorithm identified the time-series spectral trajectory as a series of sequential lines. As a result, a stack of near yearly-fitted spectral index images were obtained, and those spectral values directly taken from the fitted line were put into the forest cover estimation model as predictors. Since the first or the last year lacks part of the neighborhood information, it was difficult to detect the singular value. Therefore, we only detected forest cover change from 1987 to 2012, and neglected information from 1986 (the first year) and 2013 (the last year).

3.3. Forest Cover Modeling

We compared the result using original Landsat spectral indexes and temporal trajectory fitted spectral indexes with four types of forest cover statistic modeling algorithms. The used modeling algorithms include a stepwise linear regression (SLR) model and three machine-learning algorithms, including quantile regression neural network (QRNN), support vector machine (SVM), and regression tree random forest (RF).

SLR builds the multiple linear regressions between sub-pixel forest cover and Landsat spectral indexes iteratively using a stepwise method. For each iteration, the model is improved by adding the independent variable with the most predictive power and removing the insignificant independent variables. SLR has been used in a variety applications such as biomass estimation and wetland inundation mapping [33,34].

QRNN is the analog of linear quantile regression in the field of artificial neural network. The model uses a finite smoothing algorithm and searches for the optional left censoring, which is suitable for mixed discrete-continuous prediction (such as precipitation). The number of hidden nodes was set to four, and the hidden layer transfer function was defined as sigmoid to get the nonlinear model.

SVM attempts to determine a hyperactive plane to implement the structural risk minimization. We utilized the radial basis function (RBF) kernel with the C-SVM classifier in "caret" provided in R [35]. The particle swarm optimization (PSO) method was adopted to select the parameters. γ is the kernel parameter which defines the influence of a single training example and is set in the range of 0.01–1000. stands for the parameter for the marginal cost function, and was set in the range of 0.25–1024. All variables were scaled from −1 to 1 to speed up convergence of the regression and improve the estimation accuracy.

RF has shown good performance in vegetation height mapping [36], land cover mapping [37–39], and biomass estimation [40]. In construction of the regression trees, random forest utilized different bootstrap samples of data and changed how the regression trees are constructed as well. The node of trees is split by randomly choosing the subset of forest cover predictors and selecting the best of these. The RF algorithm is robust against over fitting and performs well relative to other classifiers [41]. Two parameters need to be settled including *numFeatures*, which means the number of features to be randomly selected at each node, and *numTrees* that means number of trees generated. We employ the suggested value of *numFeatures* as one third of number of features, and the *numTrees* was set to 500 considering the limitation of computer memory [36].

Although RF and SVM is naturally resistant to non-informative predictors, non-informative predictors can negatively affect some models [42]. We checked the importance of each variables as mean decrease in accuracy with the out of bag (OOB) samples [43]. All four methods represent similar patterns, and most of the variables play a relatively important role in each model, details can be found in the supplementary Figure S2. Then, for each regression method, the backward elimination approach was adopted, and the least important variable was thrown out until the out-of-bag prediction accuracy dropped [44]. Six out of nine variables were chosen. EVI, TCD, and TCW were discarded in further analysis.

3.4. Validation and Uncertainty Analysis

The k-fold and leave-one-out cross-validations were utilized to estimate the uncertainty of forest cover models.

The uncertainty of forest cover evaluation induced by uncertainty in predictors and model structure could be obtained from whole number of prediction with finite samples and estimated by the standard deviation of the k-fold cross-validation. In this study, ten-fold cross-validation was used. In addition, we randomly selected 5% of the entire sample as validation sample to evaluate the performance of forest cover model with scatter plots.

To assess if the forest cover model built with 2009 data was applicable to other years, the model was directly applied to the fitted Landsat spectral indexes for 2003 (as described in Section 3.2). Results were then evaluated with the Forest Management Inventory Data.

The performance of those regression models was evaluated in accordance with R^2, root mean square error (RMSEs), and mean error (ME). The optimal regression model with the lowest RMSE was employed to predict forest cover for all available years of data.

4. Results and Discussion

4.1. Performance of Forest Cover Regression Models in 2009

The sub-pixel forest cover for the LiDAR measured year in 2009 were shown in Figure 3. The result were extracted from airborne LiDAR data using the Beer's Law modified method. Non-vegetation area masked from FROM-GLC dataset were shown in white.

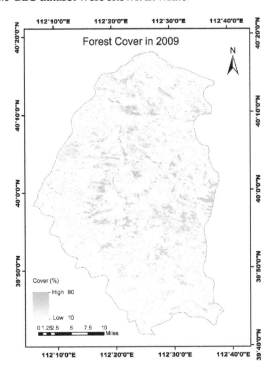

Figure 3. Sub-pixel forest cover estimated from airborne LiDAR data in 2009.

Four forest cover regression models were evaluated according to the R^2, RMSE and ME in Table 2. We compared the result with original spectral indexes (O) and temporal trajectory fitting (F).

Notably, predictors with time-series processing outperform the original spectral indexes for each regression model in varying degrees.

As can be seen from the comparison of modeling results in Table 2, the SLR model performed poor with R^2 of 0.59. SLR assumes a global linear relationship between the spectral indexes and forest cover, which may not be suitable for forest cover modeling. The use of non-linear regression may be more appropriate.

Although QRNN did not perform as well as expected in terms of its R^2 value (0.65), the model achieved a relatively low RMSE of 7.84. We assume that the result is due to the lack of efficient tuning for better model parameters.

Table 2. Modeling comparison with different algorithm in 2009. The comparison was performed with original spectral indexes (O) and temporal trajectory (T), separately. SD stands for standard derivation. The best results were marked in bold.

		\multicolumn{5}{c}{Results with Ten-Fold Cross-Validation}				
		R^2	RMSE	SD_R^2	SD_RMSE	ME
SLR	O	0.50	9.29	0.034	0.34	−0.006
	T	0.59	8.48	0.027	0.17	−0.002
QRNN	O	0.54	8.75	0.033	0.29	−0.021
	T	0.65	7.84	0.022	0.29	−0.042
SVM	O	0.68	7.47	0.034	0.37	−0.746
	T	0.73	6.08	0.024	0.32	−0.486
RF	O	0.72	7.26	0.026	0.32	0.033
	T	**0.82**	**5.19**	0.015	0.20	−0.010

With respect to the SVM method, it achieved acceptable RMSE and R^2 values, but its standard deviation and ME values were the greatest among the four methods.

Random forest, on the other hand, generated well-predicted forest cover in terms of both RMSE and R^2 values. RF regression has the ability to recognize complex non-parametric relationships. As a result, random forest is robust to the shortage of reference data. Furthermore, the impact caused by the distribution and uncertainty of training sample would be reduced because of its bootstrapping strategy [34].

Based on the ten-fold cross validation results, we further investigated the performance of RF through scatter plots in Figure 4. To show the distribution of validation data, we set aside five percent of the random sample to use as a training set. The validation was performed 10 times. The regression line revealed that the predicted results fit well with the cover calculated from LiDAR data.

Compared to the time-series fitted model in Figure 4, the RF model based on single-date Landsat data (Figure 4b) performed relatively well with an $R^2 = 0.72 \pm 0.026$ and $RMSE = 7.26\% \pm 0.320\%$. Including the temporal trajectory fitting process (Figure 4a) for the RF model substantially improved the overall performance ($R^2 = 0.82 \pm 0.015$ and $RMSE = 5.19\% \pm 0.204\%$). The fitted spectral indexes incorporating the historic information infuse the model with better prediction power and yield better results of lower bias and higher correlation, which was confirmed by Powell *et al.* [30] and Sulla-Menashe *et al.* [45]. The uncertainty assessed by standard derivation of the ten-fold cross-validation, indicates that the uncertainty of the original spectral indexes is larger by one-third than with the temporal trajectory fitting approach. The scatterplots revealed an improvement in the prediction particularly in the high and low forest cover range, increasing the dynamic range of the predictions.

Compared with a global, Landsat-based tree cover dataset [17] produced by rescaling of MODIS VCF with a piecewise linear function, our result performed well for high forest cover range. The rescaling result retained the saturation artifact of the MODIS VCF at greater than or equal to 80% tree cover, and shown negative bias at high cover [46]. Since no LiDAR data available to calibrate result in China, the rescaling result appears problematic when checked at the local scale, with the maximum forest cover of 29% in 2000 for our study area.

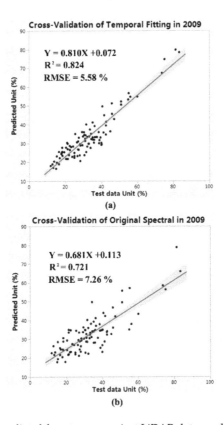

Figure 4. Scatter plots of predicted forest cover against LiDAR data results. (**a**) Validation of forest cover with temporal trajectory fitting algorithm; (**b**) validation of forest cover with original spectral indices. The blue area represents the 95% confidence intervals for the regression line.

4.2. Performance of Trajectory Landsat Imagery in Forest Cover Mapping

With respect to modeling algorithms, RF with temporal-trajectory fitted spectral indexes performs the best in terms of coefficient of determination (R^2) and prediction errors (RMSE), and RF was chosen to predict the forest cover for all Landsat imagery available each year.

The results in Figure 5 shown that the forest cover model predict forest cover along the temporal trajectory effectively. By coupling LiDAR analysis with Landsat spectral indices, this allowed us to estimate crown closure over time. In the following discussion, this model was applied to all available years to extract time-series forest cover condition and forest cover change.

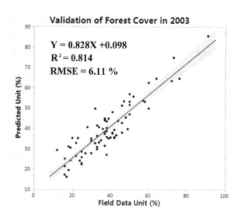

Figure 5. Scatterplot of predicted forest cover against field data in 2003.

4.3. Forest Cover Change

Forest cover changes were obtained from 1987 to 2012 as shown in Figure 6. Extrapolating to the county-level, the average forest cover represents a significant increase from the 1987 estimate of 15.5% to the 2012 estimate of 37.8%, under the combined change of natural forest and plantation. This finding agrees well with the results of Liu and Gong [47].

The plantation and native forest area were separated with visual inspection of high-resolution images in Google Earth. The plantation area represents a regular pattern while native forest is naturally spread and not thickened, which shows natural and irregular pattern. The plantation was mainly implemented in the center of Youyu county along the Cangtou River, thus largely promoting the accumulation of forest cover. As can be seen from Figure 6a,d, forest cover in these plantation areas increases dramatically. Nonetheless, forest cover in the southwest area (shown in Figure 6c) stays the same, and tree crown does not show significant growth.

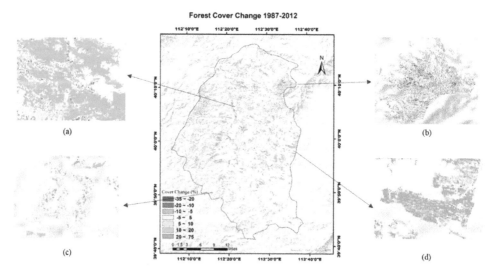

Figure 6. Forest cover change from 1987 to 2012. (a) Plantation area with forest cover increases; (b) native forest; (c) plantation area with unchanged forest cover; and (d) plantation area.

The native forest represents different pattern was shown in Figure 6b, the difference is probably due to various water-heat combinations [48].

We then examined successive forest cover trends across the time series, revealing the status and change along the time. As can be seen from Figures 7 and 8 forest cover during the designated period revealed different trends.

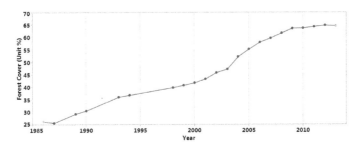

Figure 7. Annual forest cover condition and dynamic for plantation area.

For the plantation area shown in Figure 6d, forest cover was on a constant rise as illustrated in Figure 7. The dynamics of forest cover over time resembles a logistic (S-shaped) growth curve, which agree with the results of managed forest by Dewar and Roderick [49]. After 20 years of growth, the growth rate slows down when resources are limited, and forest cover reaches a stable level. This growth period is roughly consistent with the local response to the national policy of "Three North Shelter Forest Program" for Shanxi Province. For mature forest, with the stand age of 20 to 30 years for aspen [50], additional research is suggested on the mortality of trees for the next 5 to 10 years and the corresponding strategy to deal with the anticipated mortality, as in investigated by Esseen *et al.* [51], and Monserud [52] for the old-growth forest.

For the native forest area shown in Figure 6b, various patterns could be detected. This is area where crown cover is being established. Figure 8a,b, represent two different patterns of forest cover—decreasing or fluctuating. For the cover loss area, there is a slow increase before 1998, which gradually drops afterwards. Although a decease was detected, it remains relatively high in coverage. For the area whose crown cover fluctuates, additional research is suggested on the driving factor for the variation in forest cover, as is presented by Crowther *et al.* [53].

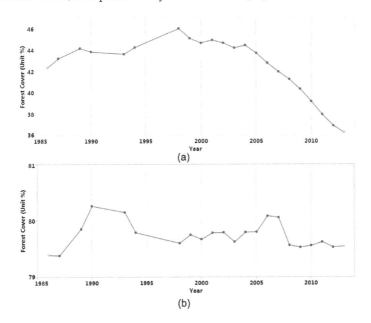

Figure 8. Forest cover dynamic for native forest area from 1987 to 2012. (**a**) Decrease; and (**b**) fluctuation.

4.4. Limitations

In many areas of China, there was a lack of Landsat imagery in the 1990s. In addition to the Landsat 7 ETM+ gaps caused by the failure of the Scan Line Corrector (SLC), cloud and its shadow make it even harder to obtain sufficient images for each year. Adding the Landsat MSS dataset would also be helpful to lengthen the temporal detection period, while the comparability introduced by system-inherent differences should be further examined.

As can be seen from Figure 7, plantation follows a logistic S-shaped curve. The logistic curve can be approximated fairly well by several straight-line segments. However, there is still room for improvement in fitting the model and in automatic parameter selection. A series of piecewise logistic functions can be used to represent intra-annual vegetation dynamics, as the previous study applied to MODIS time series data [54]. When applied to global scale, a method should be processed without setting thresholds or empirical constants [55].

The sampling strategy would violation of independence of the data [16], and sample size would influence the regression result [40]. More efforts should focused on the design of sampling.

5. Conclusions

Deriving historical and up-to-date forest cover information is important for forest management and monitoring. In this research, we developed a new approach to map time-series sub-pixel forest cover with Landsat and LiDAR data, and we further evaluated this method for deriving change information in forest cover in the Northeastern China county of Youyu.

Our results show that the inclusion of the Landtrendr temporal trajectory fitting process considerably increased the model performance versus the original spectral indices with better results of lower bias and higher correlation. In comparing regression methods, we found that random forest outperformed the SVM, NN, and SLR in quantitative forest cover estimation. A random forest model incorporating temporal trajectory fitting yields the best agreement with validation data ($R^2 = 0.82$ and $RMSE = 5.19\%$).

Our results revealed that forest cover rose from 15.5% in 1987 to 37.8% in 2012—an increase by 144.1%. Accordingly, the "Three North Shelter Forest Program" obtained impressive achievements with their afforestation program in this arid region.

More focused research is suggested to explore better sigmoid curve fitting along the time-series trajectories, aimed at better depicting forest status. Furthermore, future research should focus on the climate response to forest plantation and forest change over long periods of time, especially the mortality of trees and corresponding management planning.

Acknowledgments: This research was partially supported by Special Fund for Meteorology Scientific Research in the Public Welfare (GYHY201506010), Youth Innovation Promotion Association CAS, National Natural Science Foundation of China (Grant No. 41401484), and China Scholarship Council (Project reference number 201404910437). The authors also acknowledge the valuable discussions with Robert A. Monserud at Rocky Mountain and Pacific Northwest Research Stations, USDA Forest Service. We would also like to thank three anonymous reviewers for insightful comments and improvements to the manuscript.

Author Contributions: All authors have made significant contributions to the manuscript. Xiaoyi Wang, Huabing Huang and Peng Gong conceived and designed the experiments; Xiaoyi Wang and Huabing Huang performed the experiments; Xiaoyi Wang and Gregory S. Biging analyzed the data; Qinchuan Xin , Yanlei Chen, Jun Yang and Caixia Liu contributed in manuscript revision. Gregory S. Biging and Peng Gong contributed in the language editing.

Conflicts of Interest: The authors declare no conflict of interest.

References

1.   Bonan, G.B. Forests and climate change: Forcings, feedbacks, and the climate benefits of forests. *Science* **2008**, *320*, 1444–1449. [CrossRef] [PubMed]

2. Pan, Y.; Birdsey, R.A.; Fang, J.; Houghton, R.; Kauppi, P.E.; Kurz, W.A.; Phillips, O.L.; Shvidenko, A.; Lewis, S.L.; Canadell, J.G. A large and persistent carbon sink in the world's forests. *Science* **2011**, *333*, 988–993. [CrossRef] [PubMed]

3. Hansen, M.C.; Potapov, P.V.; Moore, R.; Hancher, M.; Turubanova, S.; Tyukavina, A.; Thau, D.; Stehman, S.; Goetz, S.; Loveland, T. High-resolution global maps of 21st-century forest cover change. *Science* **2013**, *342*, 850–853. [CrossRef] [PubMed]

4. Yang, J.; Gong, P.; Fu, R.; Zhang, M.; Chen, J.; Liang, S.; Xu, B.; Shi, J.; Dickinson, R. The role of satellite remote sensing in climate change studies. *Nat. Clim. Chang.* **2013**, *3*, 875–883. [CrossRef]

5. Tokola, T.; Löfman, S.; Erkkilä, A. Relative calibration of multitemporal landsat data for forest cover change detection. *Remote Sens. Environ.* **1999**, *68*, 1–11. [CrossRef]

6. Townshend, J.R.; Masek, J.G.; Huang, C.; Vermote, E.F.; Gao, F.; Channan, S.; Sexton, J.O.; Feng, M.; Narasimhan, R.; Kim, D. Global characterization and monitoring of forest cover using Landsat data: Opportunities and challenges. *Int. J. Digit. Earth* **2012**, *5*, 373–397. [CrossRef]

7. Schroeder, T.A.; Cohen, W.B.; Yang, Z. Patterns of forest regrowth following clearcutting in western oregon as determined from a Landsat time-series. *For. Ecol. Manag.* **2007**, *243*, 259–273. [CrossRef]

8. Kennedy, R.E.; Yang, Z.; Cohen, W.B.; Pfaff, E.; Braaten, J.; Nelson, P. Spatial and temporal patterns of forest disturbance and regrowth within the area of the northwest forest plan. *Remote Sens. Environ.* **2012**, *122*, 117–133. [CrossRef]

9. Wang, X.; Huang, H.; Gong, P.; Liu, C.; Li, C.; Li, W. Forest canopy height extraction in rugged areas with ICESAT/GLAS data. *IEEE Trans. Geosci. Remote Sens.* **2014**, *52*, 4650–4657. [CrossRef]

10. Asner, G.P.; Mascaro, J.; Muller-Landau, H.C.; Vieilledent, G.; Vaudry, R.; Rasamoelina, M.; Hall, J.S.; van Breugel, M. A universal airborne Lidar approach for tropical forest carbon mapping. *Oecologia* **2012**, *168*, 1147–1160. [CrossRef] [PubMed]

11. Chen, Q. Retrieving vegetation height of forests and woodlands over mountainous areas in the pacific coast region using satellite laser altimetry. *Remote Sens. Environ.* **2010**, *114*, 1610–1627. [CrossRef]

12. Huang, H.; Gong, P.; Cheng, X.; Clinton, N.; Li, Z. Improving measurement of forest structural parameters by co-registering of high resolution aerial imagery and low density Lidar data. *Sensors* **2009**, *9*, 1541–1558. [CrossRef] [PubMed]

13. Liu, C.; Huang, H.; Gong, P.; Wang, X.; Wang, J.; Li, W.; Li, C.; Li, Z. Joint use of ICESAT/GLAS and Landsat data in land cover classification: A case study in Henan province, China. *IEEE J. Sel. Top. Appl. Earth Obs. Remote Sens.* **2015**, *8*, 511–522. [CrossRef]

14. Huang, H.; Liu, C.; Wang, X.; Biging, G.S.; Yang, J.; Gong, P. Mapping vegetation heights in China with remotely sensed data. *ISPRS J. Photogram. Remote Sens.* **2016**, Submitted.

15. Chen, X.; Vierling, L.; Rowell, E.; DeFelice, T. Using Lidar and effective LAI data to evaluate Ikonos and Landsat 7 ETM+ vegetation cover estimates in a ponderosa pine forest. *Remote Sens. Environ.* **2004**, *91*, 14–26. [CrossRef]

16. Ahmed, O.S.; Franklin, S.E.; Wulder, M.A. Integration of Lidar and Landsat data to estimate forest canopy cover in coastal British Columbia. *Photogram. Eng. Remote Sens.* **2014**, *80*, 953–961. [CrossRef]

17. Sexton, J.O.; Song, X.-P.; Feng, M.; Noojipady, P.; Anand, A.; Huang, C.; Kim, D.-H.; Collins, K.M.; Channan, S.; DiMiceli, C. Global, 30-m resolution continuous fields of tree cover: Landsat-based rescaling of MODIS vegetation continuous fields with Lidar-based estimates of error. *Int. J. Digit. Earth* **2013**, *6*, 427–448. [CrossRef]

18. Zhou, Q.; Robson, M.; Pilesjo, P. On the ground estimation of vegetation cover in australian rangelands. *Int. J. Remote Sens.* **1998**, *19*, 1815–1820. [CrossRef]

19. Hopkinson, C.; Chasmer, L. Testing Lidar models of fractional cover across multiple forest ecozones. *Remote Sens. Environ.* **2009**, *113*, 275–288. [CrossRef]

20. Koeln, G.T.; Jones, T.; Melican, J. Geocover Lc: Generating Global Land Cover from 7600 Frames of Landsat TM Data. In Proceedings of the ASPRS 2000 Annual Conference, Washington, DC, USA, 21–26 May 2000.

21. Véga, C.; Durrieu, S. Multi-level filtering segmentation to measure individual tree parameters based on Lidar data: Application to a mountainous forest with heterogeneous stands. *Int. J. Appl. Earth Obs. Geoinf.* **2011**, *13*, 646–656. [CrossRef]

22. FAO. *Global Forest Resources Assessment 2000: Main Report*; Food and Agriculture Organization of the United Nations: Rome, Italy, 2000.

23. Smith, A.; Falkowski, M.J.; Hudak, A.T.; Evans, J.S.; Robinson, A.P.; Steele, C.M. A cross-comparison of field, spectral, and Lidar estimates of forest canopy cover. *Can. J. Remote Sens.* **2009**, *35*, 447–459. [CrossRef]

24. Trost, J.E. Statistically nonrepresentative stratified sampling: A sampling technique for qualitative studies. *Qual. Sociol.* **1986**, *9*, 54–57. [CrossRef]

25. Gao, B.-C. NDWI—A normalized difference water index for remote sensing of vegetation liquid water from space. *Remote Sens. Environ.* **1996**, *58*, 257–266. [CrossRef]

26. Jin, S.M.; Sader, S.A. Comparison of time series tasseled cap wetness and the normalized difference moisture index in detecting forest disturbances. *Remote Sens. Environ.* **2005**, *94*, 364–372. [CrossRef]

27. Kennedy, R.E.; Yang, Z.; Cohen, W.B. Detecting trends in forest disturbance and recovery using yearly landsat time series: 1. LandTrendr—Temporal segmentation algorithms. *Remote Sens. Environ.* **2010**, *114*, 2897–2910. [CrossRef]

28. Huete, A.; Didan, K.; Miura, T.; Rodriguez, E.P.; Gao, X.; Ferreira, L.G. Overview of the radiometric and biophysical performance of the MODIS vegetation indices. *Remote Sens. Environ.* **2002**, *83*, 195–213. [CrossRef]

29. Crist, E.P. A TM tasseled cap equivalent transformation for reflectance factor data. *Remote Sens. Environ.* **1985**, *17*, 301–306. [CrossRef]

30. Powell, S.L.; Cohen, W.B.; Healey, S.P.; Kennedy, R.E.; Moisen, G.G.; Pierce, K.B.; Ohmann, J.L. Quantification of live aboveground forest biomass dynamics with Landsat time-series and field inventory data: A comparison of empirical modeling approaches. *Remote Sens. Environ.* **2010**, *114*, 1053–1068. [CrossRef]

31. Liang, L.; Chen, Y.; Hawbaker, T.J.; Zhu, Z.; Gong, P. Mapping mountain pine beetle mortality through growth trend analysis of time-series Landsat data. *Remote Sens.* **2014**, *6*, 5696–5716. [CrossRef]

32. Pflugmacher, D.; Cohen, W.B.; Kennedy, R.E.; Yang, Z. Using landsat-derived disturbance and recovery history and Lidar to map forest biomass dynamics. *Remote Sens. Environ.* **2014**, *151*, 124–137. [CrossRef]

33. Zhang, Y.; Liang, S.; Sun, G. Forest biomass mapping of northeastern China using GLAS and MODIS data. *IEEE J. Sel. Top. Appl. Earth Observ. Remote Sens.* **2014**, *7*, 140–152. [CrossRef]

34. Huang, C.; Peng, Y.; Lang, M.; Yeo, I.-Y.; McCarty, G. Wetland inundation mapping and change monitoring using Landsat and airborne Lidar data. *Remote Sens. Environ.* **2014**, *141*, 231–242. [CrossRef]

35. Kuhn, M. Building predictive models in R using the caret package. *J. Stat. Softw.* **2008**, *28*, 1–26. [CrossRef]

36. Simard, M.; Pinto, N.; Fisher, J.B.; Baccini, A. Mapping forest canopy height globally with spaceborne Lidar. *J. Geophys.Res. Biogeosci* **2011**, *116*, G04021. [CrossRef]

37. Gong, P.; Wang, J.; Yu, L.; Zhao, Y.; Zhao, Y.; Liang, L.; Niu, Z.; Huang, X.; Fu, H.; Liu, S.; *et al.* Finer resolution observation and monitoring of global land cover: First mapping results with landsat TM and ETM+ data. *Int. J. Remote Sens.* **2013**, *34*, 2607–2654. [CrossRef]

38. Hao, P.; Zhan, Y.; Wang, L.; Niu, Z.; Shakir, M. Feature selection of time series MODIS data for early crop classification using random forest: A case study in Kansas, USA. *Remote Sens.* **2015**, *7*, 5347–5369. [CrossRef]

39. Hao, P.; Wang, L.; Niu, Z. Potential of multitemporal Gaofen-1 panchromatic/multispectral images for crop classification: Case study in Xinjiang uygur autonomous region, China. *J. Appl. Remote Sens.* **2015**, *9*. [CrossRef]

40. Fassnacht, F.; Hartig, F.; Latifi, H.; Berger, C.; Hernández, J.; Corvalán, P.; Koch, B. Importance of sample size, data type and prediction method for remote sensing-based estimations of aboveground forest biomass. *Remote Sens. Environ.* **2014**, *154*, 102–114. [CrossRef]

41. Fernández-Delgado, M.; Cernadas, E.; Barro, S.; Amorim, D. Do we need hundreds of classifiers to solve real world classification problems? *J. Mach. Learn. Res.* **2014**, *15*, 3133–3181.

42. Ishwaran, H. Variable importance in binary regression trees and forests. *Electron. J. Stat.* **2007**, *1*, 519–537. [CrossRef]

43. Breiman, L. *Out-of-Bag Estimation*; Technical Report; Citeseer: Berkeley, CA, USA, 1996.

44. Díaz-Uriarte, R.; De Andres, S.A. Gene selection and classification of microarray data using random forest. *BMC Bioinf.* **2006**, *7*. [CrossRef]

45. Sulla-Menashe, D.; Kennedy, R.E.; Yang, Z.; Braaten, J.; Krankina, O.N.; Friedl, M.A. Detecting forest disturbance in the Pacific northwest from MODIS time series using temporal segmentation. *Remote Sens. Environ.* **2014**, *151*, 114–123. [CrossRef]

46. Alexander, C.; Bøcher, P.K.; Arge, L.; Svenning, J.-C. Regional-scale mapping of tree cover, height and main phenological tree types using airborne laser scanning data. *Remote Sens. Environ.* **2014**, *147*, 156–172. [CrossRef]

47. Liu, S.; Gong, P. Change of surface cover greenness in China between 2000 and 2010. *Chin. Sci. Bull.* **2012**, *57*, 2835–2845. [CrossRef]

48. Niinemets, Ü. Responses of forest trees to single and multiple environmental stresses from seedlings to mature plants: Past stress history, stress interactions, tolerance and acclimation. *For. Ecol. Manag.* **2010**, *260*, 1623–1639. [CrossRef]

49. Dewar, R.C. Analytical model of carbon storage in the trees, soils, and wood products of managed forests. *Tree Physiol.* **1991**, *8*, 239–258. [CrossRef] [PubMed]

50. Gao, Y.; Yuan, Y.; Liu, S.; Wang, Y.; Liu, L. Allocation of fine root biomass and its response to nitrogen deposition in poplar plantations with different stand ages. *Chin. J. Ecol.* **2007**, *23*, 185–189.

51. Esseen, P.-A. Tree mortality patterns after experimental fragmentation of an old-growth conifer forest. *Biol. Conserv.* **1994**, *68*, 19–28. [CrossRef]

52. Monserud, R.A.; Sterba, H. Modeling individual tree mortality for austrian forest species. *For. Ecol. Manag.* **1999**, *113*, 109–123. [CrossRef]

53. Crowther, T.; Glick, H.; Covey, K.; Bettigole, C.; Maynard, D.; Thomas, S.; Smith, J.; Hintler, G.; Duguid, M.; Amatulli, G. Mapping tree density at a global scale. *Nature* **2015**, *525*, 201–205. [CrossRef] [PubMed]

54. Zhang, X.; Friedl, M.A.; Schaaf, C.B.; Strahler, A.H.; Hodges, J.C.; Gao, F.; Reed, B.C.; Huete, A. Monitoring vegetation phenology using MODIS. *Remote Sens. Environ.* **2003**, *84*, 471–475. [CrossRef]

55. Pettorelli, N.; Vik, J.O.; Mysterud, A.; Gaillard, J.-M.; Tucker, C.J.; Stenseth, N.C. Using the satellite-derived NDVI to assess ecological responses to environmental change. *Trends Ecol. Evol.* **2005**, *20*, 503–510. [CrossRef] [PubMed]

3

Application of Synthetic NDVI Time Series Blended from Landsat and MODIS Data for Grassland Biomass Estimation

Binghua Zhang [1,2,3], Li Zhang [1,2,*], Dong Xie [1,4], Xiaoli Yin [1], Chunjing Liu [1,5] and Guang Liu [1]

Academic Editors: Clement Atzberger and Prasad S. Thenkabail

[1] Key Laboratory of Digital Earth Science, Institute of Remote Sensing and Digital Earth, Chinese Academy of Sciences, No. 9 Dengzhuang South Road, Beijing 100094, China; zhangbh@radi.ac.cn (B.Z.); xiedong@gwmail.gwu.edu (D.X.); yinxiaoli6525@163.com (X.Y.); liuchunjing1127@163.com (C.L.); liuguang@radi.ac.cn (G.L.)

[2] Hainan Key Laboratory of Earth Observation, Hainan 572029, China

[3] College of Resources and Environment, University of Chinese Academy of Sciences, No. 19A Yuquan Road, Beijing 100049, China

[4] Department of Mathematics, The George Washington University, 2115 G St. NW, Washington, DC 20052, USA

[5] College of Information Science and Engineering, Shandong Agricultural University, No. 61 Daizong Road, Taian 271018, China

* Correspondence: zhangli@radi.ac.cn

Abstract: Accurate monitoring of grassland biomass at high spatial and temporal resolutions is important for the effective utilization of grasslands in ecological and agricultural applications. However, current remote sensing data cannot simultaneously provide accurate monitoring of vegetation changes with fine temporal and spatial resolutions. We used a data-fusion approach, namely the spatial and temporal adaptive reflectance fusion model (STARFM), to generate synthetic normalized difference vegetation index (NDVI) data from Moderate-Resolution Imaging Spectroradiometer (MODIS) and Landsat data sets. This provided observations at fine temporal (8-d) and medium spatial (30 m) resolutions. Based on field-sampled aboveground biomass (AGB), synthetic NDVI and support vector machine (SVM) techniques were integrated to develop an AGB estimation model (SVM-AGB) for Xilinhot in Inner Mongolia, China. Compared with model generated from MODIS-NDVI (R^2 = 0.73, root-mean-square error (RMSE) = 30.61 g/m^2), the SVM-AGB model we developed can not only ensure the accuracy of estimation (R^2 = 0.77, RMSE = 17.22 g/m^2), but also produce higher spatial (30 m) and temporal resolution (8-d) biomass maps. We then generated the time-series biomass to detect biomass anomalies for grassland regions. We found that the synthetic NDVI-derived estimations contained more details on the distribution and severity of vegetation anomalies compared with MODIS NDVI-derived AGB estimations. This is the first time that we have generated time series of grassland biomass with 30-m and 8-d intervals data through combined use of a data-fusion method and the SVM-AGB model. Our study will be useful for near real-time and accurate (improved resolutions) monitoring of grassland conditions, and the data have implications for arid and semi-arid grasslands management.

Keywords: biomass; data fusion; STARFM; MODIS; Landsat; support vector machine (SVM)

1. Introduction

Grasslands are of considerable global importance because they are important sites for biodiversity and serve as energy suppliers for mankind (*i.e.*, production of agricultural commodities such as hay and livestock); grassland conservation is a particular concern in China because of land use changes due to climate variability and the country's continued population growth [1]. Dynamic pasture growth information is of great significance for pasture capacity enhancements, the development of appropriate management strategies and rotational grazing plans, regional livestock production, and scientific evaluations of ecological benefits. In grassland ecosystems, the amount of biomass represents primary production and determines the herbivore carrying capacity [2]. It can also reflect the health status of grassland ecosystems. Therefore, accurate estimates of grassland biomass are critical for their management.

Field surveys are the most reliable method to obtain accurate grassland biomass data, but these surveys are extremely time-consuming and labor-intensive over large areas, especially in remote regions. Satellite platforms offer an effective way to collect data over large areas [3]. The normalized difference vegetation index (NDVI) is the most widely known vegetation index that can be computed from remotely sensed data (e.g., Landsat and Moderate-Resolution Imaging Spectroradiometer (MODIS) data), and the NDVI has been shown to be a sufficiently stable indicator for monitoring the intra- and inter-annual variations of vegetation greenness [4]. However, technical and financial constraints often limit the remote sensing instruments' ability to acquire data with high spatial and temporal resolutions simultaneously [5]. For example, remotely sensed images acquired from Landsat satellites have a spatial resolution of 30 m for multispectral bands. However, its long revisit cycle of 16 d, frequent cloud contamination, and other poor atmospheric conditions that are often encountered [6–8] limit the uses of Landsat data. In contrast, MODIS data can provide highly frequent (daily) observations. However, their coarse spatial resolutions ranging from 250 to 1000 m restrict their applications in relatively small and heterogeneous landscapes. Thus, integrating information from different remote sensing sensors with image fusion techniques could be a feasible and inexpensive way to better capture land surface characteristics [9].

Traditional image fusion methods such as intensity-hue-saturation (IHS) transformation [10], principal component analysis [11], and wavelet decomposition [12] are not effective for generating synthetic observations with enhanced spatiotemporal resolution [12,13]. While these traditional methods can be used to combine high spatial resolution panchromatic images and spectral information for coarse pixel images, they cannot generate time series of reflectance with both high spatial and temporal information [14]. Thus, the panchromatic images can only help to enhance the spatial resolution [15]. In order to generate reflectance data with both high spatial and temporal resolution, Gao et al. [14] developed a Spatial and Temporal Adaptive Reflectance Fusion Model (STARFM) for the prediction of daily surface reflectance through blending Landsat and MODIS data. The STARFM algorithm has been applied successfully in several studies involved with the monitoring of seasonal changes in surface features [16–19], improving the accuracy of classifications [20], and evaluating gross primary productivity (GPP) [21]. Through adopting the synthetic reflectance series generated by STARFM, Singh [22] found that the GPP derived from synthetic reflectance data could effectively capture the green-up and leaf down dates of croplands over the growing season. Bhandari et al. [23] fused MODIS and Landsat data through the use of STARFM and generated synthetic surface reflectance data to replace low-quality Landsat thematic mapper (TM) images contaminated by clouds. Consequently, they acquired a long 5-yr time series at Landsat resolution. They found that this Landsat image series, which had resolutions of 16 d and 30 m, could be effectively used to monitor vegetation phenology in broadleaf forest environments. Senf et al. [24] performed land cover classifications by detecting the phenology patterns of different vegetation with the synthetic NDVI series and found that synthetic images could be used to compensate for missing Landsat data and increase the classification accuracy compared to that with the use of single-date Landsat data. Based on the design of STARFM, several new methods and modifications have been developed, including the sparse-representational-based spatiotemporal reflectance fusion model (SPSTFM) [25], enhanced spatial and temporal adaptive reflectance fusion model (ESTARFM) [15], spatial temporal adaptive algorithm for mapping reflectance change (STAARCH) [26], spatial and temporal reflectance unmixing model (STRUM) [27], and other fusion algorithms [13,28–30]. Most of these methods need more than one image pair as inputs. However, it is very hard to acquire many scenes of TM data without cloud contamination. The STARFM is the most widely used method and can work with a single pair of fine and coarse images. Moreover, the STARFM has achieved satisfactory results in many studies [18,22,23,31,32], which have demonstrated its practicability.

Various methods have been applied with remotely sensed data for estimating grassland biomass, and these methods include multiple regression analysis [33], K-nearest neighbor [34], artificial neural network (ANN) [35], and support vector machine (SVM) techniques. Artificial neural network, because of its ability to handle complex nonlinear functions, has been successfully used in various land observation applications including biomass estimations [35–37], land classifications [38,39], and land change detection [40,41]. However, the ANN model would not produce guaranteed accuracy when there are not enough samples and it is easy to over-fit the data when there are excessive numbers of samples [42]. Support vector machine is based on the structural risk minimization principle and isgood at solving practical problems involving small numbers of training samples, nonlinearity, high numbers of dimensions, and local minima; thus, SVM is considered to be a good alternative to ANN. Several studies have demonstrated that biomass estimation based on a SVM model can provide robust results [43–45].

The objective of this study was to accurately estimate aboveground biomass (AGB) for grasslands in arid and semi-arid regions of Inner Mongolia, China. To achieve this goal, we first derived synthetic NDVI time series with high temporal and spatial resolution by blending the high-temporal MODIS data and the medium-spatial Landsat data with a data-fusion approach. We then developed different biomass models and integrated the synthetic NDVI products to acquire the optimal grassland AGB estimation model for accurate estimates of AGB (8-d intervals and 30-m resolutions). To further explore the practicability of our model, we calculated biomass anomalies with the optimal grassland AGB estimation model and further compared the results between the synthetic NDVI-derived biomass and the MODIS NDVI-derived biomass data. Our approach detailed in this study will be useful for near real-time and accurate (high spatial and temporal resolutions) monitoring of grasslands conditions. Such data have the potential to help grassland managers utilize grassland resources more effectively.

2. Study Area and Data

2.1. Study Area

Xilinhot (Figure 1) is located in the hinterland of Xilingol, and covers an area of about 15,000 km^2 (43°02′–44°52′N, 115°13′–117°06′E). The elevation in this area shows a decreasing gradient from the southeast to the northwest. This area is characterized by a typical temperate continental semi-arid climate. Winters are cold and dry, where conditions are influenced by the air flow from the Mongolian plateau, whereas summers are wet and warm, where conditions are influenced by monsoons. This area has an average annual temperature of 0.5–1.0 °C and an annual mean precipitation of 350 mm. Precipitation is often distributed unevenly in the region, and droughts occur frequently [46]. In addition, 90% of this area is dominated by temperate steppe, and the typical growing season is from May to September. Grassland biomass shows a decreasing trend from southeast to northwest that follows the elevation transition in the study area [35,47].

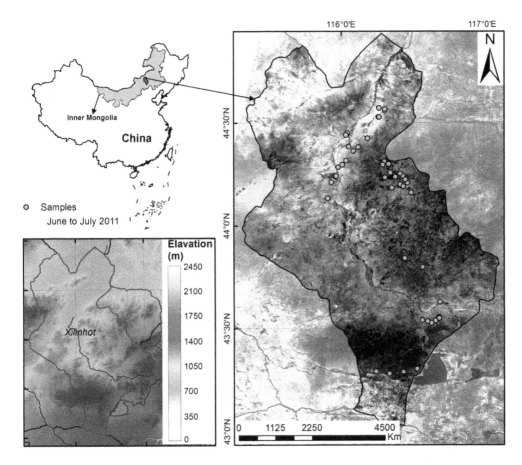

Figure 1. The location of Xilinhot. The elevation in the left-bottom corner was generated by using Shuttle Radar Topography Mission (SRTM) data (http://srtm.csi.cgiar.org/). The image on the right is the RGB (red, green, blue) composite of Landsat bands 5, 4, and 3 overlaid with field locations (orange dots) where samples were collected in 2011.

2.2. Field Data

We sampled 68 samples during the growing season of 2011 (10 June to 15 June and 20 July to 25 July), and sample locations are indicated with the orange dots in Figure 1. Distances between samples were at least 1.5 km to avoid spatial auto-correlation. The size of each sample was 300 m × 300 m to provide overlapping coverage for the corresponding MODIS pixel (250 m × 250 m). Given the high cost associated with sampling biomass in such large sample areas, and the fact that endemic grass species are not very heterogeneous within the 300 m × 300 m plots, we subsampled the large sample areas by using five 1 m × 1 m quadrats (one in the center and four about 100–150 m away from the center); the data from the subsamples were then used to represent the grassland growth conditions of the corresponding sample area. Specifically, plants in each 1 m × 1 m quadrant were clipped to the ground surface and then dried at 70 °C to obtain the dry weight. The average dry weight of the five 1 m × 1 m quadrats was calculated and used to represent the mean biomass of the 300 m × 300 m sample [48].

2.3. Remote Sensing Data

Landsat TM and MODIS data were used to generate the primary inputs for the STARFM fusion model. We obtained 12 scenes of Landsat 5 TM images (path 124/row 29, path 124/row 30, cloud cover <10%) covering the years of 2005–2011 (Table 1), and the data were processed with Landsat Ecosystem Disturbance Adaptive Processing System (LEDAPS) tools. The LEDAPS software was developed specifically for processing Landsat directional surface reflectance data through radiometric calibration, atmospheric correction, image registration, and image orthorectification [49]. The digital numbers (DNs) of the Landsat TM images were converted to reflectance values based on the atmospheric calibration procedure that was implemented by using LEDAPS [50].

The MODIS product (MOD09Q1, tile h26v04) at 250 m and 8-d intervals for the study area during the growing season (May to September) in 2005–2013 was obtained from the National Aeronautics and Space Administration's (NASA's) Earth Observing System Data and Information System (http://earthdata.nasa.gov). The MOD09Q1 product contains optimal observations during the 8-d periods given the effects of clouds, cloud shadows, and aerosol loading [51]; pixels with poor quality were excluded by using the quality assurance layer. The MODIS reflectance imagery were projected to Universal Transverse Mercator (UTM) projections and resampled to a 30-m resolution by nearest neighbor resampling to meet the projection and resolution of Landsat TM data.

Table 1. Landsat and Moderate Resolution Imaging Spectroradiometer (MODIS) image pairs for generating synthetic Normalized Difference Vegetation Index (NDVI) data during 2005–2013. Years highlighted with bold font lacked Landsat images, and therefore, the Landsat images from adjacent year (1-yr interval) or 2-yr intervals were used.

Year	Landsat TM Path = 124, Row = 29/30	MODIS (DOY) Tile = h26v04
2005	09/02/2005	241/2005
2006	09/21/2006	265/2006
2007	09/08/2007	249/2007
2008	**09/08/2007**	**249/2007**
2009	08/12/2009	225/2009
2010	08/31/2010	241/2010
2011	08/02/2011	209/2011
2012	**08/02/2011**	**209/2011**
2013	**08/02/2011**	**209/2011**

3. Methods

3.1. Procedure for Grassland AGB Estimation

The development of the AGB estimation model took place via four main steps (Figure 2).

Step I: Generation of synthetic NDVI by using STARFM. We obtained the synthetic NDVI for the AGB estimation model based on STARFM with the Landsat spatial resolution and MODIS temporal frequency. The base data pairs are shown in Table 1. The STARFM algorithm is described in more details in Section 3.2.

Step II: Development of the AGB estimation model. A total of 68 field samples were collected during June to August in 2011, and these data were used for the development and validation of the AGB model. We used the training set (46 samples) to construct four AGB estimation models, including linear regression model, power function model, exponential model, and SVM model. We then used the remaining of 22 plots (1/3 of the samples) as the testing data set to evaluate the four AGB estimation models.

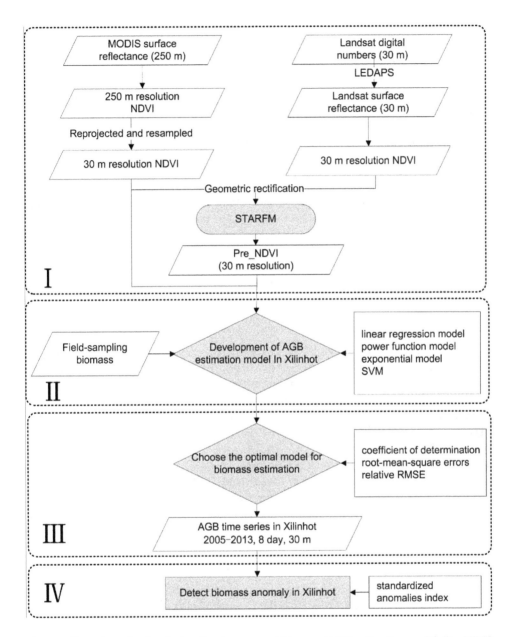

Figure 2. Procedures for developing the aboveground biomass (AGB) estimation model. Pre_NDVI represents the synthetic NDVI series.

Step III: Choose the optimal model for biomass estimation. To evaluate the model's performance, we used three statistical indices, namely, the coefficient of determination (R^2), root-mean-square error ($RMSE$), and relative RMSE ($RMSE_r$) [52]. The RMSE and $RMSE_r$ were calculated as follows:

$$RMSE = \sqrt{\sum_{i=1}^{n} (y_i - \hat{y}_i)^2 / n} \qquad (1)$$

$$RMSE_r = \left(\sqrt{\sum_{i=1}^{n} (y_i - \hat{y}_i)^2 / n/\overline{\hat{y}_i}}\right) \times 100\% \qquad (2)$$

where y_i is the observed value, $\hat{y}_i$ is the model predicted value, $\overline{\hat{y}_i}$ is the mean of the modeled values, and n is the sample number. In this study, we found that the SVM (see Section 3.3) was the optimal model for biomass estimation in this region.

Step IV: The Standardized Anomalies Index (SAI) [53] was used to detect the spatial and temporal distribution of biomass anomalies in July 2007. The SAI is defined as follows:

$$SAI_{AGB} = (AGB_i - \overline{AGB})/\sigma_{AGB} \qquad (3)$$

Here, SAI_{AGB} represents the AGB anomalies, AGB_i is the annual AGB in the *ith* year, $\overline{AGB}$ is the mean AGB for the years of 2005–2013, and σ_{AGB} is the standard deviation of the 9-yr AGB. Positive SAI values indicate that the annual AGB is larger than the 9-yr mean, while negative SAI values indicate that the annual AGB is more deficient compared to the 9-yr mean. In this study, we compared the SAI_{AGB} generated from the synthetic NDVI-derived biomass images and from the MODIS NDVI-derived biomass images to detect whether the synthetic images can help capture more details of the biomass anomalies in Xilinhot.

3.2. The STARFM Algorithm for NDVI Image Fusion

In this study, we obtained higher temporal and spatial resolution NDVI data by fusing MODIS and Landsat images with the use of the STARFM [14]. Previous studies have demonstrated the applicability of using STARFM for generating synthetic NDVI [18,22,29,54]. Among these studies, Tian *et al.* [18] and Jarihani *et al.* [54] further verified that use of the NDVI from Landsat and MODIS images as the input for STARFM can produce higher precision compared with blending reflectance data. Therefore, we used Landsat NDVI and MODIS NDVI directly as the input data for generating the synthetic NDVI.

Through establishing a linear relationship between MODIS and TM, STARFM can generate synthetic images based on an image pair of MODIS and TM NDVI data and the MODIS NDVI on the prediction date. Supposing that land cover types and system errors do not change over time, we can predict the NDVI at Landsat's spatial resolution for date (t_o) through using MODIS observations for date (t_o) and one or several pairs of Landsat and MODIS images acquired at the same date (t_k) [14]:

$$L\left(x_i, y_j, t_o\right) = M\left(x_i, y_j, t_o\right) + L\left(x_i, y_j, t_k\right) - M\left(x_i, y_j, t_k\right) \qquad (4)$$

where $L\left(x_i, y_j, t_o\right)$ is the predicted pixel (x_i, y_j) at Landsat resolution for date t_o, $M\left(x_i, y_j, t_o\right)$ is the MODIS pixel (x_i, y_j) for date t_o, $L\left(x_i, y_j, t_k\right)$ is the Landsat pixel (x_i, y_j) for date t_k, and $M\left(x_i, y_j, t_k\right)$ is the MODIS pixel (x_i, y_j) for date t_k.

However, in actual cases, the relationships between the NDVI of MODIS and TM are much more complicated. To solve this problem, STARFM uses a moving window (the size of the window was 750 m in our study) to acquire reflectance information for the neighboring pixels (750 m × 750 m), and then, it weights those pixels with a combination of spectral, temporal, and spatial differences among the MODIS and TM images as follows [14]:

$$L\left(x_{w/2}, y_{w/2}, t_o\right) = \sum_{i=1}^{w}\sum_{j=1}^{w}\sum_{k=1}^{n} W_{ijk} \times \left(M\left(x_i, y_j, t_o\right) + L\left(x_i, y_j, t_k\right) - M\left(x_i, y_j, t_k\right)\right) \qquad (5)$$

where $L\left(x_{w/2}, y_{w/2}, t_o\right)$ is the Landsat pixel $x_{w/2}, y_{w/2}$ for date (t_o), which is decided by the accumulation of adjacent pixels (x_i, y_j) multiplied by the different weights W_{ijk}. The time interval between t_0 and t_k, the NDVI value difference between Landsat and MODIS, and the relative distance among surrounding neighboring points are altogether used to determine the weights W_{ijk}.

Because of the lack of available TM images with little cloud contamination during the growing season, TM images with good quality were not available for some years. To solve this problem, we used TM and MODIS images pairs ($L\left(x_i, y_j, t_k\right)$ and $M\left(x_i, y_j, t_k\right)$) from the adjacent years (even as much as a 2-yr difference) as the input image pairs.

To examine the quality of the fusion results, we tested the accuracy of the fused images by using TM as the base image with the following three schemes: date t_0 and date t_k from the same year; date t_0 and date t_k from adjacent years; and date t_0 and date t_k with 2-yr intervals. The input and validated images are shown in Table 2.

Table 2. Description of the input images used for the validation of the three schemes for data fusion.

	Input MODIS t_k	Input Landsat t_k	Input MODIS t_0	Validation Landsat t_0
Scheme 1	09/06/2007–09/13/2007 DOY 249	09/08/2007	05/17/2007–05/24/2007 DOY 137	05/19/2007
Scheme 2	09/22/2006–09/29/2006 DOY 265	09/21/2006	05/17/2007–05/24/2007 DOY 137	05/19/2007
Scheme 3	08/29/2005–09/05/2005 DOY 241	09/02/2005	05/17/2007–05/24/2007 DOY 137	05/19/2007

We chose five statistical indices (*i.e.*, mean value, standard deviation, entropy, average gradient, and mean absolute difference) to assess the quality of the predicted synthetic NDVI products (Table 3).

Table 3. Statistics for assessment of the quality of the synthetic images.

Statistics	Formula [a]
Mean value	$\bar{p} = \sum\limits_{i=1}^{N} \sum\limits_{j=1}^{M} \dfrac{p_{(i,j)}}{N \times M}$
Standard deviation	$\sigma = \sqrt{\dfrac{1}{N \times M} \sum\limits_{i=1}^{N} \sum\limits_{j=1}^{M} \left(p_{(i,j)} - \bar{p}\right)^2}$
Entropy	$H = -\sum\limits_{j=1}^{M} \sum\limits_{i=1}^{N} P_{(i)} ln P_{(i)}$
Average gradient	$\bar{g} = \dfrac{1}{(M-1)(N-1)} \times \sum\limits_{i=1}^{M-1} \sum\limits_{j=1}^{N-1} \sqrt{\dfrac{(p(i,j) - p(i+1,j))^2 + (p(i,j) - p(i,j+1))^2}{2}}$
Mean absolute difference	$\bar{D} = \sum\limits_{i=1}^{N} \sum\limits_{j=1}^{M} \dfrac{\left|p_{(i,j)} - q_{(i,j)}\right|}{N \times M}$

[a] Here, $p_{(i,j)}$ represents the NDVI values of the ith row and jth column in image P; N represents the number of rows of the images; M is the column of the images; $P_{(i)}$ is the frequency of the pixel whose gray value was i (to calculate the entropy, values for NDVI images in our study were standardized to 0–255); $q_{(i,j)}$ represents the NDVI values of the ith row and jth column in image Q.

3.3. Biomass Estimation Model: Support Vector Machine Algorithm

In this study, we used a SVM algorithm to perform the regression analysis and generalize the biomass estimation model. The basic idea of the SVM algorithm is to find an optimal non-linear mapping ϕ and the corresponding parameters through cross-validation; we mapped the input space to the high dimensional feature space $\aleph$ and found the linear regression relationship in this space.

The SVM algorithm model can be described as follows:

$$f(x) = \langle w, \varphi(x) \rangle + b, with\ w \in \aleph, b \in R \tag{6}$$

where w and b are regressions coefficients and $\varphi(x)$ is the mapping function. R is the space of the output pattern. The flow chart for the SVM algorithm is shown in Figure 3.

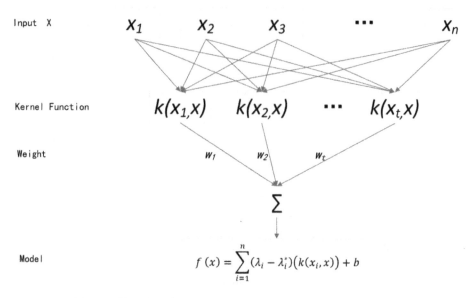

Figure 3. Flow chart for the support vector machine (SVM) algorithm.

Figure 3 is similar to a three layer neural network. The input vector for input layer nodes is $x_1, x_2, x_3 \cdots x_n$. The middle layer node for the kernel function is $k(x_i, x)$. The middle layer and output layer are connected by the weight vector, and the weight vector is $W = (w_1, w_2, \ldots, w_n)$, which is essentially the coefficients of kernel function $k(x_i, x)$. The output is the estimated AGB $f(x)$, where λ_i, λ_i^*, and b are coefficients and $k(x_i, x)$ is the kernel function. The basic kernel functions that can be used include the linear function, polynomial function, radial basis function (RBF), and sigmoid function. In this study, we decided to adopt the RBF to build the SVM-AGB model because the SVM model based on the RBF had highest accuracy; moreover, it is a recommended kernel function to acquire satisfactory results [55]. Two parameters are important for the RBF, namely, C and γ. We used the recommended grid search method and cross-validation method [55] to find the optimal parameters for the RBF. In our study, the input parameter x was the MODIS NDVI or synthetic NDVI, while $f(x)$ was the above ground biomass estimated by the MODIS NDVI or synthetic NDVI.

4. Results and Discussion

4.1. Accuracy Assessment of Synthetic NDVI Based on STARFM

Figure 4 shows the validation results for the three fusion schemes described in Section 3.2, that is, data fusion within one year (Figure 4a,d) (2007), with adjacent years (Figure 4b,e) (2007 and 2006), and with 2-yr intervals (Figure 4c,f) (2007 and 2005). Red colors represent over-estimations and blue colors indicate under-estimations for synthetic NDVI. Figure 4d–f show the scatter plots between Landsat NDVI and synthetic NDVI for the three schemes. As we only considered grasslands in Xilinhot, all other vegetation types have been masked out.

There was good agreement between the observed Landsat NDVI images and synthetic NDVI data in the three schemes. The average relative errors between observed Landsat NDVI images and synthetic data were below 10% (8.12% for Scheme 1, 7.58% for Scheme 2, and 9.41% for Scheme 3). However, the pixels near lake shores, along riverbanks (mainly in the northern and central parts of the study area), and in other sparsely vegetated areas (mainly in the southern parts of the study area) showed larger differences between synthetic NDVI and NDVI derived from observed TM. The study of Zhang *et al.* [31] also pointed out that over-estimations of synthetic images were mainly located at pixels near riverbanks and lake shores in the mid-eastern parts of New Orleans, USA, and these were due to abrupt surface changes caused by urban flooding [31]. Cloud contamination can also cause bias between observed NDVI and synthetic NDVI [16]. Although the MODIS product used in this study give the 8-day optimal observations, it can still not fully remove the atmospheric noise [56], which could affect the accuracy of the fusion result. While in most of the study area, the grass sceneries were homogenous and the three schemes showed good fusion results. However, the fusion results of Scheme 1 (within one year) and Scheme 2 (with adjacent years) were slightly better than Scheme 3, as they had higher coefficients of determination (R^2) and lower root-mean-square errors (R^2 = 0.72, RMSE = 0.021 for Scheme 1; R^2 = 0.77, RMSE = 0.024 for Scheme 2; R^2 = 0.67, RMSE = 0.037 for Scheme 3) (Figure 4).

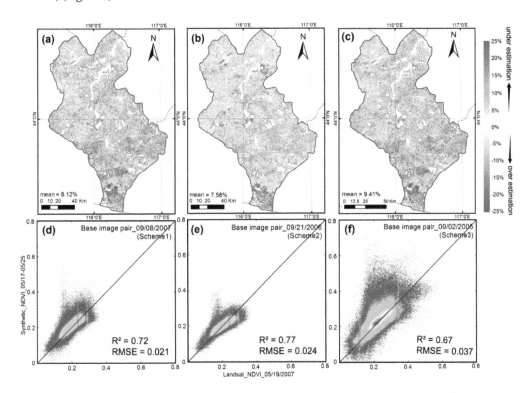

Figure 4. Difference maps and scatter plots between the observed Landsat NDVI images (taken 19 May 2007) and the synthetic NDVI images (taken during 17–24 May 2007) predicted by (Left) (**a,d**) Scheme 1; (Middle) (**b,e**) Scheme 2, and (Right) (**c,f**) Scheme 3.

We further chose five statistical indices (*i.e.*, mean value, standard deviation, entropy, average gradient, and mean absolute difference) to assess the quality of the predicted synthetic NDVI products (Table 4). There were no considerable differences between the observed and predicted NDVI images for the three schemes, and the first four indices were nearly identical. The average differences were all below 0.025. In addition, the fused images in the same year (Scheme 1) and in the adjacent years had more consistent results with observed images than those from Scheme 3, as indicated by the five indices. In another woodland area, similar coefficient of determinations (0.66–0.94) between the observed TM NDVI and predicted NDVI were found by Bhandari *et al.* [23]. Overall, our predicted NDVI images were fairly reliable and could be used for accurate estimations of AGB. In the following work, we chose different data-fusion schemes according to the availability of TM images.

Table 4. Accuracy assessment of the fused images. TM_NDVI is the observed TM NDVI image and Pre_NDVI represents the predicted NDVI images at 30-m resolution that were derived from STARFM with Schemes 1, 2, and 3. Numbers in bold represent the best fit among the three schemes.

Type	Mean	Standard Deviation	Entropy	Average Gradient	Mean Absolute Difference
TM_NDVI	0.240	0.049	3.619	0.014	/
Pre_NDVI_Scheme1	**0.244**	**0.053**	**3.653**	**0.013**	0.019
Pre_NDVI_Scheme2	0.245	**0.045**	3.566	0.012	**0.018**
Pre_NDVI_Scheme3	0.247	0.060	3.725	0.016	0.022

4.2. Prediction of Time-Series Synthetic NDVI

Based on the above analysis of the three fusion schemes, we decided to use data in the same year (except for 2008, 2012, and 2013), in adjacent years (2008, 2012), and data with 2-yr intervals (2013) (data shown in Table 1) to predict the synthetic NDVI.

We fused TM and MODIS NDVI by using the STARFM method and predicted NDVI for each 8-d intervals during 2005–2013. A total of 180 NDVI scenes (20 scenes/yr for 9 yr) were predicted. Figure 5a,b show the mean and standard deviation of MODIS NDVI and synthetic NDVI at 8-day intervals during 2005–2013, respectively. Figure 5c shows the relationship between the MODIS NDVI and synthetic NDVI results that were shown in Figure 5a,b, respectively. Figure 5d shows the relationship between the standard deviations of the MODIS NDVI and synthetic NDVI results that were shown in Figure 5a,b, respectively. Figure 5e shows the maximum NDVI values of MODIS and synthetic images during the growing seasons, which represent the highest vegetation productivity within a year.

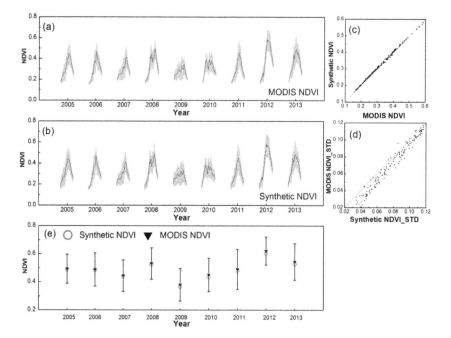

Figure 5. Temporal variations of the regional mean and standard deviation (shaded area) of NDVI for grasslands in the study areas during 2005–2013 (growing season). (a) MODIS NDVI; (b) synthetic NDVI; (c) scatter plots between MODIS NDVI and synthetic NDVI at 8-d intervals; (d) scatter plots between the standard deviation of MODIS NDVI and synthetic NDVI at 8-d intervals; (e) MODIS $NDVI_{max}$ (black) and synthetic $NDVI_{max}$ (gray). Error bars represent the standard deviations of MODIS $NDVI_{max}$ (black) and synthetic $NDVI_{max}$ (gray).

The means of the synthetic NDVI values were almost identical to those of the MODIS NDVI (Figure 5c). The predicted NDVI (30 m) therefore retained the high-frequency temporal information from the MODIS NDVI time series. The predicted images had larger standard deviations and represent larger dispersions of the imagery (Figure 5d), which indicates that the synthetic images may contain more detailed information. The differences between the standard deviation of MODIS NDVI and synthetic NDVI in this study were less than those reported in the study of Tian *et al.* [18]. In their study, these differences could be clearly observed in graphics similar to the ones in Figure 5a,b, and the shadow area of the synthetic NDVI series was much larger than the MODIS NDVI series. This can be explained by the homogeneous landscape for grasslands in Xilinhot, our study area. In contrast, in the study of Tian *et al.* [18], the study area was a mixed fragmentized landscape consisting of forests, shrublands, and residential areas.

Our results indicate that the accuracy of time-series NDVI derived from the STARFM algorithm was reliable and that the STARFM could effectively predict NDVI time series at higher spatial and temporal resolutions for the grasslands. The synthetic NDVI time series described more detailed spatial variations of NDVI at a resolution of 30 m. The temporal information from MODIS and the spatial information from Landsat were integrated in the predicted synthetic NDVI data set, and it was expected that such data can provide superior input for accurate estimations of AGB time series.

4.3. Development of the AGB Estimation Model

We developed four regression models for estimating the AGB, and these included linear regression model, power function model, exponential model, and support vector machine model (SVM-AGB). We randomly selected 46 biomass field samples and the corresponding synthetic NDVI pixels as the training set to construct the four models. We then used the other 22 samples to assess the accuracy of the four models. The accuracy assessment results indicated that for both MODIS NDVI and synthetic NDVI, the SVM-AGB model had higher accuracy than the other three models (Tables 5 and 6). We further compared the synthetic NDVI-derived SVM-AGB model and MODIS NDVI-derived SVM-AGB model and found that the accuracies of the synthetic NDVI-derived SVM-AGB model data were higher than the accuracies of MODIS NDVI-derived SVM-AGB model data, as indicated by the R^2, RMSE, and $RMSE_r$ for both the training set and testing set (Tables 5 and 6). Therefore, we finally chose to integrate the SVM-AGB model and the synthetic NDVI to predict AGB for grasslands in Xilinhot during 2005–2013.

We compared our SVM-AGB model ($R^2 = 0.77$, RMSE = 17.22 g/m^2, $RMSE_r = 24.8\%$) with other AGB estimation models in the same region, e.g., the exponential model based on 250-m MODIS NDVI ($R^2 = 0.447$) by Kawamura *et al.* [57]; the ANN model based on elevation, Landsat NDVI, and Landsat reflectance data ($R^2 = 0.82$, RMSE = 60.01 g/m^2, $RMSE_r = 40.61\%$) by Xie *et al.* [35]; and the power function model based on 250-m MODIS NDVI ($R^2 = 0.568$, RMSE = 673.88 kg/ha) by Jin *et al.* [47]. The studies by Kawamura *et al.* [57] and Jin *et al.* [47] mainly used 250-m MODIS NDVI data, and the accuracies of their models were lower than that of our estimation model ($R^2 = 0.447$ by Kawamura *et al.*; $R^2 = 0.568$ by Jin *et al.*; $R^2 = 0.77$ by our synthetic NDVI-derived SVM-AGB model). Meanwhile, in the study of Xie *et al.* [35], although their ANN model was built on Landsat NDVI and reflectance images and possessed higher accuracy, the limitations of sparse temporal data associated with Landsat images restricted their model's application to the requirements of frequent time series and dynamic monitoring. In general, our synthetic NDVI-derived SVM-AGB estimation model had both higher spatial resolutions (30 m) and temporal resolutions (8 d) than the other models in Tables 5 and 6 and showed improvements compared to other AGB estimation models generated from a single type of remotely sensed data [33,35,47,56].

Table 5. Comparison of the accuracy of different AGB estimation models based on synthetic NDVI data.

AGB Model	Regression Equation	Training Set			Testing Set		
		R^2	RMSE (g/m^2)	$RMSE_r$	R^2	RMSE (g/m^2)	$RMSE_r$
Linear regression model	$y = 2178 * x - 1140$	0.71	31.40	42.6%	0.79	26.48	34.6%
Power function model	$y = 1.044 * 10^5 * x^{12.59}$	0.68	33.62	44.6%	0.84	28.03	38.0%
Exponential model	$y = 3.902 * 10^{-4} * e^{21.61 * x}$	0.67	34.14	45.1%	0.84	28.60	38.8%
SVM-AGB	/	**0.77**	**17.22**	**24.8%**	**0.83**	**22.60**	**31.3%**

Note: x represents synthetic NDVI; y represents AGB.

Table 6. Comparison of the accuracy of different AGB estimation models based on MODIS NDVI.

AGB Model	Regression Equation	Training Set			Testing Set		
		R^2	RMSE (g/m^2)	$RMSE_r$	R^2	RMSE (g/m^2)	$RMSE_r$
Linear regression model	$y = 2010 * x - 1043$	0.66	34.13	46.3%	0.64	34.19	42.1%
Power function model	$y = 2.55 * 10^5 * x^{14.15}$	0.68	33.21	44.8%	0.69	31.23	40.9%
Exponential model	$y = 6.771 * 10^{-5} * e^{(24.71 * x)}$	0.68	33.36	44.9%	0.69	31.24	41.1%
SVM-AGB	/	**0.73**	**30.61**	**43.0%**	**0.72**	**22.89**	**37.1%**

Note: x represents MODIS NDVI; y represents AGB.

4.4. Drought Condition Monitoring with Time-Series Biomass Maps

We further used all the 68 biomass field samples and the corresponding synthetic NDVI pixels to generate synthetic NDVI-derived SVM-AGB model. Specifically, we estimated the AGB for the entire grassland area of Xilinhot, Inner Mongolia, for the growing season during 2005–2013. Figure 6 illustrates the spatial distribution of biomass in 2007 (drought year) and 2011 (non-drought year) for each 8-d intervals during the growing season. Figure 7 shows the fluctuations of the regional mean biomass in the study areas during 2005–2013. The biomass variation of the regional mean in 2011 was almost identical with the 10-yr mean (Figure 7), and therefore, biomass maps in 2011 can be used to represent the general vegetation growth conditions in Xilinhot.

Generally, biomass for May–September exhibited large spatial heterogeneity, which is of great significance for detecting grassland conditions. The AGB generally showed a decreasing trend from the southeast to the northwest. In 2011, in the first week of May (day of year (DOY): 121), the grassland biomass was generally below 20 g/m^2 and the regional mean was 16.9 g/m^2. During July, nearly half of the grassland biomass values reached 40–50 g/m^2. Grassland biomass reached its peak values during the end of July and the start of August. In the last week of July (DOY: 209), the highest value reached to 190 g/m^2 in the southeast region, while in the northwest region, the biomass was as low as 35 g/m^2. Then, biomass started to decrease during the third week of August. In the first week of September (DOY: 233), the regional mean AGB was 34.0 g/m^2, and it decreased to 26.2 g/m^2 during the last week of September.

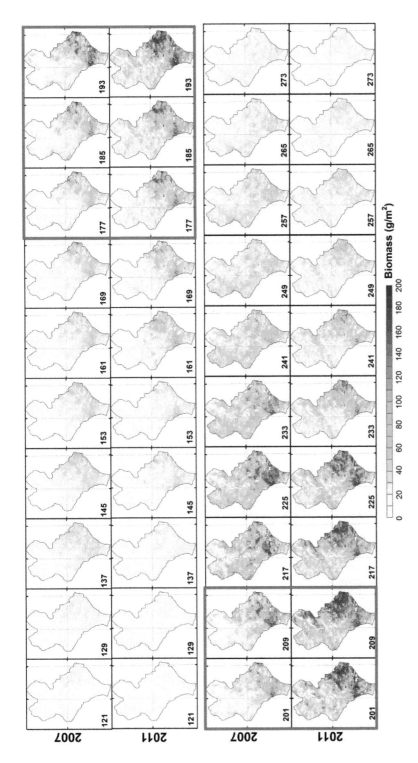

Figure 6. Spatial distribution of biomass in 2007 and 2011 for each 8-d intervals during the growing season.

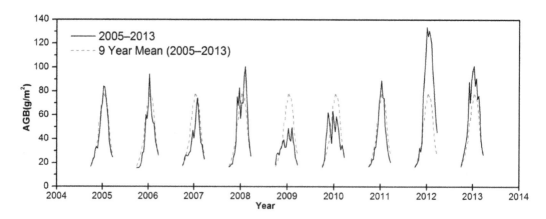

Figure 7. The 8-d intervals of the regional mean biomass estimated based on synthetic NDVI and SVM-AGB model during the growing season (May to September) of 2005–2013 in Xilinhot. Dash line represents the regional mean biomass of 9-yr mean (2005–2013).

Returning to the drought analysis, comparisons of the biomass were made between 2007 (drought year) and 2011 (non-drought year). In 2007, biomass generally exhibited temporal variations similar to the biomass variations in 2011. Importantly, during the end of June to the end of July (DOY: 177–209) (indicated by red rectangles), large areas of biomass in 2007 showed much lower biomass compared with the biomass in 2011 (Figure 6), especially in the southwestern region of Xilinhot. During the third and fourth week of July (DOY: 201 and 209) in 2007, large areas of biomass in the southwest were below 100 g/m^2, while in 2011, the biomass values reached up to 150 g/m^2. For the remainder of the growing season, no large differences were detected between the biomass maps in 2007 and 2011.

Overall, the time series of biomass during the growing season of 2005–2013 (black line) generally showed similar temporal variations with the 9-yr mean (2005–2013) (dashed line) (Figure 7). However, biomass in 2007, 2009, and 2010 showed larger negative anomalies compared with the 9-yr means largely because of the droughts in this area.

The year of 2007 was recorded as a drought year [58], and the biomass values during June and October 2007 were mostly below the multi-year mean (Figure 7). This was consistent with the findings of Yu *et al.* [59] and Hang *et al.* [60]. Yu *et al.* [61] used the Modified Grassland Index (MGI) to monitor grassland variations in Xilingol (2001–2010) and found that the MGI values in 2007 were much lower than MGI values in other years. In the study of Hang *et al.* [60], the annual and seasonal NDVI in 2007 was below that in other years (2000–2010) in Xilingol. The largest negative biomass anomaly during 2005–2013 was found during the second week of July in 2009 (51.7% less than the 9-yr mean), followed by the second week of August in 2009 (50.0% lower than the 9-yr mean) and the first week of August in 2009 (47.0% lower than the 9-year mean). Jin *et al.* [47] also found that during 2006–2012, the biomass in 2009 was the lowest and it was 31.5% lower than the 7-yr mean in Xilingol League (Xilinhot is located in the center of Xilingol League). This was mainly attributed to severe drought and insect outbreaks in this region in 2009 [62]. Biomass in 2010 showed large fluctuations. According to the national grassland report in 2010 issued by Farmer's daily [59], Xilingol League suffered from severe snow disasters during the end of 2009 and the beginning of 2010. For grasslands in Inner Mongolia, up to 32 million ha of grassland were affected [59], which could be a possible reason for the biomass fluctuations in 2010. Conversely, biomass in 2012 was much larger than that in other years. In the third week of July, the mean biomass of the study area even reached to 133.86 g/m^2, which was 88.9% higher than the 9-yr mean. Jin *et al.* [47] stated that 2012 was a prime harvest year for all of Xilingol, and the biomass value was 40% higher than the annual average (2005–2013) and twice the value in 2009.

To further explore the capability of the synthetic NDVI for detecting anomalous vegetation activity, we took the biomass during the drought year (2007) as an example and then compared the data to the 9-yr mean (2005–2013) through using SAI_{AGB}. In 2007, large vegetation anomalies were mainly found during June to the start of August (Figure 8, DOY: 153–217), which can be attributed to the rainfall shortages. Large water deficiencies were detected during the start of March to the end of June (DOY: 64–176) (Figure 8). Other studies have found that precipitation has obvious lag effects on vegetation in the study region, and the lag response time is typically 48 to 56 d [61,63]. Although precipitation in the last week of July was even more than twice the 9-yr mean, it did not relieve the stress on vegetation during this month. Biomass in the third week of July in 2007 was even 42.7% lower than the multi-year mean (Figure 8).

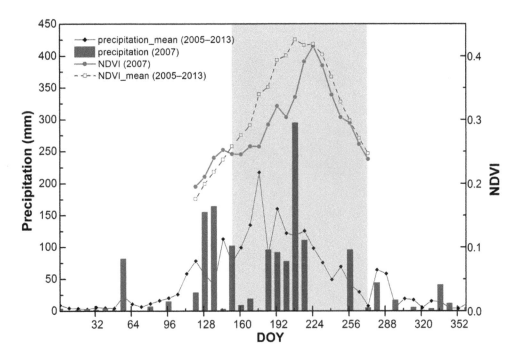

Figure 8. Time series of precipitation and predicted synthetic NDVI (8-d intervals) for Xilinhot in 2007 (precipitation data at site Xilinhot were obtained from the China meteorological data sharing service system).

Figure 9 shows the biomass anomalies in 2007 compared with the mean of 2005–2013 during July. Site 1 was located at the Maodeng Ranch, which is China's largest typical-steppe reserve [48]. Site 2 was occupied by lovely achnatherum, needlegrass, and little-leaf pea shrub [48]. The distribution of the biomass anomalies based on MODIS NDVI-derived biomass and synthetic NDVI-derived biomass data generally showed good agreement. However, the biomass derived from the synthetic NDVI image data showed more details compared with MODIS NDVI-derived biomass data. Terrain textures and roads can be clearly identified in the maps from the synthetic images. In addition, the absolute SAI of biomass maps from the synthetic NDVI images was generally larger than that of the MODIS NDVI-derived biomass images, which implies that the synthetic images can better capture the biomass anomalies than MODIS NDVI-derived biomass images.

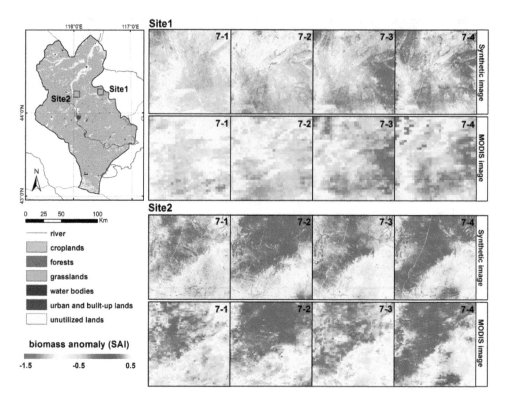

Figure 9. The Standardized Anomalies Index (SAI) indicating biomass anomalies in Xilinhot during July 2007. The land cover map in Xilinhot in the left panel was adapted from the land cover data set provided by the Data Center for Resources and Environmental Sciences, Chinese Academy of Sciences (http://www.resdc.cn). The unutilized lands represent lands that have not been used (including desert, Gobi region, saline, wetlands, bare soil).

5. Conclusions

Modeling aboveground biomass is of great importance to studies of grassland ecology and socio-economic environments in grassland regions. In order to predict biomass for grasslands with high spatial resolutions and temporal frequencies, we developed an aboveground biomass estimation model integrating synthetic NDVI data and the support vector machine technique (SVM-AGB). Our modeling method showed that the synthetic images can guarantee the accuracy of AGB estimations compared with AGB estimates from MODIS data (MODIS data: $R^2 = 0.77$ and RMSE = 17.22 g/m^2; synthetic imagery: $R^2 = 0.73$ and RMSE = 30.61 g/m^2). Importantly, the spatial-temporal resolution was improved (8 d, 30 m), which indicates that integrating the synthetic NDVI data and SVM model can produce accurate grassland AGB estimates for natural resource applications.

Based on the improved grassland AGB estimation model and synthetic NDVI, we mapped time-series AGB for grasslands of Xilinhot, Inner Mongolia, at 8-d intervals and 30-m resolutions. We successfully combined the STARFM and SVM into grassland biomass estimations. Compared with the biomass model using single data sources, our SVM-AGB model not only guarantees the accuracy of biomass estimations and achieves higher temporal frequencies, but it also improves the spatial resolution (*i.e.*, from 250 to 30 m). Thus, it can capture biomass anomalies at finer scales, which is very useful for monitoring grassland conditions and conducting disaster assessments (such as during droughts). The high temporal (8 d) and medium spatial resolution (30 m) AGB estimations should be particularly useful for dynamic monitoring of grassland AGB, animal husbandry management, and general grassland management. Our approach can also be applied to grasslands in other geographical regions.

Acknowledgments: This study was funded by the National Natural Science Foundation of China (NSFC) (Grant No. 41271372) and the National Key Technology R&D Program (Grant No. 2012BAC16B01). The authors thank STARFM developers for sharing the code, Liangyun Liu for sharing field data for this study, and Jingfeng Xiao for giving important suggestion for this study. We thank the helpful comments on our manuscript from the anonymous reviewers and associate editors.

Author Contributions: Binghua Zhang contributed to the experiments of the research, analyzed the data, and wrote the majority of the paper. Li Zhang designed the original idea of the study, supervised and designed the research work, and contributed to the construction of the manuscript structure. Dong Xie made comparison of different biomass estimation models. Xiaoli Yin contributed to data processing. Chunjing Liu and Guang Liu helped with discussion and revisions.

Conflicts of Interest: The authors declare no conflict of interest.

References

1. He, C.; Zhang, Q.; Li, Y.; Li, X.; Shi, P. Zoning grassland protection area using remote sensing and cellular automata modeling—A case study in Xilingol steppe grassland in northern China. *J. Arid Environ.* **2005**, *63*, 814–826. [CrossRef]
2. Jobbagy, E.G.; Sala, O.E. Controls of grass and shrub aboveground production in the Patagonian steppe. *Ecol. Appl.* **2000**, *10*, 541–549. [CrossRef]
3. Nordberg, M.L.; Evertson, J. Monitoring change in mountainous dry-heath vegetation at a regional scale using multitemporal Landsat TM data. *Ambio* **2003**, *32*, 502–509. [CrossRef] [PubMed]
4. Huete, A.; Didan, K.; Miura, T.; Rodriguez, E.P.; Gao, X.; Ferreira, L.G. Overview of the radiometric and biophysical performance of the MODIS vegetation indices. *Remote Sens. Environ.* **2002**, *83*, 195–213. [CrossRef]
5. Price, J.C. How unique are spectral signatures. *Remote Sens. Environ.* **1994**, *49*, 181–186. [CrossRef]
6. Asner, G.P. Cloud cover in Landsat observations of the Brazilian Amazon. *Int. J. Remote Sens.* **2001**, *22*, 3855–3862. [CrossRef]
7. Jorgensen, P.V. Determination of cloud coverage over Denmark using Landsat MSS/TM and NOAA–AVHRR. *Int. J. Remote Sens.* **2000**, *21*, 3363–3368. [CrossRef]
8. Ju, J.C.; Roy, D.P. The availability of cloud-free Landsat ETM plus data over the conterminous United States and globally. *Remote Sens. Environ.* **2008**, *112*, 1196–1211. [CrossRef]
9. Gur, E.; Zalevsky, Z. Resolution-enhanced remote sensing via multi spectral and spatial data fusion. *Int. J. Image Data Fusion* **2011**, *2*, 149–165. [CrossRef]
10. Carper, W.J.; Lillesand, T.M.; Kiefer, R.W. The use of intensity-hue-saturation transformations for merging SPOT panchromatic and multispectral image data. *Photogramm. Eng. Remote. Sens.* **1990**, *56*, 459–467.
11. Shettigara, V.K. A generalized component substitution technique for spatial enhancement of multispectral images using a higher resolution data set. *Photogramm. Eng. Remote. Sens.* **1992**, *58*, 561–567.
12. Yocky, D.A. Multiresolution wavelet decomposition image merger of Landsat thematic mapper and SPOT panchromatic data. *Photogramm. Eng. Remote. Sens.* **1996**, *62*, 1067–1074.
13. Fu, D.J.; Chen, B.Z.; Wang, J.; Zhu, X.L.; Hilker, T. An improved image fusion approach based on enhanced spatial and temporal the adaptive reflectance fusion model. *Remote Sens.* **2013**, *5*, 6346–6360. [CrossRef]
14. Gao, F.; Masek, J.; Schwaller, M.; Hall, F. On the blending of the Landsat and MODIS surface reflectance: Predicting daily Landsat surface reflectance. *IEEE Trans. Geosci. Remote Sens.* **2006**, *44*, 2207–2218.
15. Zhu, X.L.; Chen, J.; Gao, F.; Chen, X.H.; Masek, J.G. An enhanced spatial and temporal adaptive reflectance fusion model for complex heterogeneous regions. *Remote Sens. Environ.* **2010**, *114*, 2610–2623. [CrossRef]
16. Hilker, T.; Wulder, M.A.; Coops, N.C.; Seitz, N.; White, J.C.; Gao, F.; Masek, J.G.; Stenhouse, G. Generation of dense time series synthetic Landsat data through data blending with MODIS using a spatial and temporal adaptive reflectance fusion model. *Remote Sens. Environ.* **2009**, *113*, 1988–1999. [CrossRef]
17. Walker, J.J.; de Beurs, K.M.; Wynne, R.H.; Gao, F. Evaluation of Landsat and MODIS data fusion products for analysis of dryland forest phenology. *Remote Sens. Environ.* **2012**, *117*, 381–393. [CrossRef]

18. Tian, F.; Wang, Y.J.; Fensholt, R.; Wang, K.; Zhang, L.; Huang, Y. Mapping and evaluation of NDVI trends from synthetic time series obtained by blending Landsat and MODIS data around a coalfield on the Loess Plateau. *Remote Sens.* **2013**, *5*, 4255–4279. [CrossRef]

19. Schmidt, M.; Udelhoven, T.; Gill, T.; Roder, A. Long term data fusion for a dense time series analysis with MODIS and Landsat imagery in an Australian savanna. *J. Appl. Remote Sens.* **2012**, *6*, 063512.

20. Watts, J.D.; Powell, S.L.; Lawrence, R.L.; Hilker, T. Improved classification of conservation tillage adoption using high temporal and synthetic satellite imagery. *Remote Sens. Environ.* **2011**, *115*, 66–75. [CrossRef]

21. Singh, D. Generation and evaluation of gross primary productivity using Landsat data through blending with MODIS data. *Int. J. Appl. Earth Obs.* **2011**, *13*, 59–69. [CrossRef]

22. Singh, D. Evaluation of long-term NDVI time series derived from Landsat data through blending with MODIS data. *Atmosfera* **2012**, *25*, 43–63.

23. Bhandari, S.; Phinn, S.; Gill, T. Preparing Landsat image time series (LITS) for monitoring changes in vegetation phenology in Queensland, Australia. *Remote Sens.* **2012**, *4*, 1856–1886. [CrossRef]

24. Senf, C.; Leitao, P.J.; Pflugmacher, D.; van der Linden, S.; Hostert, P. Mapping land cover in complex Mediterranean landscapes using Landsat: Improved classification accuracies from integrating multi-seasonal and synthetic imagery. *Remote Sens. Environ.* **2015**, *156*, 527–536. [CrossRef]

25. Huang, B.; Song, H.H. Spatiotemporal reflectance fusion via sparse representation. *IEEE Trans. Geosci. Remote Sens.* **2012**, *50*, 3707–3716. [CrossRef]

26. Hilker, T.; Wulder, M.A.; Coops, N.C.; Linke, J.; McDermid, G.; Masek, J.G.; Gao, F.; White, J.C. A new data fusion model for high spatial and temporal-resolution mapping of forest disturbance based on Landsat and MODIS. *Remote Sens. Environ.* **2009**, *113*, 1613–1627. [CrossRef]

27. Gevaert, C.M.; García-Haro, F.J. A comparison of STARFM and an unmixing-based algorithm for Landsat and MODIS data fusion. *Remote Sens. Environ.* **2015**, *156*, 34–44. [CrossRef]

28. Shen, H.F.; Wu, P.H.; Liu, Y.L.; Ai, T.H.; Wang, Y.; Liu, X.P. A spatial and temporal reflectance fusion model considering sensor observation differences. *Int. J. Remote Sens.* **2013**, *34*, 4367–4383. [CrossRef]

29. Meng, J.H.; Du, X.; Wu, B.F. Generation of high spatial and temporal resolution NDVI and its application in crop biomass estimation. *Int. J. Digit. Earth* **2013**, *6*, 203–218. [CrossRef]

30. Huang, B.; Zhang, H.K. Spatio-temporal reflectance fusion via unmixing: Accounting for both phenological and land-cover changes. *Int. J. Remote Sens.* **2014**, *35*, 6213–6233. [CrossRef]

31. Zhang, F.; Zhu, X.L.; Liu, D.S. Blending MODIS and Landsat images for urban flood mapping. *Int. J. Remote Sens.* **2014**, *35*, 3237–3253. [CrossRef]

32. Emelyanova, I.V.; McVicar, T.R.; van Niel, T.G.; Li, L.T.; van Dijk, A.I.J.M. Assessing the accuracy of blending Landsat-MODIS surface reflectances in two landscapes with contrasting spatial and temporal dynamics: A framework for algorithm selection. *Remote Sens. Environ.* **2013**, *133*, 193–209. [CrossRef]

33. Gao, T.; Xu, B.; Yang, X.C.; Jin, Y.X.; Ma, H.L.; Li, J.Y.; Yu, H.D. Using MODIS time series data to estimate aboveground biomass and its spatio-temporal variation in Inner Mongolia's grassland between 2001 and 2011. *Int. J. Remote Sens.* **2013**, *34*, 7796–7810. [CrossRef]

34. Tomppo, E.; Nilsson, M.; Rosengren, M.; Aalto, P.; Kennedy, P. Simultaneous use of Landsat-TM and IRS-1C WIFS data in estimating large area tree stem volume and aboveground biomass. *Remote Sens. Environ.* **2002**, *82*, 156–171. [CrossRef]

35. Xie, Y.C.; Sha, Z.Y.; Yu, M.; Bai, Y.F.; Zhang, L. A comparison of two models with Landsat data for estimating above ground grassland biomass in Inner Mongolia, China. *Ecol. Model* **2009**, *220*, 1810–1818. [CrossRef]

36. Jin, Y.Q.; Liu, C. Biomass retrieval from high-dimensional active/passive remote sensing data by using artificial neural networks. *Int. J. Remote Sens.* **1997**, *18*, 971–979. [CrossRef]

37. Uno, Y.; Prasher, S.O.; Lacroix, R.; Goel, P.K.; Karimi, Y.; Viau, A.; Patel, R.M. Artificial neural networks to predict corn yield from compact airborne spectrographic imager data. *Comput. Electron. Agric.* **2005**, *47*, 149–161. [CrossRef]

38. Yool, S.R. Land cover classification in rugged areas using simulated moderate-resolution remote sensor data and an artificial neural network. *Int. J. Remote Sens.* **1998**, *19*, 85–96. [CrossRef]

39. Ban, Y.F. Synergy of multitemporal ERS-1 SAR and Landsat TM data for classification of agricultural crops. *Can. J. Remote Sens.* **2003**, *29*, 518–526. [CrossRef]

40. Erbek, F.S.; Ozkan, C.; Taberner, M. Comparison of maximum likelihood classification method with supervised artificial neural network algorithms for land use activities. *Int. J. Remote Sens.* **2004**, *25*, 1733–1748. [CrossRef]

41. Mathur, P.; Govil, R. Detecting temporal changes in satellite imagery using ANN. In Proceedings of the RAST 2005 2nd International Conference on Recent Advances in Space Technologies, Istanbul, Turkey, 9–11 June 2005; pp. 645–647.

42. Balabin, R.M.; Lomakina, E.I. Support vector machine regression (SVR/lS-SVM)—An alternative to neural networks (ANN) for analytical chemistry? Comparison of nonlinear methods on near infrared (NIR) spectroscopy data. *Analyst* **2011**, *136*, 1703–1712. [CrossRef] [PubMed]

43. Chen, G.; Hay, G.J.; Zhou, Y.L. Estimation of forest height, biomass and volume using support vector regression and segmentation from Lidar transects and Quickbird imagery. In Proceedings of the 2010 18th International Conference on Geoinformatics, Beijing, China, 18–20 June 2010; pp. 1–4.

44. Gleason, C.J.; Im, J. Forest biomass estimation from airborne LiDAR data using machine learning approaches. *Remote Sens. Environ.* **2012**, *125*, 80–91. [CrossRef]

45. Jachowski, N.R.A.; Quak, M.S.Y.; Friess, D.A.; Duangnamon, D.; Webb, E.L.; Ziegler, A.D. Mangrove biomass estimation in southwest Thailand using machine learning. *Appl. Geogr.* **2013**, *45*, 311–321. [CrossRef]

46. Tu, Y. Based on Meteorological Data of Drought Disaster Forecast Study in Pastoral Area of Xilingol League. Master's Thesis, Inner Mongolia Normal University, Hohhot, China, 2013. (In Chinese)

47. Jin, Y.X.; Yang, X.C.; Qiu, J.J.; Li, J.Y.; Gao, T.; Wu, Q.; Zhao, F.; Ma, H.L.; Yu, H.D.; Xu, B. Remote sensing-based biomass estimation and its spatio-temporal variations in temperate grassland, northern china. *Remote Sens.* **2014**, *6*, 1496–1513. [CrossRef]

48. Guan, L.L.; Liu, L.Y.; Peng, D.L.; Hu, Y.; Jiao, Q.J.; Liu, L.L. Monitoring the distribution of C3 and C4 grasses in a temperate grassland in northern China using moderate resolution imaging spectroradiometer normalized difference vegetation index trajectories. *J. Appl. Remote Sens.* **2012**, *6*. [CrossRef]

49. Masek, J.G.; Vermote, E.F.; Saleous, N.E.; Wolfe, R.; Hall, F.G.; Huemmrich, K.F.; Gao, F.; Kutler, J.; Lim, T.K. A Landsat surface reflectance dataset for North America, 1990–2000. *IEEE Geosci. Remote Sens. Lett.* **2006**, *3*, 68–72. [CrossRef]

50. Bai, Y.F.; Han, X.G.; Wu, J.G.; Chen, Z.Z.; Li, L.H. Ecosystem stability and compensatory effects in the Inner Mongolia grassland. *Nature* **2004**, *431*, 181–184. [CrossRef] [PubMed]

51. Vermonte, E.F.; Kotchenova, S.Y.; Ray, J.P. MODIS Surface Reflectance User's Guide, Version 1.4. Available online: http://modis-sr.ltdri.org/guide/MOD09_UserGuide_v1_3.pdf (accessed on 16 November 2015).

52. Makela, H.; Pekkarinen, A. Estimation of forest stand volumes by Landsat TM imagery and stand-level field-inventory data. *For. Ecol. Manag.* **2004**, *196*, 245–255. [CrossRef]

53. Katz, R.W.; Glantz, M.H. Anatomy of a rainfall index. *Mon. Weather Rev.* **1986**, *114*, 764–771. [CrossRef]

54. Jarihani, A.A.; McVicar, T.R.; van Niel, T.G.; Emelyanova, I.V.; Callow, J.N.; Johansen, K. Blending Landsat and MODIS data to generate multispectral indices: A comparison of "Index-then-blend"and "Blend-then-index" approaches. *Remote Sens.* **2014**, *6*, 9213–9238. [CrossRef]

55. Hsu, C.W.; Chang, C.C.; Lin, C.J. A Practical Guide to Support Vector Classification. Available online: https://www.csie.ntu.edu.tw/~cjlin/papers/guide/guide.pdf (accessed on 19 November 2015).

56. Rembold, F.; Atzberger, C.; Savin, I.; Rojas, O. Using low resolution satellite imagery for yield prediction and yield anomaly detection. *Remote Sens.* **2013**, *5*, 1704–1733. [CrossRef]

57. Kawamura, K.; Akiyama, T.; Yokota, H.; Tsutsumi, M.; Yasuda, T.; Watanabe, O.; Wang, S.P. Quantifying grazing intensities using geographic information systems and satellite remote sensing in the Xilingol steppe region, Inner Mongolia, China. *Agric. Ecosyst. Environ.* **2005**, *107*, 83–93. [CrossRef]

58. Chun, F.; Li, C.L.; Bao, Y.H. The wavelet analysis of average temperature and precipitation in Xilinhot during 57 years. *J. Inner Mong. Normal Univ. (Nat. Sci. Ed.)* **2013**, *42*, 47–52. (In Chinese)

59. Farmer's Daily. Available online: http://szb.farmer.com.cn/nmrb/html/2011-04/13/nw.D110000nmrb_20110413_1-03.htm?div=-1 (accessed on 4 September 2015).

60. Hang, Y.L.; Bao, G.; Bao, Y.H.; Burenjirigala; Altantuya, D. Spatiotemporal changes of vegetation coverage in Xilingol grassland and its responses to climate change during 2000–2010. *Acta Agres. Sin.* **2014**, *22*, 1194–1204. (In Chinese)

61.   Yu, H.D.; Yang, X.C.; Xu, B.; Jin, Y.X.; Gao, T.; Li, J.Y. Changes of grassland vegetation growth in Xilingol league over 10 years and analysis on the influence factors. *J. Geo-Inf. Sci.* **2013**, *15*, 270–279. (In Chinese)
62.   Xilinhaote News. Available online: http://xilinhaote.nmgnews.com.cn/system/2009/07/20/010253966.shtml (accesed on 4 September 2015).
63.   Liu, C.L.; Fan, R.H.; Wu, J.J.; Yan, F. Temporal lag of grassland vegetation growth response to precipitation in Xilinguolemeng. *Arid Land Geogr.* **2009**, *32*, 512–518. (In Chinese)

Remote Sensing Based Simple Models of GPP in Both Disturbed and Undisturbed Piñon-Juniper Woodlands in the Southwestern U.S.

Dan J. Krofcheck [1,*], **Jan U. H. Eitel** [2,3,†], **Christopher D. Lippitt** [4,†], **Lee A. Vierling** [2,3,†], **Urs Schulthess** [5,†] and **Marcy E. Litvak** [1,†]

Academic Editors: Randolph H. Wynne and Prasad S. Thenkabail

[1] Department of Biology, University of New Mexico, Albuquerque, NM 87131, USA; mlitvak@unm.edu
[2] Geospatial Laboratory for Environmental Dynamics, University of Idaho, Moscow, ID 83844, USA; jeitel@uidaho.edu (J.U.H.E.); leev@uidaho.edu (L.A.V.)
[3] McCall Outdoor Science School, University of Idaho, McCall, ID 83638, USA
[4] Department of Geography and Environmental Studies, University of New Mexico, Albuquerque, NM 87131, USA; clippitt@unm.edu
[5] CIMMYT-Bangladesh, House 10/B, Road 53, Gulshan-2, Dhaka 1213, Bangladesh; u.schulthess@cgiar.org
*   Correspondence: djkrofch@unm.edu
†   These authors contributed equally to this work.

Abstract: Remote sensing is a key technology that enables us to scale up our empirical, *in situ* measurements of carbon uptake made at the site level. In low leaf area index ecosystems typical of semi-arid regions however, many assumptions of these remote sensing approaches fall short, given the complexities of the heterogeneous landscape and frequent disturbance. Here, we investigated the utility of remote sensing data for predicting gross primary production (GPP) in piñon-juniper woodlands in New Mexico (USA). We developed a simple model hierarchy using climate drivers and satellite vegetation indices (VIs) to predict GPP, which we validated against *in situ* estimates of GPP from eddy-covariance. We tested the influence of pixel size on model fit by comparing model performance when using VIs from RapidEye (5 m) and the VIs from Landsat ETM+ (30 m). We also tested the ability of the normalized difference wetness index (NDWI) and normalized difference red edge (NDRE) to improve model fits. The best predictor of GPP at the undisturbed PJ woodland was Landsat ETM+ derived NDVI (normalized difference vegetation index), whereas at the disturbed site, the red-edge VI performed best (R^2_{adj} of 0.92 and 0.90 respectively). The RapidEye data did improve model performance, but only after we controlled for the variability in sensor view angle, which had a significant impact on the apparent cover of vegetation in our low fractional cover experimental woodland. At both sites, model performance was best either during non-stressful growth conditions, where NDVI performed best, or during severe ecosystem stress conditions (e.g., during the girdling process), where NDRE and NDWI improved model fit, suggesting the inclusion of red-edge leveraging and moisture sensitive VI in simple, data driven models can constrain GPP estimate uncertainty during periods of high ecosystem stress or disturbance.

Keywords: semi-arid; red-edge; NDWI; woody mortality

1. Introduction

Semi-arid regions and drylands together cover more than 40% of the globe. In spite of the low fractional cover of vegetation and minimal annual precipitation, the contribution of semi-arid biomes to the global uptake of carbon dioxide from the atmosphere can be significant on annual time scales [1,2]. The inter-annual variability of precipitation in these ecosystems can result in both rapid carbon stock accumulation, and subsequent turnover due to frequent disturbance (e.g., drought, insect outbreak, fire), but how these relationships alter the net storage of carbon is poorly understood (e.g., [1]). The climate in the southwestern USA has recently experienced both increased regional air temperature, and decreased precipitation, both trends which are projected to continue in the coming decades [3–7]. The combined effects of these changes in climate have increased drought severity, exacerbated ecosystem stress, and ultimately triggered widespread forest mortality across the region [8–10]. Given the rapid state transitions of vegetation in these climatically sensitive biomes, and the projected changes in climate and associated mortality for this region [11], accurate monitoring of carbon uptake dynamics in these semi-arid ecosystems is an essential part of constraining uncertainties associated with regional carbon balance.

Monitoring ecosystem carbon uptake over large geographic extents requires the use of remote sensing. Typically, remote sensing driven models of ecosystem function predict Gross Primary Productivity (GPP), the total atmospheric carbon taken up via photosynthesis (e.g., [12,13]). Satellite remote sensing provides the means to predict GPP by employing light use efficiency-based algorithms [14,15] that rely on vegetation indices (VIs), such as the normalized difference vegetation index (NDVI, [16]). This approach has been used at a variety of sites and scales, both by employing empirical (e.g., [17,18]) and light use efficiency/process driven techniques (e.g., [19]), generally using high temporal resolution, coarse spatial resolution (≥250 m) remote sensing data from sensors, such as the Moderate Resolution Imaging Spectrometer (MODIS) (e.g., [12]). To this end, the success of most space borne estimates of GPP hinge on a robust relationship between satellite estimated light interception and photosynthesis, via light use efficiency models, which generally perform well relative to *in situ* measurements from flux tower networks [20].

However, in semi-arid regions, vegetation function is constrained by water availability for the majority of the year, with periods of drought and heat interrupted briefly by spring snow melt or episodic pulses of precipitation. Consequently, in these biomes, changes in NDVI are constrained primarily by the availability of water, rather than light or temperature [21]. This results in NDVI and GPP often being decoupled in semi-arid regions due to low soil moisture, particularly where evergreen plants are present [22]. Other characteristics unique to semi-arid ecosystems that challenge the use of NDVI for characterizing changes in GPP in these biomes are highly variable precipitation patterns which trigger high inter-annual variability in GPP due to seasonal water limitations, but low variability in LAI and/or chlorophyll concentration (subsequently low variability in NDVI). Further, the low LAI (<1.5 mean LAI across landscapes, e.g., [23]) and spatially heterogeneous plant canopies typical of these systems can result in further uncertainties due to high reflectance by the soil background, thus confounding spectral signals relating to plant function (e.g., [24–26]).

One of the most universal responses to leaf stress is increasing visible reflectance [27], due to a combination of stressors which ultimately reduce chloropyll a + b, and consequently reduce the absorption of incident light [27,28]. Chlorophyll a + b strongly absorb in the red portion of the visible spectrum, resulting in saturation of the red band at low Chlorophyll a + b, and reducing its potential to track initial chlorophyll loss at the onset of stress. However, the red-edge has been shown to have a more linear response to a wide range of chlorophyll concentrations, increasing its potential as a stress indicator in vegetated systems over conventional red-NIR combinations [26,29–32]. Given the increasing availability of the red-edge waveband in commercial (e.g., RapidEye, WorldView-2, WorldView-3) sensors, and freely available data (Sentinel-2, launched June 2015), testing the potential of the red-edge waveband to improve modeled estimates of GPP in semi-arid ecosystems is becoming a feasible task.

Here, we test the ability of spectral VI's other than NDVI to model GPP in semi-arid ecosystems. We focus particularly on piñon-juniper woodlands for several reasons. First, because it is the largest biome in the Southwestern US, covering 18 million ha in New Mexico, Arizona, Colorado and Utah. Second, the changes in climate in this region have triggered a significant amount of mortality in this biome [10,33], and thus, quantifying the extent of this disturbance on productivity throughout the region is crucial to understanding how this extensive mortality has impacted both current and future carbon dynamics. Finally, we take advantage of an existing experimental manipulation in a Piñon-Juniper woodland in central New Mexico that was girdled in 2009 to simulate the widespread piñon mortality observed throughout the region. The advantage of using this experimental manipulation is that since the girdling in 2009, changes in ecosystem productivity triggered by this mortality have been continuously monitored using eddy covariance, and compared to similar measurements in a nearby intact PJ woodland that serves as a control.

Recent research [34] conducted in our experimental manipulation suggested that the red-edge employing normalized difference red-edge index (NDRE) [35] was more sensitive than NDVI to the observed initial decrease in leaf chlorophyll concentration triggered by piñon girdling in this PJ woodland, adding evidence to its potential as a stress monitoring component in GPP modeling efforts. The NDRE was determined to be more sensitive to changes in greenup of the low LAI herbaceous vegetation following mortality in this same system [22]. Based on these findings, the goal of this study was to test three specific hypotheses. The first hypothesis being that the observed variability in NDRE will allow more accurate estimation of GPP in both the disturbed and undisturbed PJ woodland in this experimental manipulation. Secondly, due to the inherent dependence of productivity on water availability in this biome, adding the normalized difference wetness index (NDWI), a VI that is sensitive to changes in foliar water content, will significantly improve the model fit by constraining the model error during the dry season. Finally, given the low vegetation cover and highly heterogeneous nature of PJ woodlands, that VIs generated from higher spatial resolution (5 m) data will provide more accurate estimates of GPP relative to traditional, moderate resolution (30 m) remote sensing data.

2. Experimental Section

2.1. Site Description

The study site includes two piñon-juniper woodlands separated by 3 km and located south of Mountainair, NM. In September of 2009, 1632 adult piñon (>7 cm diameter at breast height) in a 4 ha plot in one of the sites (hereafter referred as the girdled site; 34°26′48.54″N, 106°12′48.63″W) were mechanically girdled by severing the sapwood with a chainsaw at breast height (1.4 m) followed by the application of a 5% glyphosate solution directly applied to the cut to ensure rapid loss of conductivity and subsequent mortality of the trees. The total decoupling of the leaves with the roots following the manipulation was designed to replicate the landscape following the mortality that occurred in the response to the 1999–2002 drought [9,27]. The second PJ woodland (hereafter referred to as the control site; 34°26′18.420″N, 106°14′15.698″W) remained intact. We have continually monitored surface fluxes of carbon, water and energy at each site using tower mounted eddy covariance and associated micrometeorological sensors since 8 months prior to the girdling manipulation in February 2009.

The soil at both sites is Turkey Springs stony loam, characterized by an abundance of alluvially deposited limestone (Soil Survey Staff). The climate of the region is semi-arid, with a mean annual precipitation of 372 mm (±86.8 mm standard deviation, sd) and max, and min monthly temperatures of 19.8 °C (±0.77 °C sd) and 2.32 °C (±0.64 °C sd) respectively, over the past 20 years (PRISM Climate Group, Oregon State University). Incoming moisture to the site is largely bimodal, broken into winter snow melt from January to March and seasonal monsoon precipitation between August and October with a pronounced dry season occurring from April through July.

2.2. *Data*

2.2.1. Gross Primary Productivity

Data processing and instrumentation is identical at both sites. Eddy covariance (EC) derived surface fluxes of carbon, water and energy were measured at 10 Hz using a 3-axis sonic anemometer (CSAT-3, Campbell Scientific, Logan, UT, USA) and an open-path infrared gas analyzer (Li-7500, LiCor Biosciences; Lincoln, NE, USA). Continuous measures of net radiation, air temperature and relative humidity, soil temperature, photosynthetically active radiation, and soil moisture (volumetric water content) were made using a CNR1 (Kipp and Zonen, Delft, The Netherlands), HMP45C (Vaisala RH probe with aspirated radiation shield, Vantaa, Finland), TCAV (averaging thermocouple probes, 27 per site), up facing quantum sensors (Li-190SB, Licor Biosciences, Lincoln, NE, USA) and ECHO probes (TE 5 cm, Decagon Devices, Pullman, WA, USA, 27 per site) respectively. The fluxes were aggregated to 30 min intervals and were corrected for temperature and moisture variations (WPL, [36]) as well as frequency responses according to Massman [37]. Anemometer tilt due to terrain variability was corrected using a planar fit method. We used a friction velocity (u*) filter to reject data obtained when turbulence is low (u* less than a threshold value). Data gaps created by the u* filter, malfunctioning instruments, and rain were filled following Lasslop *et al.* [38]. We partitioned the gapfilled net ecosystem exchange (NEE) into the components of total C uptake through photosynthesis (*i.e.*, GPP) and total carbon leaving the ecosystem through both autotrophic and heterotrophic respiration (Re). We used exponential relationships between nighttime NEE and temperature following the methods of Lasslop *et al.* [38] to calculate continuous ecosystem respiration during the day. GPP was then calculated as NEE—Re [39].

We used a flux source area model [40] modified for 2 dimensions by Detto *et al.* [41], to characterize the experimental region at both sites that are measured by the flux towers. The source area model suggested that the four hectare analysis region at each site accounted for approximately 80% recovery at the tower level. In this way we ensured that the remote sensing data and the EC measurements were representing the same patch of vegetation.

2.2.2. Landsat ETM+

We used the web enabled Landsat database (WELD) web service [42] to download time series of Landsat ETM+ data from 2009 to 2011 for 36 pixels within the four hectare area measured by each flux tower. Each pixel time series was cleaned (*i.e.*, filtered to meet the following conditions) using the QA/QC data associated with WELD products [42]; only the pixels classified as containing no saturated bands and not cloudy (determined by the automatic cloud cover assessment algorithm, ACCA) [43] were retained. To remove effects of snow cover on the VI, the normalized difference snow index (NDSI) was used to remove snow covered pixels from the analysis. If more than 25% of the total analysis region for either the control or girdled site was covered by snow, the entire acquisition date was removed from the analysis. Hall *et al.* [44] suggest mapping snowy pixels as NDSI >0.4. However, we used a lower threshold of 0.2 in this study, which we validated with ground based imagery and field technician reports. The majority (82%) of the Landsat ETM+ data were acquired within ⩽3 days of a RapidEye acquisition (see Section 2.2.3), with the greatest time separation being 10 days. The Landsat ETM+ pixel time series was then linearly interpolated to daily intervals for comparison with the RapidEye data. In this manner, the same number of Landsat ETM+ VI and RapidEye VI were used in each of the models tested. Geolocation error for the Landsat ETM+ imagery was generally <0.5 pixel according to the WELD documentation.

2.2.3. RapidEye

A time series of 46 images (September 2009 to October 2011) were used with a top of atmosphere, dark object subtraction (TOA-DOS) correction applied by RapidEye, Inc. (now BlackBridge Ltd. of Berlin, Germany). The RapidEye data were reasonably registered on delivery, but to minimize variability in the spectral signature due to pixel shifts between scenes, we manually co-registered each image to a master image (September 2009) using 50 ground control points and a second order polynomial transformation. A total of 15–20 check points per image resulted in each scene being co-registered to a root mean square error (RMSE) <2.5 m (0.5 pixels). Eighteen images were excluded from the time series due to snow or cloud cover. Existing site boundary shapefiles were used to clip the images to the extent of both the control and girdled sites, coincident with the pixels extracted for the Landsat ETM+ analysis (Python Software Foundation. Python Language Reference, version 2.7). Figure 1 illustrates the distribution of RapidEye images throughout the duration of this experiment for the girdle and control site.

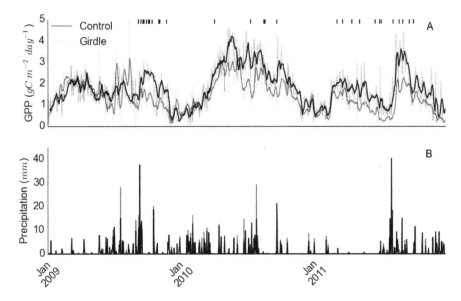

Figure 1. Time series of tower-measured gross primary production (**A**) and precipitation (**B**) at both the control (grey) and girdled (black) sites. RapidEye acquisition dates are indicated by the rug plot above the top plot.

2.2.4. Spectral Vegetation Indices

NDVI (Normalized Difference Vegetation Index), NDRE (Normalized Difference Red Edge), and NDWI (Normalized Difference Wetness Index), were calculated to inform our simple models of GPP and to test our hypotheses. NDRE and NDWI are only available using RapidEye and Landsat ETM+, respectively, and indicated by their subscripts in Table 1. The red edge here refers to a waveband on the RapidEye constellation of sensors that spans 690 to 730 nm.

$$NDVI = \frac{NIR - RED}{NIR + RED} \tag{1}$$

$$NDRE = \frac{NIR - Red\ Edge}{NIR + Red\ Edge} \tag{2}$$

$$NDWI = \frac{NIR - SWIR}{NIR + SWIR} \tag{3}$$

Table 1. Model framework description. Each model is structured as GPP ~ PAR × TA × VI where VI is represented by the corresponding Table entry above, while PAR and TA are derived from on-site instrumentation. Acronym definition: GPP (gross primary productivity), PAR (photosynthetically active radiation), TA (air temperature), VI (vegetation index), NDVI (normalized difference vegetation index), NDRE (normalized difference red-edge index), NDWI (normalized difference wetness index).

Model	VI Combination	Scale (m)
NDVI$_{LS}$	Landsat NDVI	30
NDVI$_{LSW}$	Landsat NDVI, NDWI	30
NDVI$_{RE}$	RapidEye NDVI	5
NDVI$_{REW}$	RapidEye NDVI, NDWI	5, 30
NDRE	RapidEye NDRE	5
NDRE$_W$	RapidEye NDRE, NDWI	5, 30

2.3. Statistical Analysis

We fitted six multiple linear regression models in open-source software package R 3.0.1 (R Development Core Team, 2013). Each regression model is structured as follows:

$$GPP ~ PAR × TA × VI \tag{4}$$

where PAR is photosynthetic active radiation, TA is air temperature, and VI depicts one or more vegetation index that are shown in Table 1. GPP, PAR, and TA were derived from on-site instrumentation (for GPP, see Section 2.2.1). The VI were calculated as the mean of all pixels that passed the qa/qc steps mentioned in Sections 2.2.2 and 2.2.3. Thus the spatial resolution of the driving VI was either 5 m (from RapidEye), 30 m (from Landsat ETM+), or a combination in the case of the mixed models.

Within each model group, we created a hierarchical structure of models containing all possible combinations of parameters and one first order interaction term. In this way we aimed to compare model groups by comparing the models with the best set of parameters within each group. Model performance was characterized using the root means squared error (RMSE) and the adjusted R^2 (R^2_{adj}) that adds an additional penalty for increasing model complexity:

$$R^2_{adj} = 1 - \frac{(1 - R^2)(n - 1)}{n - k - 1} \tag{5}$$

where n = the number of observations.

To identify the best model, we further calculated the Akaike information criterion (AIC_{adj}) :

$$AIC_{adj} = AIC + \frac{2k(2k + 1)}{n - k - 1} \tag{6}$$

where k = the number of parameters . We chose the AIC_{adj} to account for the relatively small number of observations and to avoid model over fitting [45]. The best model for both the control and girdle site was chosen based on the largest AIC_{adj} weight, and largest relative log likelihood, a measure of how likely a given model is actually the best.

3. Results and Discussion

3.1. Models of GPP in PJ Woodlands Had the Lowest Error at the Disturbed Site

Simple linear models driven by photosynthetically active radiation (PAR), air temperature (TA), and remotely sensed VI, summarized in Table 1, were able to explain between 80% and 90% of the variability of gross GPP in both intact and disturbed PJ woodlands (e.g., Table 2). However, the strength of this relationship was contingent on a tight coupling between the variability in VI and canopy physiological processes that affect a given index. For this reason, the model was run on individual years, as well as on all years together. Overall, the best performing model VI at the undisturbed PJ woodland was Landsat ETM+ derived NDVI, whereas at the disturbed site the red-edge VI performed best (R^2_{adj} of 0.90 and 0.92, respectively).

Table 2. Model performance statistics for only year 2009 of the experiment. See Table 1 for a description of model names. R^2_{adj} is the adjusted R^2, ΔAICc is the difference in model AICc (Akaike information criterion) relative to the lowest scoring model, RelLL is the relative log likelihood, or how likely it is that each model is actually the best, and Weights is a list of the AICc weights for each model.

Site	Sensor	Model	R^2_{adj}	ΔAICc	RelLL	Weights
Control	Landsat	NDVI$_{LS}$	0.902	0	1	0.997
	Landsat	NDVI$_{LSW}$	0.75	12.159	0.002	0.002
	RapidEye	NDVI$_{RE}$	0.813	14.072	0.001	0.001
	RapidEye	NDVI$_{REW}$	0.056	29.427	0	0
	RapidEye	NDRE	0.024	29.86	0	0
	RapidEye	NDRE$_W$	0.135	33.983	0	0
Girdle	Landsat	NDVI$_{LS}$	0.391	18.428	0	0
	Landsat	NDVI$_{LSW}$	0.439	17.684	0	0
	RapidEye	NDVI$_{RE}$	0.238	41.847	0	0
	RapidEye	NDVI$_{REW}$	0.392	39.812	0	0
	RapidEye	NDRE	0.921	0	1	1
	RapidEye	NDRE$_W$	0.361	18.851	0	0

When we drove the model hierarchy with data from all three years in the study, the collection of models performed poorly across the board (Figure 2, Table 3), with the Landsat ETM+ based models performed best (R^2_{adj} = 0.40 at control, R^2_{adj} = 0.50 at girdle). We believe this to be due to the coupling strength between changes in canopy VI and canopy function during as a function of interannual variability in climate. As an example, the fit statistics from both the Landsat ETM+ and RapidEye driven models improved significantly when a single year (e.g., 2009) was used to drive the regression (Figure 3), dramatically increasing the fit of the simple Landsat ETM+ NDVI model at the control site (R^2_{adj} from 0.40 to 0.90) and increased the fit of the RapidEye NDRE model at the girdled site (R^2_{adj} from 0.14 to 0.92) (Table 2, discussed in following sections). At the control site, Fall 2009 was very productive, with a strong monsoon bringing sufficient moisture to the ecosystem, and consequently the coupling of Landsat ETM+ NDVI, PAR, and air temperature explained almost 90% of the variability seen in GPP. In the case of girdled site during the same period, the manipulation in piñon mortality drove both the rapid decrease in VI, and the sudden decrease in canopy uptake of carbon. Both during this period of significant canopy stress and throughout the entire analysis period, the best model fits were seen at the girdled site, possibly due to the drastic decreases in LAI and subsequent increased coupling between the VI and GPP, previously documented as a potential effect of the manipulation itself [22].

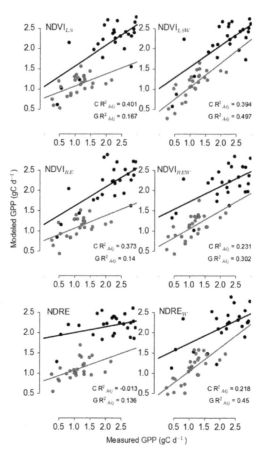

Figure 2. One to one plots of tower derived (measured) and predicted (modeled) gross primary productivity for the control (C, black) and girdled (G, grey) sites for all years of the study. See Table 1 for a description of model names.

Table 3. Model performance statistics for all years of data. See Table 1 for a description of model names. R^2_{adj} is the adjusted R^2, $\Delta AICc$ is the difference in model AICc relative to the lowest scoring model, RelLL is the relative log likelihood, or how likely it is that each model is actually the best, and Weights is a list of the AICc weights for each model.

Site	Sensor	Model	R^2_{adj}	$\Delta AICc$	RelLL	Weights
Control	Landsat	$NDVI_{LS}$	0.401	0.836	0.658	0.352
	Landsat	$NDVI_{LSW}$	0.394	3.518	0.172	0.092
	RapidEye	$NDVI_{RE}$	0.373	0	1	0.534
	RapidEye	$NDVI_{REW}$	0.231	7.608	0.022	0.012
	RapidEye	NDRE	−0.013	12.934	0.002	0.001
	RapidEye	$NDRE_W$	0.218	8.042	0.018	0.01
Girdle	Landsat	$NDVI_{LS}$	0.167	10.991	0.004	0.004
	Landsat	$NDVI_{LSW}$	0.497	0	1	0.899
	RapidEye	$NDVI_{RE}$	0.14	13.979	0.001	0.001
	RapidEye	$NDVI_{REW}$	0.302	8.552	0.014	0.012
	RapidEye	NDRE	0.136	11.945	0.003	0.002
	RapidEye	$NDRE_W$	0.45	4.803	0.091	0.081

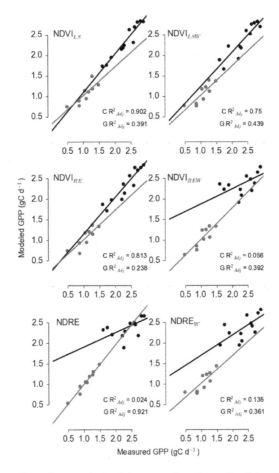

Figure 3. One to one plots of tower derived (measured) and predicted (modeled) gross primary productivity for the control (C, black) and girdled (G, grey) sites for 2009 only. See Table 1 for a description of model names.

3.2. NDRE and NDWI Reduced Model Error only during Periods of Significant Stress

In our linear model hierarchy, including NDRE did not significantly improve model performance relative to models that used only NDVI (Table 2). Previous work in this region suggests NDRE is more responsive to decreases in canopy chlorophyll concentration relative to NDVI, yet following a lag period (roughly 2 weeks), the NDVI performed equally well in quantifying foliar concentration of chlorophyll [34]. Given that our time series spanned multiple years, with rapid decreases in canopy chlorophyll occurring only at the girdled site in 2009, we ran the same model structures on only data from 2009, which captured the girdling process (Figure 3). Over this smaller time period of significant canopy stress, the NDRE dramatically improved model performance at the girdled site (R^2_{adj} from 0.136 for NDVI to 0.92 for NDRE, Table 2). However, we did not observe significant improvement in NDRE driven model fit at the control site, which is consistent with the idea that NDRE explains variability in canopy function best during early periods of rapid canopy change.

The incorporation of Landsat ETM+-derived NDWI into the model structure did not improve model fit in either Landsat ETM+ or RapidEye based models when the entire time series was used in the regression (Figure 2, Table 3). NDWI did however slightly reduce prediction error in both Landsat ETM+ and RapidEye NDVI based models at the girdled site, in 2009 alone (Figure 3, Table 2). While the inclusion of the moisture index slightly improved model performance, the added complexity was penalized fairly heavily given the small number of observations used to parameterize the model.

Previous work has shown that including Landsat ETM+ moisture indices can improve estimates of productivity in some dryland ecosystems [21,46], although it appears to be less effective in sparse canopy systems [47]. Our data suggest NDWI may only improve estimates of productivity in semi-arid coniferous woodlands during rapid changes in canopy physiological status, such as during canopy mortality, yet do not provide sufficient evidence to justify increasing the complexity of the model structure in this case. Moisture sensitive VI are less effective at representing canopy water content during periods of low to moderate drought stress [48,49], suggesting that GPP predictions during periods of the year when water is less limiting may not benefit from the inclusion of a canopy moisture VI as much as during periods of drought stress. This is evident at the girdle site during 2011, a year which was characterized by significant drought across the southwestern US. Model performance improved at the girdle site with the inclusion of Landsat ETM+ NDWI. However, the increased complexity of the model and our small sample size limit the generalizability of the finding that NDWI may constrain model uncertainty in these systems during drought periods (Table 4).

Table 4. Model performance statistics for only year 2011 of the experiment, a significant drought year. See Table 1 for a description of model names. R^2_{adj} is the adjusted R^2, ΔAICc is the difference in model AICc relative to the lowest scoring model, RelLL is the relative log likelihood, or how likely it is that each model is actually the best, and Weights is a list of the AICc weights for each model.

Site	Sensor	Model	R^2_{adj}	ΔAICc	RelLL	Weights
Control	Landsat	NDVI$_{LS}$	0.815	0	1	0.899
	Landsat	NDVI$_{LSW}$	0.698	4.414	0.11	0.099
	RapidEye	NDVI$_{RE}$	0.795	22.327	0	0
	RapidEye	NDVI$_{REW}$	0.292	12.08	0.002	0.002
	RapidEye	NDRE	−0.042	15.56	0	0
	RapidEye	NDRE$_W$	0.5	30.359	0	0
Girdle	Landsat	NDVI$_{LS}$	0.592	4.36	0.113	0.04
	Landsat	NDVI$_{LSW}$	0.677	1.575	0.455	0.161
	RapidEye	NDVI$_{RE}$	0.711	0.234	0.89	0.315
	RapidEye	NDVI$_{REW}$	0.662	2.097	0.35	0.124
	RapidEye	NDRE	0.717	0	1	0.354
	RapidEye	NDRE$_W$	0.672	8.709	0.013	0.005

3.3. Inconsistency in Sensor View Imposed Significant Variability RapidEye VI Model Performance

Our results from the entire 3 year period do not support the hypothesis that VI generated from higher spatial resolution data should be a better predictor of ecosystem carbon uptake (Table 2). In fact, using higher spatial resolution data from RapidEye surprisingly decreased model performance at the control site in some cases, and did not significantly improve model performance at the girdled site. Some of this can be explained by the inconsistent scene to scene sensor view angle in our RapidEye time series, which influenced the apparent fractional cover of vegetation (Figure 4A). In an effort to test whether or not the variability in view angle was imposing noise into the RapidEye data, we binned the analysis by sensor view angles <7 °C off nadir, given the majority of our data were collected at more than 7 degrees off nadir (60%, Figure 4B). Limiting the analysis to scenes with absolute sensor view angle <7 degrees significantly improved model fits in NDVI RapidEye mods (at the girdled site (R^2_{adj} increased from 0.14 to 0.77), yet decreased model performance at the control site; Table 5, Figure 5), potentially due to the changing apparent fraction of under-story and inter-canopy vegetation as a function of sensor view angle, which has a more pronounced impact at the girdle site relative to the control due to the lower fractional cover of overstory vegetation post manipulation. The inclusion of view angle into the model hierarchy as a parameter resulted in model over-fitting given the small sample size. The high inter-annual variability in system state (e.g., monsoonal precipitation or severe drought) at these sites further confounded our ability to account for view angle effects, due to the coupling between canopy function and the remotely derived VI for some dates being much tighter than others (see Section 2.1).

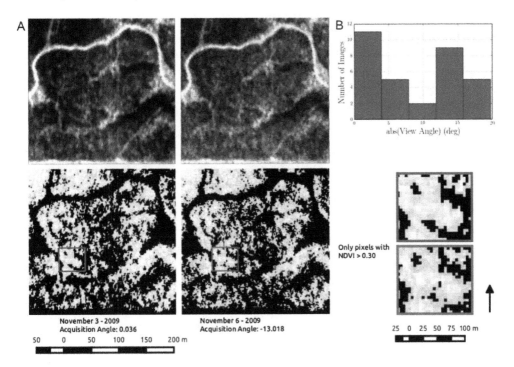

Figure 4. (A) RapidEye false color composites of the PJ control site (**top**) for two image dates 3 days apart in 2009, and the corresponding fraction of pixels with NDVI >0.30 (**bottom**), with detailed subsets on the far right; (**B**) Histogram of view angles that comprised the entire RapidEye time series. View angles are represented as absolute values from nadir.

Table 5. Model performance statistics for RapidEye sensor view angles less than 7 degrees off nadir. See Table 1 for a description of model names. R^2_{adj} is the adjusted R^2, ΔAICc is the difference in model AICc relative to the lowest scoring model, RelLL is the relative log likelihood, or how likely it is that each model is actually the best, and Weights is a list of the AICc weights for each model.

Site	Sensor	Model	R^2_{adj}	ΔAICc	RelLL	Weights
Control	Landsat	$NDVI_{LS}$	0.172	0.134	0.935	0.348
	Landsat	$NDVI_{LSW}$	0.103	5.782	0.056	0.021
	RapidEye	$NDVI_{RE}$	0.078	6.113	0.047	0.018
	RapidEye	$NDVI_{REW}$	−0.133	8.584	0.014	0.005
	RapidEye	NDRE	0.446	0	1	0.372
	RapidEye	$NDRE_W$	0.665	0.91	0.634	0.236
Girdle	Landsat	$NDVI_{LS}$	−0.248	9.964	0.007	0.003
	Landsat	$NDVI_{LSW}$	0.475	0.433	0.805	0.317
	RapidEye	$NDVI_{RE}$	0.777	0	1	0.394
	RapidEye	$NDVI_{REW}$	0.206	4.985	0.083	0.033
	RapidEye	NDRE	0.087	0.894	0.64	0.252
	RapidEye	$NDRE_W$	0.437	10.209	0.006	0.002

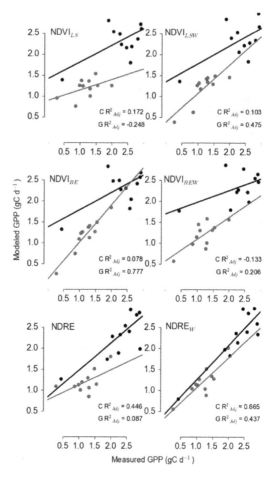

Figure 5. One to one plots of tower derived (measured) and predicted (modeled) gross primary productivity for the control (C, black) and girdled (G, grey) sites for RapidEye acquisition angles <7 degrees off nadir. See Table 1 for a description of model names.

Previous work indicates the effects of sensor view angle on the fractional cover of various feature types (e.g., [50], pertaining to composites derived from the advanced very high radiometric resolution spectrometer (AVHRR)), and generally near-nadir view angles are a selection criteria for multi-image or multi-sensor analyses (e.g., [51]). Given the increasing amount and complexity of remote sensing data at the disposal of terrestrial scientists, multi-resolution and time series data assimilation approaches to developing temporally and spatially resolved remote sensing products will need to address the confounding effects of view angle on apparent vegetation cover, especially in sparse canopy systems.

3.4. Implications for Regional Remote Sensing Based Estimations of GPP in PJ Woodlands

The results of our case study suggest that remote sensing driven, simple linear models of GPP have the potential to accurately describe patterns in regional carbon uptake, using locally measured PAR and air temperature as covariates. In this study, we chose to use Landsat ETM+ data, given its ease of access via the WELD portal. The highly heterogeneous composition of PJ woodlands, and the small scale of our manipulation experiment (200 m × 200 m), precluded the use of MODIS sized pixels (≥250 m × 250 m), and given the heterogeneous patterns of mortality and subsequent recovery typical of disturbed PJ woodlands (e.g., [9,22,27]), Landsat scale or finer resolution (≤30 m) products are of the appropriate spatial scale to resolve the patchy, heterogeneous patterns of mortality typical in disturbed PJ woodlands. In spite of the sensor view angle complications described in Section 3.3, limiting image acquisition angle variability impacted model performance in some cases, suggesting that increased spatial resolution data may have a role in constraining GPP predictions in these heterogeneous woodlands in the future, yet we were unable to fully test the hypothesis due to inadequate sample sizes when considering the variation in sensor view angle.

The inclusion of the red-edge leveraging NDRE from RapidEye only improved model performance during periods of significant stress, such as during the selective mortality of piñon pine that we imposed on the girdled site in Fall 2009, or during periods of severe drought such as in 2011 (see Section 3.2). Given the imminent transition in climate projected for the southwestern US, and the already evident impacts on piñon mortality regionally, GPP prediction efforts in PJ woodlands should benefit from the incorporation of NDRE during mortality. Given recent launch of Sentinel-2, which can leverage the red-edge and infrared wavebands required to compute NDRE and NDWI respectively, the potential to incorporate NDRE into near-future regional models of GPP as a consequence of piñon mortality may significantly constrain the current uncertainty in estimating PJ woodland ecosystem function. Further, while NDWI did not significantly contribute to model performance in this study, the measured productivity of semi-arid plant canopies is strongly a function of available water, which ultimately affects the coupling between canopy VI and GPP (e.g., [22]), and consequently the predictive power of VI driven models of GPP in these regions. Previous work has shown NDWI to correlate reasonably well with foliar water content, and water potential in piñon pine, but not juniper [52], suggesting further that the inclusion of these VI to empirical modeling attempts should be perhaps constrained to piñon dominated pixels, requiring higher resolution imagery. Remotely sensed regional estimates of surface soil water content from the soon to launch Soil Moisture Active Passive sensor (SMAP), may provide an alternative to constrain the inter-annual variability in the VI ~ GPP relationship in semi-arid PJ woodlands.

4. Conclusions

Our results are promising in that we can use simple linear models to estimate GPP in both disturbed and undisturbed PJ woodlands driven by remotely sensed datasets. While structurally sensitive, NDVI is more informative to models of GPP than NDRE except during periods of extreme stress or disturbance. Similarly, we only saw a significant improvement in model performance using NDWI at the girdled site, during the manipulation event that took place in the Fall of 2009. Finally, the use of the RapidEye data did slightly improve estimates of GPP in both the control and girdled sites relative to Landsat ETM+, however this was only true when we reduced the variability in scene to scene sensor view angle in our RapidEye time series. This apparent advantage of the RapidEye data may be due to a combination of factors, including spatial resolution (5 m pixels *vs.* 30 m pixels) and spectral sensitivity of the sensor. While this may not play a strong role in more homogeneous, closed canopy systems, sensor view angle in this study often imposed more variability on NDVI than natural seasonal variability. Consequently, we recommend that remote sensing efforts to model VI sensitive processes in heterogeneous, low fractional cover systems, place high constraints on acquisition angles for time series, or bin analyses by viewing angle to minimize the potential confounding effects.

We recognize that the temporally resolved RapidEye data set we utilized for this study is not a common commodity and currently carries with it a large cost. Using red-edge data added sensor and illumination geometry complexity, but did improve estimates of GPP during periods of ecosystem stress despite it. Our results suggest high resolution, red-edge employing platforms will potentially be very useful for resolving changes in canopy function during periods of rapid disturbance or recovery where LAI may be changing slowly in relation to chlorophyll content. The recently launched Sentinel-2 satellite mission will allow this to be tested on a broader scale by providing greater spatial and temporal resolution than Landsat, as well as the ability to calculate NDRE and NDWI, and be freely available. Secondly, the upcoming soil moisture active passive sensor (SMAP) may provide either direct measurements, or modeled estimates of soil moisture, providing further predictive power to estimate carbon uptake rates in semi-arid ecosystems.

Acknowledgments: This research was supported primarily by the Office of Science (BER), US Department of Energy. We also acknowledge the funding from grant/cooperative agreement number 08HQGR0157 from the United States Geological Survey (USGS) via a subaward from AmericaView, the National Science Foundation through funds to the Sevilleta LTER site and NM EPSCoR, in addition to the funding from the USFS for support of the flux towers, and from the New Mexico Space Grant Consortium (NMSGC) for the partial funding of satellite data acquisitions. Its contents are solely the responsibility of the authors and do not necessarily represent the official views of the Office of Science, US Department of Energy. Finally, many thanks to Greg Maurer for his work in processing the eddy-covariance data and sound advice throughout the project.

Author Contributions: Dan Krofcheck conducted the primary data analysis and contributed in the eddy covariance instrumentation management, as well as crafting the manuscript. Jan Eitel, Christopher Lippitt, and Lee Vierling advised on and contributed to the data analysis, as well as provided remote sensing expertice, and significant editing of the manuscript. Urs Schulthess contributed to the RapidEye data processing, initial data mining, as well as editing the manuscript. Marcy Litvak provided all of the eddy-covariance data, advised on research question generation, and reshaping of the manuscript.

Conflicts of Interest: The authors declare no conflict of interest.

References

1. Poulter, B.; Frank, D.; Ciais, P.; Myneni, R.B.; Andela, N.; Bi, J.; Broquet, G.; Canadell, J.G.; Chevallier, F.; Liu, Y.Y.; et al. Contribution of semi-arid ecosystems to interannual variability of the global carbon cycle. *Nature* **2014**, *509*, 600–603. [CrossRef] [PubMed]

2. Ahlstrom, A.; Raupach, M.R.; Schurgers, G.; Smith, B.; Arneth, A.; Jung, M.; Reichstein, M.; Canadell, J.G.; Friedlingstein, P.; Jain, A.K.; et al. The dominant role of semi-arid ecosystems in the trend and variability of the land CO_2 sink. *Science* **2015**, *348*, 895–899. [CrossRef] [PubMed]

3. Intergovernmental Panel on Climate Change (IPCC). *Climate Change 2007: Impacts, Adaptation, and Vulnerability*; Contribution of Working Group II to the Fourth Assessment Report of the Intergovernmental Panel on Climate Change; IPCC: Cambridge, UK, 2007.

4. Seager, R.; Ting, M.; Held, I.; Kushnir, Y.; Lu, J.; Vecchi, G.; Huang, H.; Harnik, N.; Leetmaa, A.; Lau, N.; et al. Model projections of an imminent transition to a more arid climate in southwestern North America. *Science* **2007**, *316*, 1181–1184. [CrossRef] [PubMed]

5. Seager, R.; Naik, N.; Vogel, L. Does global warming cause intensified interannual hydroclimate variability? *J. Clim.* **2012**, *25*, 3355–3372. [CrossRef]

6. Intergovernmental Panel on Climate Change (IPCC). *Climate Change 2014: Impacts, Adaptation, and Vulnerability. Part B: Regional Aspects. Contribution of Working Group II to the Fifth Assessment Report of the Intergovernmental Panel on Climate Change*; Barros, V.R., Field, C.B., Dokken, D.J., Mastrandrea, M.D., Mach, K.J., Bilir, T.E., Chatterjee, M., Ebi, K.L., Estrada, Y.O., Genova, R.C., Eds.; Cambridge University Press: Cambridge, UK, 2014; p. 688.

7. Garfin, G.; Franco, G.; Blanco, H.; Comrie, A.; Gonzalez, P.; Piechota, T.; Smyth, R.; Waskom, R. *Southwest: Climate Change Impacts in the United States: The Third National Climate Assessment*; Melillo, J.M., Richmond, T.C., Yohe, G.W., Eds.; USA Global Change Research Program: Washington, DC, USA, 2014; pp. 462–486.

8. Allen, C.D.; Macalady, A.K.; Chenchouni, H.; Bachelet, D.; McDowell, N.; Vennetier, M.; Kitzberger, T.; Rigling, A.; Breshears, D.D.; Gonzalez, P.; et al. A global overview of drought and heat-induced tree mortality reveals emerging climate change risks for forests. *For. Ecol. Manag.* **2010**, *259*, 660–684. [CrossRef]

9. Breshears, D.D.; Cobb, N.S.; Rich, P.M.; Price, K.P.; Allen, C.D.; Balice, R.G.; Romme, W.H.; Kastens, J.H.; Floyd, M.L.; Belnap, J.; et al. Regional vegetation die-off in response to global-change-type drought. *Proc. Natl. Acad. Sci. USA* **2005**, *102*, 15144–15148. [CrossRef] [PubMed]

10. Clifford, M.; Royer, P.; Cobb, N. Precipitation thresholds and drought-induced tree die-off: Insights from patterns of Pinus edulis mortality along an environmental stress gradient. *New Phytol.* **2013**, *200*, 413–421. [CrossRef] [PubMed]

11. Williams, A.; Allen, C.; Macalady, A. Temperature as a potent driver of regional forest drought stress and tree mortality. *Nat. Clim.* **2012**, *3*, 8–13. [CrossRef]

12. Running, S.W.; Baldocchi, D.D.; Turner, D.P.; Gower, S.T.; Bakwin, P.S.; Hibbard, K.A. A global terrestrial monitoring network integrating tower fluxes, flask sampling, ecosystem modeling and eos satellite data. *Remote Sens. Environ.* **1999**, *127*, 108–127. [CrossRef]

13. Zhao, M.; Running, S.; Heinsch, F.A. MODIS-derived terrestrial primary production. In *Land Remote Sensing and Global Environmental Change*; Ramachandran, B., Justice, C.O., Abrams, M.J., Eds.; Springer New York: New York, NY, USA, 2011; Volume 11, pp. 635–660.

14. Monteith, J.L. Solar radiation and productivity in tropical ecosystems. *J. Appl. Ecol.* **1972**, *9*, 747–766. [CrossRef]

15. Monteith, J.L.; Moss, C.J. Climate and the efficiency of crop production in britain and discussion. *Philos. Trans. R. Soc. B Biol. Sci.* **1977**, *281*, 277–294. [CrossRef]

16. Tucker, C. Red and photographic infrared linear combinations for monitoring vegetation. *Remote Sens. Environ.* **1979**, *8*, 127–150. [CrossRef]

17. Sims, D.; Rahman, A.; Cordova, V.; Elmasri, B.; Baldocchi, D.; Bolstad, P.; Flanagan, L.; Goldstein, A.; Hollinger, D.; Misson, L. A new model of gross primary productivity for North American ecosystems based solely on the enhanced vegetation index and land surface temperature from MODIS. *Remote Sens. Environ.* **2008**, *112*, 1633–1646. [CrossRef]

18. Sims, D.A.; Brzostek, E.R.; Rahman, A.F.; Dragoni, D.; Phillips, R.P. An improved approach for remotely sensing water stress impacts on forest C uptake. *Glob. Chang. Biol.* **2014**, *20*, 2856–2866. [CrossRef] [PubMed]

19. Yuan, W.; Liu, S.; Yu, G.; Bonnefond, J.-M.; Chen, J.; Davis, K.; Desai, R.R.; Goldstein, A.H.; Gianelle, D.; Rossi, F.; et al. Global estimates of evapotranspiration and gross primary production based on MODIS and global meteorology data. *Remote Sens. Environ.* **2010**, *114*, 1416–1431. [CrossRef]

20. Verma, M.; Friedl, M.A.; Richardson, A.D.; Kiely, G.; Cescatti, A.; Law, B.E.; Wohlfahrt, G.; Gielen, B.; Roupsard, O.; Moors, E.J.; et al. Remote sensing of annual terrestrial gross primary productivity from MODIS: An assessment using the FLUXNET La Thuile data set. *Biogeosciences* **2014**, *11*, 2185–2200. [CrossRef]

21. Fensholt, R.; Sandholt, I.; Rasmussen, M.S.; Stisen, S.; Diouf, A. Evaluation of satellite based primary production modeling in the semi-arid Sahel. *Remote Sens. Environ.* **2006**, *105*, 173–188. [CrossRef]

22. Krofcheck, D.J.; Eitel, J.U.H.; Vierling, L.A.; Schulthess, U.; Hilton, T.M.; Dettweiler-Robinson, E.; Pendleton, R.; Litvak, M.E. Detecting mortality induced structural and functional changes in a piñon-juniper woodland using Landsat and RapidEye time series. *Remote Sens. Environ.* **2014**, *151*, 102–113. [CrossRef]

23. Anderson-Teixeira, K.J.; Delong, J.P.; Fox, A.M.; Brese, D.A.; Litvak, M.E. Differential responses of production and respiration to temperature and moisture drive the carbon balance across a climatic gradient in New Mexico. *Glob. Chang. Biol.* **2011**, *17*, 410–424. [CrossRef]

24. Huete, A.R.; Jackson, R.D.; Post, D.F. Spectral response of a plant canopy with different soil backgrounds. *Remote Sens. Environ.* **1985**, *17*, 37–53. [CrossRef]

25. Huete, A.R. A soil-adjusted vegetation index (SAVI). *Remote Sens. Environ.* **1988**, *25*, 53–70. [CrossRef]

26. Eitel, J.U.H.; Long, D.S.; Gessler, P.E.; Hunt, E.R.; Brown, D.J. Sensitivity of ground-based remote sensing estimates of wheat chlorophyll content to variation in soil reflectance. *Soil Sci. Soc. Am. J.* **2009**, *73*, 1715–1723. [CrossRef]

27. Carter, G.A.; Knapp, A.K. Leaf optical properties in higher plants: Linking spectral characteristics to stress and chlorophyll concentration. *Am. J. Bot.* **2001**, *88*, 677–684. [CrossRef] [PubMed]

28. Hendry, G.A.F.; Houghton, J.D.; Brown, S.B. The degradation of chlorophyll—A biological enigma. *New Phytol.* **1987**, *107*, 255–302. [CrossRef]

29. Carter, G.A.; Miller, R.L. Early detection of plant stress by digital imaging within narrow stress-sensitive wavebands. *Remote Sens. Environ.* **1994**, *50*, 295–302. [CrossRef]

30. Eitel, J.U.H.; Long, D.S.; Gessler, P.E.; Smith, A.M.S. Using *in-situ* measurements to evaluate the new RapidEye satellite series for prediction of wheat nitrogen status. *Int. J. Remote Sens.* **2007**, *28*, 4183–4190. [CrossRef]

31. Eitel, J.U.H.; Long, D.S.; Gessler, P.E.; Hunt, E.R. Combined spectral index to improve ground-based estimates of nitrogen status in dryland wheat. *Agron. J.* **2008**, *100*, 1694–1702. [CrossRef]

32. Eitel, J.U.H.; Keefe, R.F.; Long, D.S.; Davis, A.S.; Vierling, L.A. Active ground optical remote sensing for improved monitoring of seedling stress in nurseries. *Sensors* **2010**, *10*, 2843–2850. [CrossRef] [PubMed]

33. Drought induced Tree Mortality and Ensuing Bark Beetle Outbreaks in Southwestern Pinyon-Juniper Woodlands. Available online: http://citeseerx.ist.psu.edu/viewdoc/download?doi=10.1.1.165.4291&rep= rep1&type=pdf#page=49 (accessed on 17 December 2015).

34. Eitel, J.U.H.; Vierling, L.A.; Litvak, M.E.; Long, D.S.; Schulthess, U.; Ager, A.A.; Krofcheck, D.J.; Stoscheck, L. Broadband, red-edge information from satellites improves early stress detection in a New Mexico conifer woodland. *Remote Sens. Environ.* **2011**, *115*, 3640–3646. [CrossRef]

35. Gitelson, A.A.; Merzlyak, M.N. Quantifying estimation of chlorophyll using reflectance spectra. *J. Photochem. Photobiol. B Biol.* **1994**, *22*, 247–252. [CrossRef]

36. Webb, E.K.; Pearman, G.I.; Leuning, R. Correction of flux measurements for density effects due to heat and water vapour transfer. *Quart. J. R. Met. Soc.* **1980**, *106*, 85–100. [CrossRef]

37. Massman, W.J. A simple method for estimating frequency response corrections for eddy covariance systems. *Agric. For. Meteorol.* **2000**, *104*, 185–198. [CrossRef]

38. Lasslop, G.; Reichstein, M.; Papale, D.; Richardson, A.D.; Arneth, A.; Barr, A.; Stoy, P.; Wohlfahrt, G. Separation of net ecosystem exchange into assimilation and respiration using a light response curve approach: Critical issues and global evaluation. *Glob. Chang. Biol.* **2010**, *16*, 187–208. [CrossRef]

39. Flanagan, L.; Wever, L.; Carlson, P. Seasonal and interannual variation in carbon dioxide exchange and carbon balance in a northern temperate grassland. *Glob. Chang. Biol.* **2002**, *8*, 599–615. [CrossRef]

40. Hsieh, C.; Katul, G.; Chi, T. An approximate analytical model for footprint estimation of scalar fluxes in thermally stratified atmospheric flows. *Adv. Water Resour.* **2000**, *23*, 765–772. [CrossRef]

41. Detto, M.; Montaldo, N.; Albertson, J.D.; Mancini, M.; Katul, G. Soil moisture and vegetation controls on evapotranspiration in a heterogeneous Mediterranean ecosystem on Sardinia, Italy. *Water Resour. Res.* **2006**. [CrossRef]

42. Roy, D.P.; Ju, J.; Kline, K.; Scaramuzza, P.L.; Kovalskyy, V.; Hansen, M.; Loveland, T.R.; Vermote, E.; Zhang, C. Web-enabled Landsat Data (WELD): Landsat ETM+ composited mosaics of the conterminous United States. *Remote Sens. Environ.* **2010**, *114*, 35–49. [CrossRef]

43. Irish, R.I.; Barker, J.L.; Goward, S.N.; Arvidson, T. Characterization of the Landsat-7 ETM+ automated cloud-cover assessment (ACCA) algorithm. *Photogramm. Eng. Remote Sens.* **2006**, *72*, 1179–1188. [CrossRef]

44. Hall, D.K.; Foster, J.L.; Verbyla, D.L.; Klein, A.G. Assessment of snow-cover mapping accuracy in a variety of vegetation-cover densities in central Alaska. *Remote Sens. Environ.* **1998**, *66*, 129–137. [CrossRef]

45. Bedrick, E.J.; Tsai, C.L. Model selection for multivariate regression in small samples. *Biometrics* **1994**, *50*, 226–331. [CrossRef]

46. Leuning, R.; Cleugh, H.A.; Zegelin, S.J.; Hughes, D. Carbon and water fluxes over a temperate Eucalyptus forest and a tropical wet/dry savanna in Australia: Measurements and comparison with MODIS remote sensing estimates. *Agric. For. Meteorol.* **2005**, *129*, 151–173. [CrossRef]

47. Gu, Y.; Hunt, E.; Wardlow, B.; Basara, J.B.; Brown, J.F.; Verdin, J.P. Evaluation of MODIS NDVI and NDWI for vegetation drought monitoring using Oklahoma Mesonet soil moisture data. *Geophys. Res. Lett.* **2008**, *35*, 1–5. [CrossRef]

48. Hunt, E.R.; Rock, B.N. Detection of changes in leaf water content using near- and middle-infrared reflectance. *Remote Sens. Environ.* **1989**, *30*, 43–54.

49. Eitel, J.U.H.; Gessler, P.E.; Smith, A.M.S.; Robberecht, R. Suitability of existing and novel spectral indices to remotely detect water stress in *Populus* spp. *For. Ecol. Manag.* **2006**, *229*, 170–182. [CrossRef]

50. Stoms, D.M.; Bueno, M.J.; Davis, F.W. Viewing geometry of AVHRR image composites derived using multiple criteria. *Photogramm. Eng. Remote Sens.* **1997**, *63*, 681–689.

51. Abuzar, M.; Sheffield, K.; Whitfield, D.; O'Connell, M.; McAllister, A. Comparing inter-sensor NDVI for the analysis of horticulture crops in south-eastern Australia. *Am. J. Remote Sens.* **2014**, *2*, 1–9. [CrossRef]

52. Stimson, H.C.; Breshears, D.D.; Ustin, S.L.; Kefauver, S.C. Spectral sensing of foliar water conditions in two co-occurring conifer species: Pinus edulis and Juniperus monosperma. *Remote Sens. Environ.* **2005**, *96*, 108–118. [CrossRef]

Scaling up Ecological Measurements of Coral Reefs Using Semi-Automated Field Image Collection and Analysis

Manuel González-Rivero [1,2,*], Oscar Beijbom [1,3], Alberto Rodriguez-Ramirez [1], Tadzio Holtrop [1,4], Yeray González-Marrero [1,5], Anjani Ganase [1,2,6], Chris Roelfsema [7], Stuart Phinn [7] and Ove Hoegh-Guldberg [1,2,6]

Academic Editors: Stuart Phinn, Chris Roelfsema, Xiaofeng Li and Prasad S. Thenkabail

[1] Global Change Institute, The University of Queensland, St Lucia, QLD 4072, Australia; obeijbom@eecs.berkeley.edu (O.B.); alberto.rodriguez@uq.edu.au (A.R.-R.); tadzio.holtrop@gmail.com (T.H.); yeraygma@gmail.com (Y.G.-M.); anjani.ganase@uq.net.au (A.G.); oveh@uq.edu.au (O.H.-G.)
[2] Australian Research Council Centre of Excellence for Coral Reef Studies, St Lucia, QLD 4072, Australia
[3] Department of Electrical Engineering & Computer Sciences, University of California, Berkeley, CA 94709, USA
[4] Institute for Biodiversity and Ecosystem Dynamics (IBED), University of Amsterdam, P.O. Box 94248, Amsterdam 1090 GE, The Netherlands
[5] Department of Biology, Ghent University, Ghent 9000, Belgium
[6] School of Biological Sciences, The University of Queensland, St Lucia, QLD 4072, Australia
[7] Remote Sensing Research Centre, School of Geography, Planning and Environmental Management, The University of Queensland, St Lucia, QLD 4072, Australia; c.roelfsema@uq.edu.au (C.R.); s.phinn@uq.edu.au (S.P.)
[*] Correspondence: m.gonzalezrivero@uq.edu.au

Abstract: Ecological measurements in marine settings are often constrained in space and time, with spatial heterogeneity obscuring broader generalisations. While advances in remote sensing, integrative modelling and meta-analysis enable generalisations from field observations, there is an underlying need for high-resolution, standardised and geo-referenced field data. Here, we evaluate a new approach aimed at optimising data collection and analysis to assess broad-scale patterns of coral reef community composition using automatically annotated underwater imagery, captured along 2 km transects. We validate this approach by investigating its ability to detect spatial (e.g., across regions) and temporal (e.g., over years) change, and by comparing automated annotation errors to those of multiple human annotators. Our results indicate that change of coral reef benthos can be captured at high resolution both spatially and temporally, with an average error below 5%, among key benthic groups. Cover estimation errors using automated annotation varied between 2% and 12%, slightly larger than human errors (which varied between 1% and 7%), but small enough to detect significant changes among dominant groups. Overall, this approach allows a rapid collection of *in-situ* observations at larger spatial scales (km) than previously possible, and provides a pathway to link, calibrate, and validate broader analyses across even larger spatial scales (10–10,000 km^2).

Keywords: XL Catlin Seaview Survey; coral reefs; monitoring; support vector machine

1. Introduction

Understanding the underlying drivers and causal factors determining the existence and sustainability of coral reefs has been propelled by the rapid degradation of these ecosystems [1,2]. These issues include Crown-of-Thorns outbreaks [3], coral bleaching and mortality [4,5], and damage from tropical storms [6], in addition to the impacts of sedimentation, nutrient run-off, and pollution [7]. Given the large footprint and cumulative effect of perturbations caused by these stressors, understanding their net effect on coral reef communities requires system-wide analysis, and generates research and management demand for broad-scale (>10,000 km^2), standardised data sets. This is especially important given the rapid ocean warming and acidification, effects of which are predicted to produce large-scale changes that are beginning to occur across the planet at regional and global scales [8,9].

While meta-analysis can be an efficient way to synthesise underwater assessment efforts and generalise within and throughout regions [10–14], variability across spatial scales, multiple observers, metrics and methodologies can pose serious challenges for broad generalisations [1,14]. Alternatively, integrative and multi-disciplinary approaches using extensive field observations, optical remote-sensing datasets (satellite- and aerial-derived products) and modelling tools have the potential to enable investigators to scale up field observations in order to understand processes driving change in coral reefs [6,15–17]. Irrespective of the approach used to understand reef functioning and change, there is a fundamental need for broad-scale and standardised field data to accurately record and understand reefs under transition in order to provide informed management advice in a timely manner [1,2].

Optical remote sensing provides broad-scale aerial coverage of coral reef systems (e.g., 10,000 km^2) with every pixel assessed relative to its habitat composition and with pixel size determining the level of benthic detail mapped, resulting in 100% coverage of the study area [18]. However, optical remote-sensing products do not provide sufficient detail and reliability when compared to field-based measurements [19]. Conversely, field-based observations, while critical for the calibration of remote-sensing imagery and validation of the resulting maps [17,20], typically cover only small areas (<1%) of the study site [21].

Over the past three decades, underwater photography and videography has become increasingly accessible and is now widely used for monitoring coral reef benthic communities. Recent advances in digital photographic technology have enabled more efficient ways of obtaining observations and collecting data on the state of coral reef ecosystems [17,22,23]. Underwater vehicles, as well as diver-acquired methods [24–27], have also extended the capability of capturing large volumes of photographic records. Such is the case of the approach we evaluate here, the XL Catlin Seaview Survey (CSS), a method aimed at evaluating spatial and temporal patterns of benthic community structure in coral reefs using high-resolution imagery collected across linear transects (~2 km in length) by a customised underwater diver propulsion vehicle [24]. Insofar as the necessity for field data persists, the underlying challenge is shifting from a focus on the generation of information to a focus on the capacity of new tools to decode such information into meaningful metrics, which can extend our understanding of how coral reefs are impacted by a rapidly changing environment.

The challenge of rapid and accurate analysis of large volumes of images has led to productive collaborations between marine and computer science. While automated image analysis is extensively used in satellite image analysis [28] and plankton ecology [29], its application to coral reef systems is relatively new. Such methods, which typically rely on machine learning to map visual attributes of images to semantic classes, are enabling marine scientists to extract useful ecological data from photographic records [30–32] at speeds significantly faster than manual methods [24]. Furthermore, the development of photographic sensors and computer vision methods has enabled integration of new approaches in coral reef ecology to quantify a range of other metrics relevant to the discipline (e.g., reef terrain complexity [33,34] and fish abundance [35]).

Previous studies have revealed the potential of automated image annotation to rapidly optimise data mining from underwater imagery and generate reliable ecological metrics relative to coral reefs [30,36–38]. While there is high fidelity between human and automated annotations, the latter tend to introduce a level of variability or "noise" to the benthic coverage estimations, perhaps attributed to changes in image quality over time and space (e.g., light, water clarity and distance of the camera from the substrate, *etc.*) [24,30]. This raises the question of whether such methodologies as those used in the CSS produce viable outcomes for detecting and monitoring reef-scale changes in benthic composition over time and at large spatial extensions (>10,000 km^2)?

Here, we validate the application of automated image analysis on field imagery collected by the CSS [24] with a central aim of scaling up underwater observations of coral reefs. In particular, we explored and analysed the error introduced by automated image classification when it is used to estimate changes in benthic community composition across large spatial (regional) and temporal scales (years). Using imagery collected at multiple sites across 49 reefs from the Great Barrier Reef (GBR) and Coral Sea Commonwealth Marine Reserve (CSCMR), we discuss the fidelity of automated and human estimations in the context of variability introduced by multiple observers, in order to understand, describe and offer potential applications and limitations of this technology.

2. Material and Methods

2.1. Study Site

Data were collected in 2012 and 2014, and extracted from a total of 107 transects, comprised of 126,700 images and totalling 325,000 m^2 of area surveyed across the outer reefs of the GBR and three main atolls in the CSCMR: the Flinders, Holmes and Osprey reef atolls (Figure 1). A subset of these transects was resurveyed in 2014 in the northern GBR (Figure 1), following the impact of category 5 Tropical Cyclone Ita. The re-survey data were used to evaluate the ability of the technique to detect temporal changes.

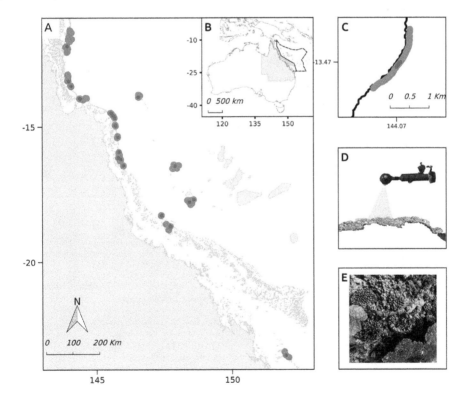

Figure 1. (**A**) Survey locations of the XL Catlin Seaview Survey in the (**B**) Great Barrier Reef (solid line) and Coral Sea Commonwealth Marine Reserve (dotted line), Queensland, Australia. Survey transects are shown in blue, while test sites are highlighted in red; (**C**) A detail of a specific transect is shown, where the sampling unit is depicted in red; (**D**) Images are collected along each 2-km transect using a customised diver propulsion vehicle; (**E**) Capturing underwater imagery of the reef benthos every three seconds.

2.2. Field Image Collection

A customised diver propulsion vehicle with a camera system mounted on it (named SVII, Figure 1D), consisting of three synchronised Cannon 5D-MkII cameras, was used to survey the fore-reef habitats in the outer reefs of the GBR and CSCMR. Images (21 Mp resolution) were collected every three seconds, approximately every 2 m, following a linear transect, averaging 1.8 km in length, along the 10 m depth contour (Figure 1C–E). Images were geo-referenced using a surface GPS unit tethered to the diver [24]. An on-board tablet computer enabled the diver to control camera settings (exposure and shutter speed) according to light conditions. Depth and altitude of the camera relative to the surface and the reef substrate were logged at half-second intervals using a Micron Tritech transponder (altitude) and pressure sensor (depth). This meta-data allowed for selection of imagery within a particular depth and altitude range (9–10 m depth and 0.5–2 m altitude) needed to maintain consistency and address variable environmental conditions, as well as ensuring a spatial resolution for each image of approximately 10 pixels$\cdot$cm^{-1} (ratio between the number of pixels contained in the diagonal of the image and its estimated size in centimetres). Using the geometry of the lens and altitude values, this pixel to centimetre ratio was calculated to crop the image to standardised 1m^2 photo quadrants. Details are provided in González-Rivero et al. [24].

2.3. Image Analysis

2.3.1. Label Set for Benthic Categories

A label set of 19 functional categories was established (Table 1). These categories were chosen for their functional relevance to coral reef ecosystems and their ability to be reliably identified from images by human annotators [38]. Four broad groups represent the main benthic components of coral reefs in the GBR and CSCMR: "Hard Corals", "Soft Corals", "Algae", and "Others". Hard corals comprise 11 functional groups classified based on a combination of taxonomy (i.e., family) and colony shape (i.e., branching, massive, encrusting, plating, and tabular). These groups were derived, modified and simplified from existing classification schemes [39,40]. Soft corals were classified and represented by three main functional groups: (1) Alcyoniidae soft corals, in particular the dominant genera; (2) sea fans and plumes from the family Gorgoniidae; and (3) other soft corals.

Table 1. Label set defining of the benthic categories employed for the classification of coral reefs benthos in the Great Barrier Reef (GBR) and Coral Sea Commonwealth Marine Reserve (CSCMR), Australia.

Group	Short Name	Taxonomic Description	Overall Functional Attributes	Ref.
Hard Coral	Acr. branching	Family Acroporidae, branching morphology (excluding hispidose type branching).	Major reef framework builders. Competitive life-history strategy: fast-growing species, spawning reproduction, high susceptibility to thermal stress (bleaching) and wave action. Provide habitat to a range of other reef-dwelling species.	[41–47]
	Acr. hispidose	Family Acroporidae, hispidose morphology	Competitive life-history strategy: fast-growing species, spawning reproduction, high susceptibility to thermal stress (bleaching) and wave action.	[44–47]
	Acr. other	Other corals from the family Acroporidae (e.g., Isopora)	Brooding reproduction, severe/high susceptibility to thermal stress, and low/moderate susceptibility to wave action.	[44,45,47,48]
	Acr. encrusting	Family Acroporidae, plate and encrusting morphologies	Major reef framework builders. Competitive and generalist life-history strategies: fast/moderate growth rates, spawning reproduction, high/severe susceptibility to thermal stress, moderate to low susceptibility to wave action.	[43–46,48]

Table 1. *Cont.*

Group	Short Name	Taxonomic Description	Overall Functional Attributes	Ref.
	Acr. Tabular	Family Acroporidae, table, corymbose and digitate morphologies	Major reef framework builders. Competitive life-history strategy: fast-growing species, spawning reproduction, high/severe susceptibility to thermal stress (bleaching), and high to low susceptibility to wave action. High to moderate susceptibility to disease outbreaks. Largely contribute to reef structural complexity.	[43–46,49,50]
	Massive meandroid	Families Favidae and Mussidae, massive and meandroid morphologies	Major reef framework builders. Stress-tolerant life history: slow-growing species, spawning reproduction, moderate susceptibility to thermal stress, and low susceptibility to wave action.	[43–46,48]
	Other corals	Other hard coral including all other groups not represented by the other coral categories of this label set.	Mixed attributes. Low/moderate susceptibility to thermal stress.	[51]
	Pocillopora	Family Pocilloporidae	Reef framework builders. Competitive and weedy life-history strategies: early colonisers in reef succession trajectories, fast-growing species, brooding reproduction, highly susceptible to thermal stress, but moderate resistant to wave action. High prevalence of coral diseases.	[43–46,48,51,52]
	Por. branching	Family Poritidae, branching morphology	Weedy life-history strategy: spawning reproduction, high/moderate susceptibility to thermal stress.	[45,46,53]
	Por. encrusting	Family Poritidae, encrusting morphology	Brooding reproduction. Low/moderate susceptible to thermal stress.	[51,54]
	Por. massive	Family Poritidae, massive morphology	Major reef framework builders. Stress-tolerant life history: slow-growing species, spawning reproduction, low/moderate susceptibility to thermal stress, and low susceptibility to wave action.	[43–46,48]
Algae	CCA	Crustose Coralline Algae	Major reef framework builders and cementers. Provide key contribution to coral reef primary production. Facilitation of coral recruitment.	[55–58]
	Macroalgae	Macroalgae. All genera.	Key contribution to coral reef primary production. Provide food source and habitat to a range of other reef dwelling species. Critical role during phase shifts.	[59–62]
	Turf	Multi-specific algal assemblage of 1 cm or less in height	Provide key contribution to coral reef primary production. Nitrogen fixation. Provide food source and habitat to a range of other reef dwelling species.	[63–66]
Others	Sand	Unconsolidated reef sediment	Not applicable (N/A)	N/A
	Other Invert.	Other sessile invertebrates	Mixed attributes.	N/A
Soft Coral	Alc. Soft coral	Soft coral, family Alcyoniidae, genera Lobophytum and Sarcophytum.	An important contributor to reef's structural complexity and biodiversity, providing habitat to a range of other reef-dwelling species. Slow-growing and suspension feeders. Spawning and asexual propagation. Deplete large amounts of suspended particulate matter.	[67–70]
	Gorg. Soft coral	Sea fans and plumes	Provide habitat to a range of other reef-dwelling species. Spawning and brooding reproduction. Suspension feeders.	[67–69]
	Other Soft coral	Other Soft corals	Mixed attributes.	N/A

The main algae groups were categorised according to their functional relevance: (1) Crustose Coralline Algae (CCA); (2) Macroalgae; and (3) Turf Algae. The latter is considered a grazed assemblage of algae species of up to 1 cm in height. The remaining group, categorised as "Other", consisted of sand and other benthic invertebrates (Table 1).

2.3.2. Point Annotations

All manual image annotations were conducted using point-sampling methods, *sensu* the Coral Point Count Method [71], adapted to GBR species and functional groups (Table 1). In this method a number of points are overlaid over the image at random locations, and the substrate types at each point location are assigned to one of the labels (Table 1). We used the point annotation tool of CoralNet for all manual annotation work [72]. A summary of the images and point annotations used in this work is provided in Table 2.

Table 2. Summary statistics of the annotation effort employed in this study for training the machine and validating the automated image estimations.

Description	# Images	# Points· Image^{-1}	# Sites
Training of automated annotator	1237	100	N/A
Spatial error	1042	40	41
Temporal error	335	40	7
Inter-observer variability	124	40	5

2.3.3. Automated Estimation of Benthic Composition

In order to automatically estimate benthic composition from collected imagery, we used a machine learning method, Support Vector Machine (SVM) [73], to automatically classify or identify benthic substrate categories from images based on a training provided by human annotators. In general terms, SVM are supervised classification models with associated learning algorithms that recognise patterns from data (in our case, visual parameters described below) to discretise categories assigned *a priori* (Table 1). Given a set of training examples, each one marked for belonging to a given category, a SVM training algorithm builds a classifier that assigns new examples into one category or another.

A total of 1237 images were randomly drawn from the pool of 126,700 images and used as training data for the automated annotator. Training images were annotated by a human annotator at 100-point locations per image (Table 2), where the substrate beneath each point was identified to the taxonomic resolution described herein (Table 1). The goal of the automated annotation method was to learn from the human annotations and to automatically analyse the remaining 125,463 images (representing approximately 1% of the total number of images in the dataset). Although the automated annotation method can technically annotate every single pixel in every single remaining image, here we followed the standard point sampling protocol and only created automated annotations at 100 randomly selected points per image. The reasoning for this was twofold: (1) to ensure straight-forward comparisons to percentage cover estimates from the human annotations; and (2) because studies indicate that the estimation error is small if the number of points per image is sufficiently large (>10) [71].

We adopted the automated annotation method of Beijbom *et al.* [30,38]. In this method, the visual parameters of texture and colour descriptors are first encoded as 24-dimensional vectors for all pixels in all images. This information is then summarised into a 250×250 pixel neighbourhood [38] (which roughly corresponds to 25×25 cm of substrate) around each point of interest (e.g., a labelled point in the trained data or a point in the unlabelled images) using Fisher Encoding [74], resulting in a 1920 dimensional "feature" vector. The feature vectors extracted from the training data are used to train a linear SVM and the trained SVM is finally used to annotate entire datasets from 2012 and 2014, including the "validation set" discussed below. We use the publicly available code [75] to extract the features and liblinear [76] to train the SVM, all performed within a MATLAB environment (MATLAB 2014a, MathWorks, Inc., Natick, MA, USA). Feature extraction required ~20 s per image, SVM training required ~1 h for training and <1 s per image with 100 points per image, which approximately translates to an automated analysis rate of 170 images per hour for benthic coverage.

2.4. Validation of Automated Analysis of Field Imagery

In order to validate automated annotations, we performed a cross-validation exercise between machine and human estimations of benthic cover. Specifically, we evaluated two types of errors for detecting patterns in benthic composition using automated image annotation: (1) error variability across a large spatial extent (GBR and CSCMR), accounting for inter-reef variability in composition, appearance and light conditions; and (2) error for detecting temporal change in a given reef. These analyses are contextualised by comparing machine-introduced variability (error) to human inter-observer variability, which enabled the determination of thresholds for pattern detection using automated annotation. These results provide an important benchmark regarding the opportunities and limitations of rapid, broad-scale and automated assessment tools.

2.4.1. Evaluation Sites

Coral reefs are heterogeneous ecosystems composed of spatial patterns that range from the scale of an individual to that of an ecosystem [77,78]. At the broadest spatial scales, physical processes are the dominant determinant of spatial structures, whereas biological processes override the effects of environmental processes at fine spatial scales [79,80]. Here, we use spatial autocorrelation analysis to set a scale of reference to depict assemblages of communities on the reef where behaviour is more predictable. Without a set reference, we are left with a spectrum of spatial scales where, at one end, we are describing the noise in the data (spatial units smaller than the spatial range of the pattern), while at the other end we are failing to detect patterns because multiple assemblages are aggregated into a single spatial unit (spatial units larger than the spatial range of the pattern).

Therefore, it was necessary to identify a minimum sampling unit from the kilometre-scale transects derived from the CSS in order to assess the ability of the machine to capture benthic community composition patterns across time and space in a replicable and predictable fashion. To do so, we determined a standard scale for replicable and predictable patterns of community composition by investigating the multivariate spatial autocorrelation of benthic coverage (Appendix). Using this approach, we were able to standardise a sampling unit (hereafter: "site") from the 2 km transects collected by the CSS to 70 m subsections which best account for local heterogeneity of a given reef (Figure 1), and we ensured replicability by maximising the detection of change across space and time (details provided in the Appendix).

2.4.2. Estimating Error across Space

A total of 41 sites containing a total of 1042 images were randomly selected across the GBR and CSCMR transects from 2012 and 2014. These images are hereafter collectively denoted "validation set", and were selected to eliminate overlap with the set used to train the machine learning algorithm (Figure 1, Table 2). All images in the validation set were manually annotated at 40 points per image using the random point methodology detailed above. The relative cover of 19 benthic categories (Table 1) was averaged for all the images present within each site, and the error estimated as the absolute difference between the machine and a human annotator (Equation (1)):

$$S_{k,z} = \left| c_{i,k,z} - c_{j,k,z} \right| \tag{1}$$

where S is the spatial error of automated annotation across sites (z) for a given benthic category or label (k), c represents the mean percentage cover of label k at site z, as estimated by the human (i) or the automated (j) annotator.

The distribution of the absolute errors followed a non-normal and skewed distribution, thus the geometric mean was calculated to represent the average error among sites for each label category. Non-parametric bootstrap simulations were used to estimate the 95% confidence intervals around the mean [81].

While S (Equation (1)) provides the necessary details for understanding the limitations of the methods for each category, a broad analysis on the differences between automated and manual estimations was provided using a community approach. Specifically, Diversity (Shannon-Weiner Index—H′), Evenness (Pielou index—J′), and Community Dissimilarity (Bray-Curtis index) indices were used to describe the variability in the automated estimates compared to manual estimates from a compositional perspective.

Diversity indices measure the community biodiversity by considering the number of species, or groups, and their proportional abundance. While diversity is central to the compositional structure of the community, comparative analysis may be difficult because diversity indices combine both the number of species (richness) and their dominance (evenness). Therefore, we also included an index for evenness (Pielou index—J′) derived from the Shannon Index, which is commonly used in ecology [82], to evaluate the capacity of machine annotations to capture the species contribution to the community composition. Bray-Curtis dissimilarity (Equation (2)) was used as a summary metric to quantify the pair-wise compositional dissimilarity between automatic and manual estimations, based on the percentage cover of each benthic category.

$$BC_{k,z} = 100 \frac{\sum_{k=1}^{n} \left| c_{i,k,z} - c_{j,k,z} \right|}{\sum_{k=1}^{n} \left| c_{i,k,z} + c_{j,k,z} \right|} \tag{2}$$

where BC represents the Bray-Curtis dissimilarity and c is the relative abundance of each benthic category (k) for a given site (z) where percentage cover has been estimated manually (i) and automatically (j). Using this metric, a value of zero means a complete congruence between manual and automated estimations of the community composition, while a value of 100 means complete dissimilarity.

2.4.3. Estimating Error in Detecting Change over Time

Following a major cyclone impact on the GBR in 2014, a subset of 40 transects from 2012 were re-surveyed in 2014. Using this dataset, we evaluated whether temporal changes in benthic composition can be resolved using automated image annotations. For this purpose, seven sites were selected within a gradient range of impact assessed at being between 0 and 40% loss in coral cover. Images contained within these sites for 2012 and 2014 surveys were automatically annotated and contrasted against human annotations. For each site, the benthic coverage for each label category was averaged and the error of detecting change was calculated as the absolute difference between the automated and human estimates of change for each of the 19 labels (Equation (3)):

$$TE_{k,z} = \left| \Delta c_{i,k,z} - \Delta c_{j,k,z} \right| \tag{3}$$

where TE is the temporal error of machine annotation in detecting the change for each label (k) at a given site (z), and Δc is the difference of mean percent cover between 2012 and 2014 (change after the cyclone) estimated by both the human (i) and the automated (j) annotators for each label and site.

As with the spatial error, the distribution of the temporal error followed non-normal and rather skewed distributions. Therefore, the geometric mean and bootstrap 95% confidence interval were estimated for each label category [81].

2.4.4. Inter-Observer Variability

In order to maximise the sampling size to evaluate the error introduced by automated annotation, previous error estimates were conducted using one human annotator as a reference. However, underlying variability across annotators, when identifying benthic substrate [23], can compromise automated estimates and hinder the application of this tool. On the other hand, machine estimations that resemble or even improve human error can serve as an ideal tool. While it is expected that human observations have higher levels of agreement than automated annotations, differences among human annotators across the label set can help contextualise the capacity of automated annotation in resolving different labels being assigned. To explore inter-observer variability further, we used three different human annotators with similar ecological backgrounds and expertise to annotate five randomly selected sites in order to enable estimation of variability among human annotators. We assumed the average percent cover estimates for each label across the three human annotators to be a close approximation to the true abundance of benthic groups than any individual estimates. Human estimates in conjunction with automated estimates for each label were compared against the average estimation among human annotators for each sampling unit. Hence, the error estimation was calculated as the absolute difference between each annotator (including the automated annotator) and the average estimation for each label. The geometric mean of the difference and the bootstrap 95% confidence intervals across sampling units were calculated as the error followed a non-normal distribution.

3. Results

3.1. Estimating Error across Space

On average, automated image annotation introduced an absolute error of 2.5% ± 0.2% for all 19 benthic categories and across 41 randomly selected sites in the GBR and CSCMR. This means machine predictions of benthic cover were, on average, 2.5% different from the estimations provided by human annotators. Furthermore, differences were observed across different benthic categories (Figure 2). Within the algal functional group, Turf algae estimates showed the largest error: 10.9% ± 2.5%. Broad categories, such as "Other Soft Corals" and "Turf algae", which include a number of species with different archetypes and morphologies, generally exhibited the largest automated annotation errors. Conversely, and despite the large phenotypic plasticity as well as the number of species grouped into these categories, a high fidelity between the automated and human annotations was recorded.

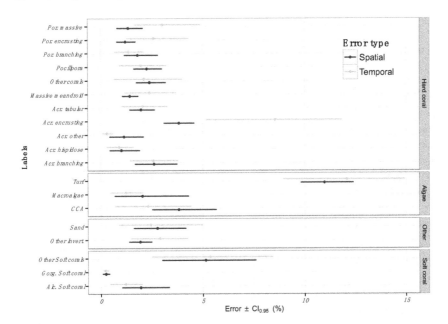

Figure 2. Spatial (S) and temporal (T) errors from automated benthic estimations across the GBR and CSCMR. Errors are presented for each one of the 19 labels presented in Table 1, where the short name for each label is detailed, and aggregated by functional groups ("Hard Corals", "Algae", "Other" and "Soft Coral"). Points represent the mean machine error and error bars represent the machine error margins.

Spatial patterns of benthic community structure across GBR and CSCMR sites, described by indices of diversity and evenness, were captured by automated estimations (Figure 3, Linear Regression, $R^2 = 0.5$, $p < 0.05$). A relatively disperse relationship between automated and human estimates of diversity and evenness was observed, showing increased variability when compared to more consistent patterns observed for the spatial error (Figure 3). Using the Bray-Curtis distance as an overall metric of dissimilarity between automated and human annotations at a community level, automated estimations differed by 24.5% ± 2.9% from human estimates. As a comparison, the Bray-Curtis distance among 41 sampling units for the GBR and CSCMR atolls for manually annotation averaged 37.7% ± 0.8%.

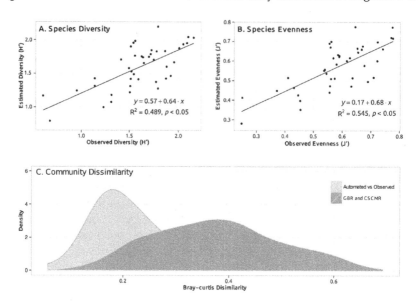

Figure 3. Comparisons of automated and manually estimated benthic composition among test sites, using: (**A**) Diversity index of Shannon; (**B**) Evenness index of Pielou; and (**C**) Community dissimilarity index of Bray-Curtis. The first two panels (**A,B**) represent a correlation plot of estimated (by the machine) and observed (by human annotator) values of diversity and evenness for each test site. The continuous line in both plots represents a fitted linear regression model to evaluate the relationship between observed and estimated values. Linear fit, Coefficient of Determination (R^2) and significance (*p*-value) are shown in the graph. In panel C, the range of community dissimilarity values between automatically and manually estimated benthic composition for each test site is presented in a density histogram (light grey). Overall dissimilarity among all test sites (manually estimated only) across the GBR and CSCMR (dark grey) is included for context.

3.2. Estimating Error in Detecting Change over Time

Comparisons of survey data prior to and after a major cyclone impact on the Northern GBR (Tropical Cyclone Ita) demonstrated that automated predictions of change in benthic communities had an error of 4.2% ± 3.3% (Figure 2). When examining each benthic category in detail, we found differences in the amount of error introduced by the automated annotation among the labels, similar to the differences found in spatial errors (Figure 2). Within the coral functional group, the temporal error averaged 2.5% ± 0.6% and remained below 5% for most of the labels. The exception was the group of hard corals from the family *Acroporidae* with an encrusting morphology (*Acr. encrusting*), where the estimated error was 8.6% ± 3.9%. In regard to the algae groups, turf algae again showed the largest error (12.1% ± 2.8%). The error for other substrate types, including "Soft Corals", remained below 5% with the exception of the broad category of soft corals, "Other Soft Corals", which included the largest diversity of soft corals species (5.3% ± 2.8%).

3.3. Inter-Observer Variability

Human annotations were consistently more precise than machine annotations, with human error ranging between 1% and 7% among categories (Figure 4). However, similar to automated annotation errors, considerable variation was noted across labels, with turf algae showing the highest error among human annotators 4.8% ± 0.78%. The turf algae errors were also the largest for the automated annotator. Note that the error estimates from machine annotations are greater than in Figure 2. This is most likely due to a lower sampling size for this exercise (41 vs. 5 sites). Therefore, our results were interpreted based on the relative difference between machine estimation errors and inter-observer variability.

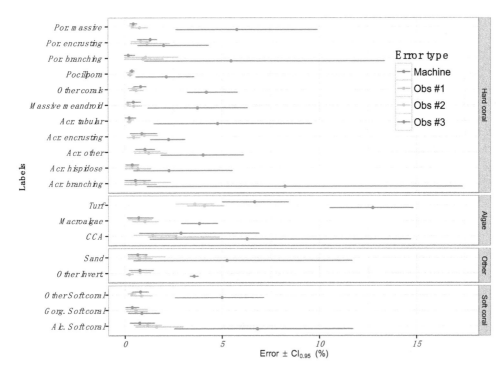

Figure 4. Inter-observer variability in the estimation of benthic composition for a subset of sites ($n = 5$) compared against the error introduced by the automated annotator. Error is presented for each one of the 19 labels presented in Table 1 and aggregated by functional groups ("Hard Corals", "Algae", "Other" and "Soft Coral"). Points represent the mean machine error and error bars represent the machine error margins.

4. Discussion

Overall, the percentage cover estimated from automated annotations captured spatial and temporal patterns of benthic community composition across the GBR and CSCMR, but with higher quantification errors the than inter-observer variability among human annotators. Our study reveals that machine estimates can measure changes in benthic composition, over time and space, with a minimum detection threshold ranging from 2% to 12% among 19 benthic categories. Using this approach, the method described here has the capacity to gather ecological metrics at kilometre scales with consistently low errors. The generation of standardised, high-definition and spatially sound datasets, via the methods described here, presents an opportunity to fill key data gaps (e.g., stock assessment, biodiversity, temporal trends) [17] and to track and understand the functional attributes of coral reef systems across broad temporal (e.g., years to decades) and spatial (e.g., >10,000 km) scales. This methodology is limited by a narrower taxonomic precision when compared to many smaller and more controlled photographic or *in-situ* surveys. These are clearly trade-offs that need to be considered in applying "hands-on" *versus* semi-automated survey technologies [22,23,30]. Advances and further improvements in image capturing and analysis tools are likely to reduce this limitation over time and are further discussed herein.

4.1. Error across Space

Mean estimations of the spatial error, introduced by the automated analysis, varied among categories and averaged 2.5%. These errors increased with the functional aggregation of communities (e.g., diversity and dissimilarity indices). Overall, the relative impact on this error of the data interpretations will depend on the relative abundance of the organisms, taxonomic resolution and the ecological relevance of the variability recorded. While we observed a relatively low estimation error, the noise introduced by the automated analysis may lead to misinterpretations of rare categories for which the average abundance is similar to the quantification errors estimated here (2%–3%). The impact of automated analysis errors on the assessment of more dominant benthic groups, on the other hand, is less pronounced. For example, the mean abundance of hard corals in the GBR and CSMCR ranged from 21% to 31%, in accordance with other studies [50,83,84], and the error of automated estimations averaged to <5%. The errors reported here fall within the expected values for these sites considering the complexity of their respective substrates.

Such variability in the representation of rare and dominant categories is carried over in community structure metrics, where indices of diversity and evenness are sensitive to the abundance of rare categories [85]. Hence, greater variability in the automated estimations of diversity and evenness was observed, although the ability of the machine to capture spatial trends was preserved. Community structure estimation errors can be summarised by measuring the dissimilarity in the community assemblage estimated by the machine when compared to a human annotator, following a traditional approach in community ecology [86]. In this study, we observed that automated estimations of benthic composition differed, on average, from the reference (human estimations) by 25% (Bray-Curtis dissimilarity). However, large heterogeneity of benthic community assemblages has been found in the GBR and CSCMR [83,87], where dissimilarity metrics of benthic assemblages typically range from 40% to 60% [84], and concurs with these results as observed in the Bray-Curtis dissimilarity of reference sites across the GBR and CSCMR (Figuew 3C). Therefore, while the method described here provides advantages for broad-scale community assessments, restrictions may apply to fine comparisons of benthic assemblages where automated annotation errors may obscure subtle differences among communities.

4.2. Error in Detecting Change over Time

The automated annotator estimated the temporal change of benthic composition with similar levels to the spatial error recorded (2%–12%, mean among labels). As with intra-year comparisons, the noise added by the automated annotator for temporal change relative to community composition may affect the interpretation of subtle changes, suggesting the applications of this approach as more suitable for mid-range temporal scales (years or decades) as opposed to subtle inter-seasonal fluctuations. As a reference, coral cover has decreased in the GBR by as much as 25%–30% over the past three decades [6,50], while less representative coral species can fluctuate around 5% [83]. Our results suggest the detection of subtle temporal variations of coral categories (<5%) may be hindered by the noise of automated estimations. Nonetheless, this approach has the capacity to provide detailed and broad-scale information on significant temporal changes (>2%–12%, depending on the categories) in coral reef benthic communities, thus providing data required for assessing causes and consequences of accelerated coral reef degradation over large extents [1].

4.3. Sources of Error

Errors introduced by the automated estimation of percentage cover of benthic groups across the GBR and CSCMR (spatial and temporal errors) can be attributed to two methodological caveats:

(a) image quality as a result of variability in reef appearance, underwater light irradiance and distance of the camera to the substrate; and (b) complexity of the label set, where groups enclosing many species with high morphological variation and phenotypic plasticity introduce a large variability due to overlapping visual features that challenges classification [30]. Since the imaging technique does not use artificial light, underwater light irradiance and reef light reflection pose imaging challenges. To compensate for this, fish-eye lenses, high-dynamic range cameras, and flexibility to adjust the camera ISO (International Standards Organization) settings on the fly, optimise the amount of light captured by the camera [24]. An on-board altitude sensor records the distance from the camera to the substrate, which enables the selection of only those images taken within a range from 0.5 to 2 m above the substrate to maintain a fixed resolution for the imagery [24].

Taxonomic resolution and morphological plasticity within the label set can also affect the capacity of automated and manual methods to accurately estimate composition and abundance [23,30]. Quantification of benthic composition from underwater images is limited by the level of taxonomic identification that can be resolved [23,24,30], whereas high-taxonomic resolution (e.g., species level) requires quantifying micro-scale morphology and internal structures of the reef organisms. In addition, species and groups of species exhibiting large morphological variations [88] may have visual attributes or features that overlap among groups, therefore hindering the capacity of automated classification to accurately resolve these classes. Furthermore, depending on the taxonomic aims of the study, taxonomically challenging categories are more prone to human errors, which are carried across to machine estimations from training data sets [22,23,30]. Therefore, the classification reference or label set needs to be designed in such a way that it conveys the taxonomic resolution which is functionally relevant for the intended study while acknowledging the taxonomic limitations of underwater image analysis, both manually and automatically [30,40].

4.4. Future Directions

While we acknowledge a higher taxonomic resolution can be achieved from underwater images [22,23,40], here we used a conservative approach. In this study, we amplified the benthic categorization compared to a previous study [24] whilst maintaining a relatively low number of taxonomic groups categorised by their functional traits (Table 1) and ensuring minimal overlap of visual features among labels. Further research is needed to evaluate the feasibility of expanding the resolution of taxonomic identifications and the capacity of automated methods to discern among labels.

Looking forward, our results indicate that there is room for improvement and the errors reported here can be further reduced by two orthogonal developments. The first involves recent development in the field of automated image analysis using deep Convolutional Neural Networks (CNNs) which have dramatically increased the classification accuracy for a wide range of images [89,90]. The second involves the opportunity to complement the RGB camera used here with multi-spectral or fluorescence cameras [91]. In this case, collecting additional spectral information could improve the ability to detect additional spectral signals, thereby improving precision and accuracy when distinguishing visually overlapping categories.

5. Conclusions

Overall, automated quantification of the relative abundance of coral reef functional groups, over time and space in the GBR and CSCMR, resulted in a relatively low error (2%–12% among labels) when compared to the human variability of benthic abundance estimations by multiple human annotators (1%–7% among labels). While limitations apply to the interpretation of this semi-automatically generated data, overall spatio-temporal changes in benthic community structure can be captured by this technology. Acknowledging the limitations previously discussed, this approach allows the scaling up of coral reef assessments by capturing underwater-water observations over large extensions (2 km transects) and processing such observations (images) at much faster rates than manual image analysis (about 170 image· h^{-1}). Data generated in this way can also contribute to more geographically comprehensive and integrative approaches (e.g., [27,92]) to improve assessment of the functionality and temporal variability for reef systems at regional scales (10–10,000 km) and provide important and informed management advice [93].

Acknowledgments: The study was conducted as part of the "XL Catlin Seaview Survey" funded by XL Catlin, with support from Global Change Institute (Ove Hoegh-Guldberg. and Manuel González-Rivero). Funding was also provided by the Australian Research Council (ARC Laureate to Ove Hoegh-Guldberg), an Early Career Grant at the University of Queensland (Manuel González-Rivero), the Waitt Foundation (Manuel González-Rivero and Ove Hoegh-Guldberg) and the EMBC+ programme at the Ghent University (Yeray González-Marrero). The authors would like to thank the contribution from XL Catlin Seaview Survey, Global Change Institute, Underwater Earth and Waitt Foundation teams for their essential support in field operations. Special thanks to Abbie Taylor and Eugenia Sampayo for their support in annotations of a large number of images for this exercise, and to Sara Naylor, Rachael Hazell and the four anonymous reviewers for their editorial comments on the manuscript.

Author Contributions: Manuel González-Rivero and Oscar Beijbom conceived and designed the study and analyses. Manuel González-Rivero, Oscar Beijbom, Alberto Rodriguez-Ramirez, Anjani Ganase and Yeray González-Marrero contributed towards image processing and analysis. Manuel González-Rivero, Oscar Beijbom, Tadzio Holtrop conducted data analysis and summary. Manuel González-Rivero, Oscar Beijbom, Alberto Rodriguez-Ramirez, Chris Roelfsema, Stuart Phinn and Ove Hoegh-Guldberg contributed to the writing of the manuscript.

Conflicts of Interest: The authors declare no conflict of interest.

Appendix

Defining Sampling Units by Structure Functions

Spatial autocorrelation is a common phenomenon in ecosystems and underlines the first law of geography: "Everything is related to everything else, but near things are more related than distant things" [94]. The structure function defined by this spatial correlation results from both environmental (induced) processes and biotic (inherent) processes interacting on the community [95,96]. Here we described the relation between the spatial structure of each transect and the inter-image distance by means of structure functions [93,94] where autocorrelation is quantified by the Mantel statistic Equation (A1) [95]:

$$Z = \sum_{i=1 \wedge i \neq j}^{n} \sum_{j=1 \wedge j \neq i}^{n} w_{ij} x_{ij} \tag{A1}$$

where Z_m is the mantel statistic; w_{ij} is Euclidean distance (geographic) matrix among n images per transect; and x_{ij} is Hellinger distance (community) matrix among images. Hellinger distance was calculated as the Euclidean distance of Hellinger transformed benthic coverage per image [97]. To compute a Mantel correlogram, the matrix w_{ij} was converted to a connectivity distance class matrix. A spline correlogram was then used to describe the structure function of the Mantel index as a function of continuous distance, allowing (1) a continuous estimate of the spatial covariance function instead of a discretised approximation like correlograms; and (2) a confidence envelope calculated around the spline using a bootstrap algorithm [98]. Large-scale spatial trends on the data were minimised, or de-trended, by identifying significant spatial trends associated with environmental gradients using polynomial regressions of all variables (benthic cover of 19 categories) on the X-Y coordinates [99]. For the de-trending of data, residuals from selected polynomial regression models (step-wise selection) were used to construct the correlograms. The structure of spatial autocorrelation was calculated individually for each surveyed transect, and the x-intercept of the spline and bootstrap confidence interval envelop was used to estimate the spatial range of benthic communities.

On average, the scale of patterns of benthic assemblages in the GBR and CSMR varied between 30 and 170 m along each transect (Figure A1). Within this range, the spatial autocorrelation was the highest and community composition was more homogeneous, therefore representing an ideal sampling size to better describe the attributes of community assembly. Here we use the upper limit of the confidence interval around the mean across 107 transects (70 m) as the size threshold of the sampling size for evaluating the performance of automated annotations in describing the benthic structure of coral reefs in the GBR and the Coral Sea (Figure A2).

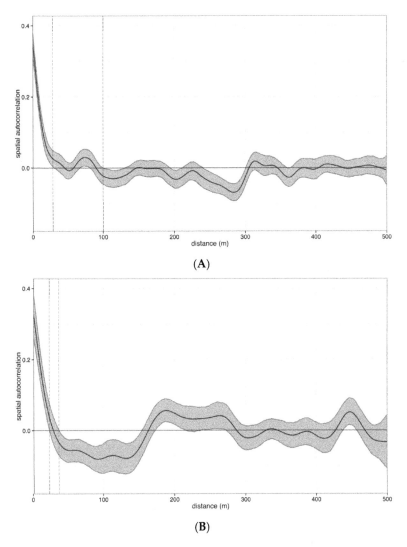

Figure A1. Examples of multivariate spatial autocorrelation of benthic communities in the GBR and CSCMR. These correlograms show the spatial autocorrelation structure of two transects as an example of the spatial structure of reef benthic assemblages with (**A**) large and (**B**) small spatial range. Autocorrelation is calculated by Mantel correlation index of spatially de-trended data as a function of linear distance between sample points. The line represents the fitted spline after 1000 permutations of the data and the shaded polygon shows the standard deviation of these permutations. Vertical dotted lines represent the range of the standard deviation where the splines intercept the x axis, or the range of distance between sampling points from where spatial autocorrelation is minimal (spatial range).

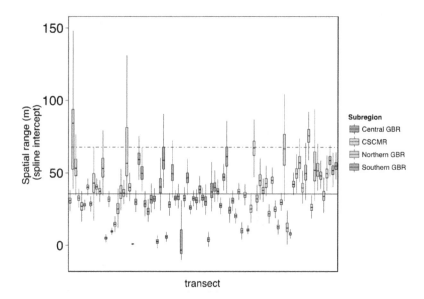

Figure A2. Spatial range (patch size) of benthic community assemblages across the GBR and CSCMR using 107 transects automatically annotated (see main manuscript for details). Here, the spatial range has been defined by the x-intercept of spatial autocorrelation splines, the point where the correlation between community distance and geographic distance becomes minimal. Each box-plot represents the outputs of 1000 bootstrap simulations of the autocorrelation spline for each transect, colour-coded according to the geographic location. The horizontal solid line shows the overall mean of the spline intercept and the dotted line represents the upper 95% percentile across the dataset.

References

1. Hughes, T.P.; Graham, N.A.J.; Jackson, J.B.C.; Mumby, P.J.; Steneck, R.S. Rising to the challenge of sustaining coral reef resilience. *Trends Ecol. Evol.* **2010**, *25*, 633–642. [CrossRef] [PubMed]

2. Knowlton, N.; Jackson, J.B.C. Shifting baselines, local impacts, and global change on coral reefs. *PLoS Biol.* **2008**, *6*. [CrossRef] [PubMed]

3. Fabricius, K.E.; Okaji, K.; De'ath, G. Three lines of evidence to link outbreaks of the crown-of-thorns seastar acanthaster planci to the release of larval food limitation. *Coral Reefs* **2010**, *29*, 593–605. [CrossRef]

4. Baker, A.C.; Glynn, P.W.; Riegl, B. Climate change and coral reef bleaching: An ecological assessment of long-term impacts, recovery trends and future outlook. *Estuar. Coast. Shelf Sci.* **2008**, *80*, 435–471. [CrossRef]

5. Hoegh-Guldberg, O. Climate change, coral bleaching and the future of the world's coral reefs. *Mar. Freshw. Res.* **1999**, *50*, 839–866. [CrossRef]

6. De'ath, G.; Fabricius, K.E.; Sweatman, H.; Puotinen, M. The 27-year decline of coral cover on the great barrier reef and its causes. *Proc. Natl. Acad. Sci. USA* **2012**, *109*, 17995–17999. [CrossRef] [PubMed]

7. Brodie, J.; Fabricius, K.; De'ath, G.; Okaji, K. Are increased nutrient inputs responsible for more outbreaks of crown-of-thorns starfish? An appraisal of the evidence. *Mar. Pollut. Bull.* **2005**, *51*, 266–278. [CrossRef] [PubMed]

8. Hoegh-Guldberg, O.; Cai, R.; Poloczanska, E.S.; Brewer, P.G.; Sundby, S.; Hilmi, K.; Fabry, V.J.; Jung, S. The ocean. In *Climate Change 2014: Impacts, Adaptation, and Vulnerability. Part B: Regional Aspects*; Barros, V.R., Field, C.B., Dokken, D.J., Mastrandrea, M.D., Mach, K.J., Bilir, T.E., Chatterjee, M., Ebi, K.L., Estrada, Y.O., Genova, R.C., *et al.* Eds.; Cambridge University Press: Cambridge, UK, 2014; pp. 1655–1734.

9. Gattuso, J.P.; Magnan, A.; Billé, R.; Cheung, W.; Howes, E.; Joos, F.; Allemand, D.; Bopp, L.; Cooley, S.; Eakin, C. Contrasting futures for ocean and society from different anthropogenic CO_2 emissions scenarios. *Science* **2015**, *349*. [CrossRef] [PubMed]

10. Gardner, T.A.; Cote, I.M.; Gill, J.A.; Grant, A.; Watkinson, A.R. Long-term region-wide declines in caribbean corals. *Science* **2003**, *301*, 958–960. [CrossRef] [PubMed]

11. Schutte, V.G.; Selig, E.R.; Bruno, J.F. Regional spatio-temporal trends in caribbean coral reef benthic communities. *Mar. Ecol. Prog. Ser.* **2010**, *402*, 115–122. [CrossRef]

12. Hughes, T.; Baird, A.; Dinsdale, E.; Harriott, V.; Moltschaniwskyj, N.; Pratchett, M.; Tanner, J.; Willis, B. Detecting regional variation using meta-analysis and large-scale sampling: Latitudinal patterns in recruitment. *Ecology* **2002**, *83*, 436–451. [CrossRef]

13. Parmesan, C.; Yohe, G. A globally coherent fingerprint of climate change impacts across natural systems. *Nature* **2003**, *421*, 37–42. [CrossRef] [PubMed]

14. Côté, I.; Gill, J.; Gardner, T.; Watkinson, A. Measuring coral reef decline through meta-analyses. *Philos. Trans. R. Soc. B Biol. Sci.* **2005**, *360*, 385–395. [CrossRef] [PubMed]

15. Scopélitis, J.; Andréfouët, S.; Largouët, C. Modelling coral reef habitat trajectories: Evaluation of an integrated timed automata and remote sensing approach. *Ecol. Model.* **2007**, *205*, 59–80. [CrossRef]

16. Mumby, P.J.; Hastings, A.; Edwards, H.J. Thresholds and the resilience of caribbean coral reefs. *Nature* **2007**, *450*, 98–101. [CrossRef] [PubMed]

17. Roelfsema, C.; Phinn, S. Integrating field data with high spatial resolution multispectral satellite imagery for calibration and validation of coral reef benthic community maps. *J. Appl. Remote Sens.* **2010**, *4*. [CrossRef]

18. Mumby, P.J.; Skirving, W.; Strong, A.E.; Hardy, J.T.; LeDrew, E.F.; Hochberg, E.J.; Stumpf, R.P.; David, L.T. Remote sensing of coral reefs and their physical environment. *Mar. Pollut. Bull.* **2004**, *48*, 219–228. [CrossRef] [PubMed]

19. Hedley, J.D.; Roelfsema, C.M.; Phinn, S.R.; Mumby, P.J. Environmental and sensor limitations in optical remote sensing of coral reefs: Implications for monitoring and sensor design. *Remote Sens.* **2012**, *4*, 271–302. [CrossRef]

20. Wooldridge, S.; Done, T. Learning to predict large-scale coral bleaching from past events: A bayesian approach using remotely sensed data, *in-situ* data, and environmental proxies. *Coral Reefs* **2004**, *23*, 96–108.

21. Roelfsema, C.; Phinn, S.; Udy, N.; Maxwell, P. An integrated field and remote sensing approach for mapping seagrass cover, moreton bay, australia. *J. Spat. Sci.* **2009**, *54*, 45–62. [CrossRef]

22. Carleton, J.H.; Done, T.J. Quantitative video sampling of coral reef benthos: Large-scale application. *Coral Reefs* **1995**, *14*, 35–46. [CrossRef]

23. Ninio, R.; Delean, J.; Osborne, K.; Sweatman, H. Estimating cover of benthic organisms from underwater video images: Variability associated with multiple observers. *Mar. Ecol. Prog. Ser.* **2003**, *265*, 107–116. [CrossRef]

24. González-Rivero, M.; Bongaerts, P.; Beijbom, O.; Pizarro, O.; Friedman, A.; Rodriguez-Ramirez, A.; Upcroft, B.; Laffoley, D.; Kline, D.; Bailhache, C.; *et al.* The Catlin Seaview Survey—Kilometre-scale seascape assessment, and monitoring of coral reef ecosystems. *Aquat. Conserv. Mar. Freshw. Ecosyst.* **2014**, *24*, 184–198. [CrossRef]

25. Williams, S.B.; Pizarro, O.; Webster, J.M.; Beaman, R.J.; Mahon, I.; Johnson-Roberson, M.; Bridge, T.C. Autonomous underwater vehicle–assisted surveying of drowned reefs on the shelf edge of the Great Barrier Reef, Australia. *J. Field Robot.* **2010**, *27*, 675–697. [CrossRef]

26. Armstrong, R.A.; Singh, H.; Torres, J.; Nemeth, R.S.; Can, A.; Roman, C.; Eustice, R.; Riggs, L.; Garcia-Moliner, G. Characterizing the deep insular shelf coral reef habitat of the hind bank marine conservation district (US Virgin islands) using the seabed autonomous underwater vehicle. *Cont. Shelf Res.* **2006**, *26*, 194–205. [CrossRef]

27. Roelfsema, C.; Lyons, M.; Dunbabin, M.; Kovacs, E.M.; Phinn, S. Integrating field survey data with satellite image data to improve shallow water seagrass maps: The role of auv and snorkeller surveys? *Remote Sens. Lett.* **2015**, *6*, 135–144. [CrossRef]

28. Mountrakis, G.; Im, J.; Ogole, C. Support vector machines in remote sensing: A review. *ISPRS J. Photogram. Remote Sens.* **2011**, *66*, 247–259. [CrossRef]

29. Culverhouse, P.F.; Williams, R.; Benfield, M.; Flood, P.R.; Sell, A.F.; Mazzocchi, M.G.; Buttino, I.; Sieracki, M. Automatic image analysis of plankton: Future perspectives. *Mar. Ecol. Prog. Ser.* **2006**, *312*, 297–309. [CrossRef]

30. Beijbom, O.; Edmunds, P.J.; Roelfsema, C.; Smith, J.; Kline, D.I.; Neal, B.P.; Dunlap, M.J.; Moriarty, V.; Fan, T.-Y.; Tan, C.-J. Towards automated annotation of benthic survey images: Variability of human experts and operational modes of automation. *PLoS ONE* **2015**, *10*. [CrossRef] [PubMed]

31. Gleason, A.C.R.; Reid, R.P.; Voss, K.J. Automated classification of underwater multispectral imagery for coral reef monitoring. In Proceedings of the 2007 OCEANS, Vancouver, BC, USA, 29 September–4 October 2007; pp. 1–8.

32. Marcos, M.S.A.; Soriano, M.; Saloma, C. Classification of coral reef images from underwater video using neural networks. *Opt. Express* **2005**, *13*, 8766–8771. [CrossRef] [PubMed]

33. Friedman, A.; Pizarro, O.; Williams, S.B. Rugosity, slope and aspect from bathymetric stereo image reconstructions. In Proceedings of the 2010 IEEE OCEANS, Sydney, NSW, Australia, 24–27 May 2010; pp. 1–9.

34. Leon, J.X.; Roelfsema, C.M.; Saunders, M.I.; Phinn, S.R. Measuring coral reef terrain roughness using "Structure-from-Motion" close-range photogrammetry. *Geomorphology* **2015**, *242*, 21–28. [CrossRef]

35. Boom, B.J.; He, J.; Palazzo, S.; Huang, P.X.; Beyan, C.; Chou, H.M.; Lin, F.P.; Spampinato, C.; Fisher, R.B. A research tool for long-term and continuous analysis of fish assemblage in coral-reefs using underwater camera footage. *Ecol. Inform.* **2014**, *23*, 83–97. [CrossRef]

36. Shihavuddin, A.; Gracias, N.; Garcia, R.; Gleason, A.C.; Gintert, B. Image-based coral reef classification and thematic mapping. *Remote Sens.* **2013**, *5*, 1809–1841. [CrossRef]

37. Pizarro, O.; Rigby, P.; Johnson-Roberson, M.; Williams, S.B.; Colquhoun, J. Towards image-based marine habitat classification. In Proceedings of the 2008 OCEANS, Quebec City, QC, Canada, 15–18 September 2008; pp. 1–7.

38. Beijbom, O.; Edmunds, P.J.; Kline, D.I.; Mitchell, B.G.; Kriegman, D. Automated annotation of coral reef survey images. In Proceedings of the 2012 IEEE Conference on Computer Vision and Pattern Recognition (CVPR), Providence, RI, USA, 16–21 June 2012; pp. 1170–1177.

39. Wallace, C. *Staghorn Corals of the World: A Revision of the Genus Acropora*; CSIRO Publishing: Clayton, VIC, Australia, 1999.

40. Althaus, F.; Hill, N.; Ferrari, R.; Edwards, L.; Przeslawski, R.; Schönberg, C.H.L.; Stuart-Smith, R.; Barrett, N.; Edgar, G.; Colquhoun, J.; *et al.* A standardised vocabulary for identifying benthic biota and substrata from underwater imagery: The catami classification scheme. *PLoS ONE* **2015**, *10*. [CrossRef] [PubMed]

41. Graham, N.A.J.; Nash, K.L. The importance of structural complexity in coral reef ecosystems. *Coral Reefs* **2013**, *32*, 315–326. [CrossRef]

42. Vytopil, E.; Willis, B. Epifaunal community structure in *Acropora* spp. (Scleractinia) on the Great Barrier Reef: Implications of coral morphology and habitat complexity. *Coral Reefs* **2001**, *20*, 281–288.

43. Van Woesik, R.; Done, T.J. Coral communities and reef growth in the southern great barrier reef. *Coral Reefs* **1997**, *16*, 103–115. [CrossRef]

44. Marshall, P.A.; Baird, A.H. Bleaching of corals on the great barrier reef: Differential susceptibilities among taxa. *Coral Reefs* **2000**, *19*, 155–163. [CrossRef]

45. Richmond, R.H.; Hunter, C.L. Reproduction and recruitment of corals: Comparisons among the Caribbean, the Tropical Pacific, and the Red Sea. *Mar. Ecol. Prog. Ser. Oldendorf.* **1990**, *60*, 185–203. [CrossRef]

46. Darling, E.S.; Alvarez-Filip, L.; Oliver, T.A.; McClanahan, T.R.; Côté, I.M. Evaluating life-history strategies of reef corals from species traits. *Ecol. Lett.* **2012**, *15*, 1378–1386. [CrossRef] [PubMed]

47. Madin, J. Mechanical limitations of reef corals during hydrodynamic disturbances. *Coral Reefs* **2005**, *24*, 630–635. [CrossRef]

48. Hughes, T.P.; Connell, J.H. Multiple stressors on coral reefs: A long-term perspective. *Limnol. Oceanogr.* **1999**, *44*, 932–940. [CrossRef]

49. Madin, J.S.; Baird, A.H.; Dornelas, M.; Connolly, S.R. Mechanical vulnerability explains size-dependent mortality of reef corals. *Ecol. Lett.* **2014**, *17*, 1008–1015. [CrossRef] [PubMed]

50. Bruno, J.F.; Selig, E.R. Regional decline of coral cover in the indo-pacific: Timing, extent, and subregional comparisons. *PLoS ONE* **2007**, *2*. [CrossRef] [PubMed]

51. Wooldridge, S. Differential thermal bleaching susceptibilities amongst coral taxa: Re-posing the role of the host. *Coral Reefs* **2014**, *33*, 15–27. [CrossRef]

52. Willis, B.L.; Page, C.A.; Dinsdale, E.A. Coral disease on the Great Barrier Reef. In *Coral Health and Disease*; Springer: Berlin, Germany, 2014; pp. 69–104.

53. Guest, J.R.; Baird, A.H.; Maynard, J.A.; Muttaqin, E.; Edwards, A.J.; Campbell, S.J.; Yewdall, K.; Affendi, Y.A.; Chou, L.M. Contrasting patterns of coral bleaching susceptibility in 2010 suggest an adaptive response to thermal stress. *PLoS ONE* **2012**, *7*. [CrossRef] [PubMed]

54. Kolinski, S.P.; Cox, E.F. An update on modes and timing of gamete and planula release in Hawaiian scleractinian corals with implications for conservation and management. *Pac. Sci.* **2003**, *57*, 17–27. [CrossRef]

55. Chisholm, J.R. Primary productivity of reef-building crustose coralline algae. *Limnol. Oceanogr.* **2003**, *48*, 1376–1387.

56. Chisholm, J.R.M. Calcification by crustose coralline algae on the northern great barrier reef, Australia. *Limnol. Oceanogr.* **2000**, *45*, 1476–1484. [CrossRef]

57. Harrington, L.; Fabricius, K.; De'Ath, G.; Negri, A. Recognition and selection of settlement substrata determine post-settlement survival in corals. *Ecology* **2004**, *85*, 3428–3437. [CrossRef]

58. Heyward, A.J.; Negri, A.P. Natural inducers for coral larval metamorphosis. *Coral Reefs* **1999**, *18*, 273–279. [CrossRef]

59. Diaz-Pulido, G. Macroalgae. In *The Great Barrier Reef: Biology, Environment and Management*; Hutchings, P., Kingsford, M.J., Hoegh-Guldberg, O., Eds.; CSIRO Publishing: Clayton, VIC, Australia, 2008; pp. 145–155.

60. Done, T.J. Effects of tropical cyclone waves on ecological and geomorphological structures on the Great Barrier Reef. *Cont. Shelf Res.* **1992**, *12*, 859–872. [CrossRef]

61. Hughes, T.P. Catastrophes, phase-shifts, and large-scale degradation of a caribbean coral reef. *Science* **1994**, *265*, 1547–1551. [CrossRef] [PubMed]

62. Schaffelke, B.; Klumpp, D. Biomass and productivity of tropical macroalgae on three nearshore fringing reefs in the central Great Barrier Reef, Australia. *Bot. Mar.* **1997**, *40*, 373–384. [CrossRef]

63. Hatcher, B.; Larkum, A. An experimental analysis of factors controlling the standing crop of the epilithic algal community on a coral reef. *J. Exp. Mar. Biol. Ecol.* **1983**, *69*, 61–84. [CrossRef]

64. Hatcher, B.G. Coral reef primary productivity: A beggar's banquet. *Trends Ecol. Evol.* **1988**, *3*, 106–111. [CrossRef]

65. Klumpp, D.D.; McKinnon, D.A. Community structure, biomass and productivity of epilithic algal communities on the Great Barrier reef: Dynamics at different spatial scales. *Mar. Ecol. Prog. Ser.* **1992**, *86*, 77–89. [CrossRef]

66. Larkum, A.; Kennedy, I.; Muller, W. Nitrogen fixation on a coral reef. *Mar. Biol.* **1988**, *98*, 143–155. [CrossRef]

67. Alderslade, P.P.; Fabricius, K.K. Octocorals. In *The Great Barrier Reef: Biology, Environment and Management*; Hutchings, P., Kingsford, M.J., Hoegh-Guldberg, O., Eds.; CSIRO Publishing: Clayton, VIC, Australia, 2008; pp. 222–245.

68. Fabricius, K.K.; Alderslade, P.P. *Soft Corals and Sea Fans: A Comprehensive Guide to the Tropical Shallow Water Genera of the Central-West Pacific, the Indian Ocean and the Red Sea*; Australian Institute of Marine Science (AIMS): Cape Ferguson, QLD, Australia, 2001.

69. Kahng, S.E.; Benayahu, Y.; Lasker, H.R. Sexual reproduction in octocorals. *Mar. Ecol. Prog. Ser.* **2011**, *443*, 265–283. [CrossRef]

70. Van Oppen, M.J.H.; Mieog, J.C.; Sanchez, C.A.; Fabricius, K.E. Diversity of algal endosymbionts (zooxanthellae) in octocorals: The roles of geography and host relationships. *Mol. Ecol.* **2005**, *14*, 2403–2417. [CrossRef] [PubMed]

71. Pante, E.; Dustan, P. Getting to the point: Accuracy of point count in monitoring ecosystem change. *J. Mar. Biol.* **2012**, *2012*. [CrossRef]

72. Beijbom, O.; Chan, S.; Sampat, D.; Hu, A.; Sandvik, J.; Kriegman, D.; Belongie, S.; Kline, D.I.; Treibitz, T.; Neal, B.; *et al.* Coralnet. Available online: http://www.coralnet.ucsd.edu (accessed on 1 June 2014).

73. Cortes, C.; Vapnik, V. Support-vector networks. *Mach. Learn.* **1995**, *20*, 273–297. [CrossRef]

74. Fan, R.-E.; Chang, K.-W.; Hsieh, C.-J.; Wang, X.-R.; Lin, C.-J. Liblinear: A library for large linear classification. *J. Mach. Learn. Res.* **2008**, *9*, 1871–1874.

75. Beijbom, O.; Edmunds, P.J.; Kline, D.I.; Mitchell, B.G.; Kriegman, D. Moorea Labeled Corals. Available online: http://vision.ucsd.edu/content/moorea-labeled-corals (accessed on 15 October 2014).

76. Liblinear—A Library for Large Linear Classification. Available oneline: https://www.csie.ntu.edu.tw/~cjlin/liblinear/ (accessed on 20 October 2014).

77. Levin, S.A. The problem of pattern and scale in ecology: The robert H. Macarthur award lecture. *Ecology* **1992**, *73*, 1943–1967. [CrossRef]

78. Habeeb, R.L.; Johnson, C.R.; Wotherspoon, S.; Mumby, P.J. Optimal scales to observe habitat dynamics: A coral reef example. *Ecol. Appl.* **2007**, *17*, 641–647. [CrossRef] [PubMed]

79. Rietkerk, M.; van de Koppel, J. Regular pattern formation in real ecosystems. *Trends Ecol. Evol.* **2008**, *23*, 169–175. [CrossRef] [PubMed]

80. Wiens, J.A. Spatial scaling in ecology. *Funct. Ecol.* **1989**, *3*, 385–397. [CrossRef]

81. Wickham, H. *Ggplot2: Elegant Graphics for Data Analysis*; Springer: Berlin, Germany, 2009.

82. Magurran, A.E. Measuring biological diversity. *Afr. J. Aquat. Sci.* **2004**, *29*, 285–286.

83. Ninio, R.; Meekan, M. Spatial patterns in benthic communities and the dynamics of a mosaic ecosystem on the great barrier reef, australia. *Coral Reefs* **2002**, *21*, 95–104. [CrossRef]

84. Jupiter, S.; Roff, G.; Marion, G.; Henderson, M.; Schrameyer, V.; McCulloch, M.; Hoegh-Guldberg, O. Linkages between coral assemblages and coral proxies of terrestrial exposure along a cross-shelf gradient on the southern great barrier reef. *Coral Reefs* **2008**, *27*, 887–903. [CrossRef]

85. Colwell, R.K. Biodiversity: Concepts, patterns, and measurement. In *The Princeton Guide to Ecology*; Princeton University Press: Princeton, NJ, USA, 2009; pp. 257–263.

86. Faith, D.P.; Minchin, P.R.; Belbin, L. Compositional dissimilarity as a robust measure of ecological distance. *Vegetatio* **1987**, *69*, 57–68. [CrossRef]

87. Done, T.J. Patterns in the distribution of coral communities across the central great barrier reef. *Coral Reefs* **1982**, *1*, 95–107. [CrossRef]

88. Todd, P.A. Morphological plasticity in scleractinian corals. *Biol. Rev.* **2008**, *83*, 315–337. [CrossRef] [PubMed]

89. Krizhevsky, A.; Sutskever, I.; Hinton, G.E. Imagenet classification with deep convolutional neural networks. In Proceedings of the 26th Annual Conference on Neural Information Processing Systems, Lake Tahoe, NV, USA, 3–6 December 2012; pp. 1097–1105.

90. Girshick, R.; Donahue, J.; Darrell, T.; Malik, J. Rich feature hierarchies for accurate object detection and semantic segmentation. In Proceedings of the 2014 IEEE Conference on Computer Vision and Pattern Recognition (CVPR), Columbus, OH, USA, 23–28 June 2014; pp. 580–587.

91. Treibitz, T.; Neal, B.P.; Kline, D.I.; Beijbom, O.; Roberts, P.L.; Mitchell, B.G.; Kriegman, D. Wide field-of-view fluorescence imaging of coral reefs. *Sci. Rep.* **2015**, *5*, 1–9. [CrossRef] [PubMed]

92. Mumby, P.J. Stratifying herbivore fisheries by habitat to avoid ecosystem overfishing of coral reefs. *Aquac. Fish. Fish Sci.* **2014**. [CrossRef]

93. Beger, M.; McGowan, J.; Treml, E.A.; Green, A.L.; White, A.T.; Wolff, N.H.; Klein, C.J.; Mumby, P.J.; Possingham, H.P. Integrating regional conservation priorities for multiple objectives into national policy. *Nat. Commun.* **2015**, *6*. [CrossRef] [PubMed]

94. Tobler, W.R. A computer movie simulating urban growth in the Detroit Region. *Econ. Geogr.* **1970**, *46*, 234–240. [CrossRef]

95. Dale, M.R.; Fortin, M.J. *Spatial Analysis: A Guide for Ecologists*; Cambridge University Press: Cambridge, UK, 2014.

96. Fortin, M.J.; Dale, M.R. Spatial autocorrelation in ecological studies: A legacy of solutions and myths. *Geogr. Anal.* **2009**, *41*, 392–397. [CrossRef]

97. Legendre, P.; Gallagher, E.D. Ecologically meaningful transformations for ordination of species data. *Oecologia* **2001**, *129*, 271–280. [CrossRef]

98. BjØrnstad, O.N.; Falck, W. Nonparametric spatial covariance functions: Estimation and testing. *Environ. Ecol. Stat.* **2001**, *8*, 53–70. [CrossRef]

99. Borcard, D.; Gillet, F.; Legendre, P. *Numerical Ecology with R*; Springer: Berlin, Germany, 2011.

Spatially-Explicit Testing of a General Aboveground Carbon Density Estimation Model in a Western Amazonian Forest Using Airborne LiDAR

Patricio Xavier Molina [1,2,*], Gregory P. Asner [3], Mercedes Farjas Abadía [2], Juan Carlos Ojeda Manrique [2], Luis Alberto Sánchez Diez [2] and Renato Valencia [4]

Academic Editors: Nicolas Baghdadi and Prasad S. Thenkabail

[1] Gestión de Investigación y Desarrollo, Instituto Geográfico Militar, Seniergues E4-676 y Gral, Telmo Paz y Miño, El Dorado 170403, Quito, Ecuador
[2] Technical University of Madrid (UPM), C/Ramiro de Maeztu, 7, Madrid 28040, Spain; m.farjas@upm.es (M.F.A.); juancarlos.ojeda@upm.es (J.C.O.M.); luisalberto.sanchez@upm.es (L.A.S.D.)
[3] Department of Global Ecology, Carnegie Institution for Science, 260 Panama Street, Stanford, CA 94305, USA; gpa@carnegiescience.edu
[4] Laboratorio de Ecología de Plantas, Escuela de Ciencias Biológicas, Pontificia Universidad Católica del Ecuador, Apartado 17-01-2184, Quito, Ecuador; lrvalencia@puce.edu.ec
*   Correspondence: xavier.molina@mail.igm.gob.ec or px.molina@alumnos.upm.es

Abstract: Mapping aboveground carbon density in tropical forests can support CO_2 emission monitoring and provide benefits for national resource management. Although LiDAR technology has been shown to be useful for assessing carbon density patterns, the accuracy and generality of calibrations of LiDAR-based aboveground carbon density (ACD) predictions with those obtained from field inventory techniques should be intensified in order to advance tropical forest carbon mapping. Here we present results from the application of a general ACD estimation model applied with small-footprint LiDAR data and field-based estimates of a 50-ha forest plot in Ecuador's Yasuní National Park. Subplots used for calibration and validation of the general LiDAR equation were selected based on analysis of topographic position and spatial distribution of aboveground carbon stocks. The results showed that stratification of plot locations based on topography can improve the calibration and application of ACD estimation using airborne LiDAR ($R^2 = 0.94$, RMSE = 5.81 Mg·C·ha^{-1}, BIAS = 0.59). These results strongly suggest that a general LiDAR-based approach can be used for mapping aboveground carbon stocks in western lowland Amazonian forests.

Keywords: aboveground carbon density; biomass; Ecuador; LiDAR; topographic features; tropical rainforest

1. Introduction

Tropical forests are important carbon and biodiversity reserves, and characterizing the spatial distribution of their aboveground carbon density (ACD; units of Mg of carbon per hectare or Mg·C·ha^{-1}) is a prerequisite for understanding the spatial and temporal dynamics of the terrestrial carbon cycle. Accurate estimations of ACD and any changes in carbon stocks due to human activities are required in order to reduce emissions from deforestation and forest degradation (REDD+), and so contribute to the efforts being made to mitigate climate change [1]. Tropical forests hold large stores of carbon, yet uncertainty remains regarding their precise contribution to the global carbon cycle and how it is distributed in space and time [2,3]. In the last ten years, estimating carbon capture reserves in tropical forests has evolved from activities based mainly on inventories carried out in the field [4] to approaches based on airborne and spaceborne remote sensing [5–9].

Modern forest ecology and management applications require accurate maps so that their dynamics, biodiversity and carbon content can be tracked through time, especially in ecologically fragile and/or inaccessible regions, as is the case of many tropical forests. Several studies on estimating ACD have been carried out using different approaches [10–12]. The most widely used LiDAR-based approaches to ACD prediction are based on regression models that combine LiDAR metrics with field estimations of carbon stocks in forest plots. The derived model is then used to assess ACD across larger geographic areas. In the last five years, studies to estimate ACD in tropical forests have been performed using a plot-aggregate allometric approach [13–15].

The plot-aggregate allometry approach [16,17] for estimating ACD from airborne LiDAR has provided estimates that are comparable in predictive power to locally-calibrated models. This approach is based on a simplified general model showing that dry tree biomass, and its carbon content, roughly ~48% of dry biomass by weight [18], can be estimated from LiDAR-derived top-of-canopy height (TCH), basal area and wood density information. This approach has the potential to reduce the time required to calibrate airborne LiDAR data; however, it requires testing in new regions.

Although broad-scale mapping is based primarily on remote sensing data, the accuracy of resulting forest carbon stock estimates depends critically on the quality of field measurements and calibration procedures [19]. A careful quantification of local spatial variability and spatial structure in ACD should be useful for remote sensing calibration efforts. Reported errors associated with LiDAR-based carbon maps range from 17 to more than 40 Mg·C·ha^{-1} (RMSE) in the tropics [20]. These errors apply to the calibration step (*i.e.*, the ability of LiDAR to predict the carbon density of a set of field inventory plots as assessed by a regression model), and not necessarily to carbon maps produced by such regression.

The size of field plots is an important design parameter in forest inventory using LiDAR, as it has the potential to either reduce or inflate the impact of edge effects [21] and co-registration error on ACD estimates [22]. Disagreement in protocol between LiDAR and field observations—namely the effects of bisecting tree crowns in LiDAR data *versus* calling a tree "in" or "out" of the plot in field data—decreases to a manageable level when field plots reach 1 ha in size [20]. The spatial variability in ACD is large for standard plot sizes (e.g., 0.1, 0.25, 1 ha), averaging 46.3% for replicate 0.1 ha subplots within a single large plot, and 16.6% for 1 ha subplots [19]. Furthermore, large parcels are needed for dynamic forest studies in order to include all the important populations of most of the species present, especially in tropical forests [23]. Errors in estimating forest structural attributes tend to decline and highest coefficient of determination (R^2) values are reached by combining large plot sizes [21] and high density LiDAR data [24]. A review of studies published on estimating ACD using LiDAR [25] concludes that there is an uncertainty level between airborne and field-based ACD estimates of around 10% when plot size is approximately 1 ha. Plot sizes of around 1 ha are usually considered to give sufficiently accurate results in forest biomass estimation [26].

Marvin *et al.* [27] found that an average sample size of 44 plots of 1 ha in the lowland Peruvian Amazon are needed to reliably (*i.e.*, a probability of 0.9) estimate ACD to within 90% of the actual landscape-scale (10^2–10^4 ha) mean value. The results obtained in [19] show the importance of keeping the topography in mind, and suggest that sampling should be stratified by topographic position (e.g., ridges, valleys and slopes), especially when the estimations involve a terrain-based approach.

The accuracy of carbon stock estimations also depends on reliable allometric models being available in order to estimate AGB from forest inventories [28–30]. Estimating biomass in tropical forests is limited by the available information on the allometry of tropical tree species. When it turns out to be impossible to obtain allometric relationships for a specific area of interest, pre-existing allometric equations are normally used. Due to the large number of studies in which these relationships are documented [31], it is important to identify the most suitable equations [32].

The objective of this study was to calibrate and evaluate the use of a general LiDAR-based approach for estimating ACD in a western lowland Amazonian forest of Ecuador, where little work has been undertaken in the past. We used geo-statistical modeling techniques and LiDAR-derived topographic features to locate plots for fitting and validating the general ACD estimation model.

2. Data and Methods

2.1. Study Area

The study area is located at 0°41'S and 76°24'W, to the south of the Tiputini River, in the Yasuní National Park (PNY) of Ecuador (Figure 1). This park contains a high concentration of western lowland Amazonian species, making it one of the highest biodiversity regions in the world [33]. Due to its wealth of natural life, in 1995 the Catholic University of Ecuador, together with the Smithsonian Tropical Research Institute and the University of Aarhus, selected an area of 50 ha in the northwest of the Yasuní National Park to study the distribution and dynamics of species. The first census of the west 25 ha showed that 1104 species were coexisting in this area [34], which represent more species of trees than in North America. The Yasuní plot is part of a global network of 61 large permanent plots associated to the Center for Tropical Forest Science (CTFS) [35].

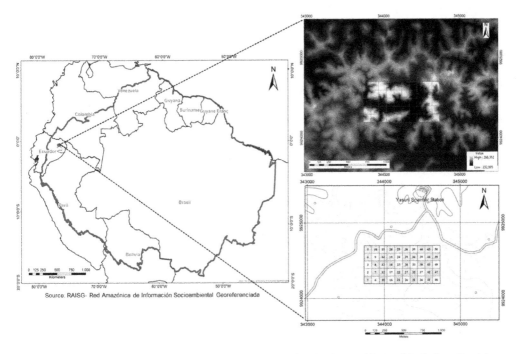

Figure 1. (Left) Location of the study area (red point) within the Yasuní National Park (in yellow) in the western Amazonian Basin. **(Upper right)** Digital Terrain Model of the 50-ha forest plot (centered rectangle). **(Lower right)** A regular grid system of 1-ha subplots used for calibration (red) and validation (green) of the airborne LiDAR model for aboveground carbon density estimation.

The plot is 50 ha in size, with an average topographic gradient of 13%, and an elevation range of 215–249 m above sea level. The study area is crossed by a valley that divides it into three groups of small hills. The southwest portion of the plot contains a 0.48-ha patch of secondary forest corresponding to a heliport, abandoned around 1987, during oil exploration. The plot contains two types of characteristic and dominant topographical environments: valleys and ridges. The valleys include several small permanent streams and swamps associated with a depression to the east of the study area. A swampy area, topographically recessed, contains water throughout the year. Both the northern and southern boundaries consist of low ridges with some steep-sided gullies. The lowest zones between the ridges contain grey and brown sedimentary deposits. Most of the soil in the Yasuní National Park is classified as Ultisol, but the swamps and floodplains are dominated by Histosols, which are prone to flooding [36]. The plot is located in lowland evergreen forest of the western Amazon region.

2.2. Field Data

2.2.1. The Forest Census Data

The first forest census was carried out from 1995 to 1997 in the west 25 ha of the plot, and after that, several censuses were carried out in different portions of the 50-ha plot. Trees were measured and tagged following a standard method used by the global network of large forest plots [37]. In the first census, all trees ≥1 cm were measured, and 1104 morphospecies were identified among 152,353 individuals. Several species of understory treelets in the genera *Matisia* and *Rinorea* dominated the forest, while important canopy species were *Iriartea deltoidea* and *Eschweilera coriacea*. The swampy area occupying 1 ha in the eastern half of the plot is most notably different. The palm *Mauritia flexuosa* (Palmae), a *Sapium* sp. (*Euphorbiaceae*), and several species of *Piper* (*Piperaceae*) are found only in the swamp [36–38]. The canopy is 15–30 m tall, and some emergent trees reach 40–50 m in height.

The first complete forest census of the 50-ha plot was conducted from 2002 to 2006. Detailed information on taxonomic and ecological characteristics of tree species in the 50-ha plot can be found in [39]. For this study, measures of stem diameters at breast height (D) for all individuals in the 50-ha plot and wood density data (wood specific gravity) were used to compute ACD. Wood density data were taken from the literature or obtained from direct measurements in and around the plot. Tree height was unavailable in these censuses.

2.2.2. Field-Based Aboveground Biomass Estimation

It has often argued that local allometric equations should be constructed in as many sites and for as many species as possible. However, the extreme diversity of the species in Yasuní National Park prevents the allometry of specific species being developed, so that generalist relationships are usually applied. We estimated AGB using a new allometric model [30] shown in Equation (1), which is used in cases in which tree height is not known and includes variables such as trunk diameter, wood density, and the bioclimatic stress variable (E). The value of the E parameter for Yasuní National Park is -0.0228121.

$$AGB_i = \exp\left[-1.803 - 0.976E + 0.976\ln(\rho) + 2.673\ln(D) - 0.0299\left[\ln(D)\right]^2\right] \tag{1}$$

where D represents trunk diameter at breast height in cm, and ρ is wood density of each tree in g·cm^{-3}. The quantity of AGB (in Mg·ha^{-1}) of all trees in the entire plot was calculated from the data obtained in the census. Within each 1-ha subplot, AGB was calculated by summing AGB estimates for all trees whose stems were located within the subplot and expressing this on a per ha-basis (Equation (2)). A summary of the field plot characteristic is presented in Table 1.

$$AGB_{SUBPLOT} = \sum_{i=1}^{n} AGB_i \tag{2}$$

Table 1. Characteristics of 1-ha field plots.

Characteristic	Range	Mean
N [a] (ha^{-1})	4680–7430	6012.68
D [b] (cm)	1–185	19.30
BA [c] (m$^2 \cdot$ ha^{-1})	25.38–39.78	32.84
AGB [d] (Mg$\cdot$ ha^{-1})	161.47–339.39	250.25

[a] number of trees; [b] diameter at breast height (1.3 m); [c] basal area; [d] aboveground biomass.

2.2.3. Georeferencing of the Field Plots

Any errors in identification or imprecision of the estimated variables are often attributed to discrepancies between the information on the reference plots obtained on the ground and that obtained from LiDAR [22,40]. To avoid such discrepancies, a planimetric survey was carried out to precisely locate the coordinates of the plot's corners, and to correlate the data obtained from the forest census with that obtained from the airborne LiDAR system. The Horizontal Reference System used in the survey was the Geocentric Reference System for the Americas (SIRGAS–ECUADOR) [41], which is compatible with GNSS satellite positioning system. Four geodetic survey markers using GNSS technology and formed the baselines for the planimetric survey. Observations were made from the Y-NPF geodetic survey marker, which belongs to Ecuador's GPS RENAGE network. The coordinates of the geodetic survey markers were fixed by the static phase differential GPS positioning method using TRIMBLE R8 dual frequency receivers which can measure baselines up to 200 km long with an accuracy of $\pm$0.005 m + 1.0 ppm. Horizontal precision was set at <0.050 m + 1.0 ppm, and vertical precision at <0.100 m + 1.0 ppm. The basic geodetic network was then used to carry out the observations required to calculate the coordinates of the four corners of the plot (Table 2) using a TRIMBLE S3 total station with a precision of 2″.

Table 2. Coordinates of the four corners of the plot (UTM 18S).

Plot Corner	East (m)	North (m)	Precision (m)	Ellipsoidal. Elev.
NW	343,737.794	9,924,696.411	0.050	249.31
SW	343,735.023	9,924,196.086	0.050	252.79
NE	344,733.958	9,924,695.236	0.050	244.89
SE	344,734.085	9,924,195.180	0.050	252.07

2.3. LiDAR Data

2.3.1. Data Collection

The LiDAR data were acquired in May 2014 from a Cessna 172 Skyhawk aircraft, which can cover a large area at low altitudes and low speeds. The LiDAR sensor used in the airborne platform was an Optech ALTM Gemini. Flight data and configurations are given below in Table 3:

Table 3. Flight data and LiDAR configuration.

Flight Data		LiDAR Configuration	
Height above ground (m)	781.25	Pulse frequency repetition (Khz)	166
Distance between lines of flight (m)	203.89	Scanning frequency (Hz)	40
Overlapping	50%	Scan angle /FOV	$\pm$15
Speed (m/s)	56.6	Nominal density of pulses per m^2	5.08
Flight lines	16	Sweep width (m)	407.78
		Number of returns	Up to 4
		Laser beam divergence (mrad) (IFOV)	0.8
		Space between points (m)	0.24
		Density of points per m^2	19.4

2.3.2. LiDAR Data Processing

Estimating the tree canopy height using LiDAR data relies on an accurate representation of the ground surface in a Digital Terrain Model (DTM). Any errors in the DTM will propagate and affect the accuracy of the derived vegetation metrics [26,42,43] and canopy height models (CHM) [44,45]. The DTM (Figure 2a) was obtained by applying the procedure described in [17]. The vertical accuracy of the DTM was assessed using GNSS measurements for georeferencing the 50-ha plot. The LiDAR data were normalized at ground level and gridded into 1-ha subplots using LAStools [46].

The canopy height model (CHM) (Figure 2c) was obtained as the difference between the digital surface model (DSM) and the DTM [47]. In each 1-ha subplot, the average of all 0.5 m CHM pixels was used to estimate mean subplot top-of-canopy height (TCH).

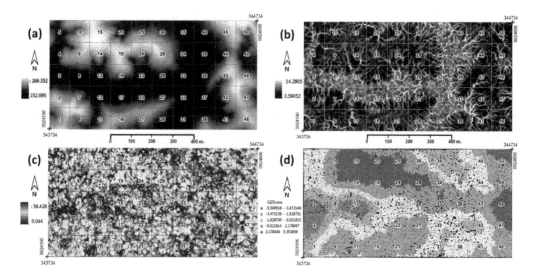

Figure 2. Variables used to select subplots for fitting and validating the general model. (**a**) Digital Terrain Model at 5 m spatial resolution; (**b**) Compound Topographic Index at 5 m spatial resolution; (**c**) Canopy Height Model at 0.5 m spatial resolution; and (**d**) Hot Spot Analysis of ACD distribution.

2.4. Data Analysis

2.4.1. Selection of Subplots for Fitting and Validating the General Model

LiDAR is capable of characterizing both terrain and vegetation structure. However, LiDAR-based DTM variables have been rarely used to plan plot locations in ACD calibration schemes. In the present study, when selecting the subplots for fitting and validating the model, we considered the spatial distribution of ACD and topographic position (e.g., valleys, ridges and slopes) in the 50-ha plot. The 1-ha subplot size was necessary to encompass substantial populations of most tree species in the community.

The sample design was a regular grid of 1-ha subplots, allowing us to capture spatial variation in ACD and forest structure throughout the study area. The X, Y coordinates of the southwest corner obtained in the planimetric survey were used as the starting point, so that the subplots and all trees whose stems were located within the subplots were geo-referenced. Subplots were numbered from 1 to 50 starting from the lower left-hand corner.

The DTM was used to calculate a topographic wetness index using the Compound Topographic Index (CTI) tool [48] in ArcGIS 10.1 (ESRI, Redlands, CA, USA: Environmental Systems Research Institute) (Figure 2b). The CTI is a function of both the slope and the upstream contributing area per unit width orthogonal to the flow direction [49]. Higher CTI values represent water accumulation (potential wetland formation), and lower CTI values represent dryness or steep places where water would not likely accumulate based on topography [50]. Zonal Statistics for each 1-ha subplot was summarized using Spatial Analyst tools for DTM and CTI. Pearson's correlation coefficient was calculated to analyze relationships between field-ACD and mean elevation ($r = 0.6794$) and mean CTI ($r = -0.5902$) in each 1-ha subplots. The Hot Spot Analysis tool (Getis-Ord Gi * statistic) in spatial statistics tools in ArcGIS, were used to locate the patterns of biomass distribution in the plot (Figure 2d). This tool identifies clusters or statistically significant spatial clusters of high values (hot spots) and low values (cold spots) that provide important insights into the underlying processes that produce spatial patterns. An incremental spatial autocorrelation tool was used to test for spatial autocorrelation within distance bands, measuring the intensity of spatial clustering for each distance.

The sampling was then stratified by topographic position: valley, slope and ridges. These three topographic positions were defined after evaluating DTM and CTI zonal statistics in each 1 ha subplot. A valley was defined as all 1-ha subplots with mean elevation <239.10 m and mean CTI >6.60. The ridge was defined as all 1-ha subplots with mean elevation >248.20 m and mean CTI <6.12. The remaining 1-ha subplots were defined as slope (Table 4).

Table 4. Summary Statistics for the LiDAR plot 1-ha grid of study area.

Topographic Position	Number of Plots	Mean (SD)			
		LiDAR TCH (m)	Basal Area ($m^2 \cdot ha^{-1}$)	Wood Density ($g \cdot cm^{-3}$)	AGB_{field} ($Mg \cdot ha^{-1}$)
Valley	18	20.60 (1.8)	24.37 (2.5)	0.557 (0.01)	198.24 (29.5)
Ridge	16	23.50 (1.1)	29.90 (2.6)	0.574 (0.01)	265.39 (33.3)
Slope	15	21.47 (1.9)	27.41 (3.1)	0.563 (0.01)	228.30 (37.06)
R_{SF}	1	22.74	30.76	0.471	199.40

R_{SF} = Ridge with remnant of secondary forest; TCH = Top of canopy height.

We selected 66% of the sample for the fitting of the model (32 subplots) and 34% for validating the model (18 subplots) using random sampling in every topographic position (Figure 1). Subplot #2, which contained secondary forest (0.48 ha), was used for model validation. In the data exploratory analysis, two atypical subplots were identified (36, 50) as having the largest trees in the area. Both subplots presented the highest values of coefficient of variation in elevation and variance. These were left out of the fitting process but were included for model validation. Only trees with a diameter at breast height (dbh) of ⩾10 cm were considered when fitting the model. Previous study on AGB estimation by habitats in 25-ha of the plot, reported that trees <10 cm dbh contributes ~5% in ridge and ~7% in valley of total AGB [51].

2.4.2. LiDAR Model Application

The most widely used LiDAR-based approaches to ACD prediction are based on regression models that correlate LiDAR metrics with field estimations of biomass in forest plots. The models are obtained from a statistical analysis to ensure consistency, mathematical rigor and predictive power. We used the plot-aggregate allometric approach [17] (Equation (3)).

$$ACD = aTCH^{b1}BA^{b2}WD_{BA}^{b3} \tag{3}$$

where ACD is the AGB obtained from Equation (2) and multiplied by 0.48, TCH is the top of canopy height obtained by LiDAR, BA is the basal area (*i.e.*, cross-sectional area of all stems), estimated using individual tree diameter and WD_{BA} is basal area-weighted wood density taken from the literature or obtained from direct measurements in and around the plot. Equation (3) was fitted using multiple linear regression on ln-transformed subplot level data for ACD, TCH, BA and WD_{BA} at 1-ha subplots in the form:

$$\ln(ACD) = \ln a + b_1 \ln(TCH) + b_2 \ln(BA) + b_3 \ln(WD_{BA}) \tag{4}$$

Model fitting and diagnostics were performed with R Commander Software [52]. After fitting the model we assessed the following issues [53]: (i) normal residuals; (ii) homoscedastity (constant variance); (iii) linear relationship; (iv) presence of atypical observations; and (v) absence of colinearity. The model estimated satisfies all the conditions. A rigorous model validation process requires that the results be verified with a sample different from the one used to build it. We therefore validated the model by applying 18 subplots selected for this purpose.

3. Results

We back-transformed the final model, since we were interested in the ACD parameter per hectare and not its natural logarithm. The model was multiplied by a correction factor (CF) to account for the back-transformation of the regression error [54]. The correction factor is given by $CF = e^{MSE/2}$ where MSE is the mean square error of the regression model. In this case it is equal to 1.00044. The equation thus became (Equation (5)):

$$ACD = 2.15813 * TCH^{0.14015} BA^{1.2292} WD^{0.9839} \tag{5}$$

When we applied the resulting model (Equation (5)) to the validation plots, we obtained the results as shown in Figure 3. The LiDAR-based ACD equation accurately and precisely predicted field plot-based ACD in eighteen plots (Figure 3a). The low bias (0.59 Mg·C·ha^{-1}) and RMSE (5.81 Mg·C·ha^{-1}), along with an adjusted R^2 of 0.94, validates the use of plot-aggregate allometric approach for estimating ACD in the study area. The final model was spatially sensitive to ACD variation in valley, ridge and slope areas (Figure 3b–d), which were the main habitats in the zone.

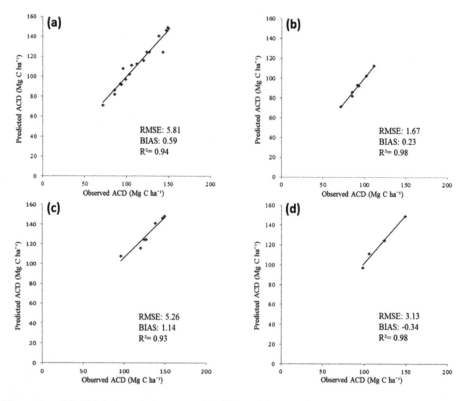

Figure 3. (a) Validation of the general ACD model in eighteen independently 1-ha plots. The performance of the general model is shown for topographic position at (b) valley; (c) ridge including plot with remnant of secondary forest; and (d) slope. RMSE = root mean squared error, BIAS = mean errors.

Our results strongly suggest that the spatial arrangement of forest carbon stocks, based on topographic location, influences the accuracy of estimates of ACD in western lowland Amazonian mature forest achieving an overall accuracy of 5% at 1-ha resolution (RMSE = 5.81 Mg·C·ha^{-1} relative to forest carbon density of ~120 Mg·C·ha^{-1}). The model validation for ACD estimation in ridge positions (Figure 3c) indicated higher bias and RMSE compared with the other topographic positions because these areas include the plot with remnant of secondary forest. Secondary forest growth is higher than adjacent mature forest on ridges where the growth is minimal [51], affecting the estimation of the model due to the time lag between field census and the LiDAR survey (RMSE = 2.33, bias = 0.74 and R^2 = 0.97 without secondary forest).

For the fitting of the model, only trees with stem diameter $\geq$10 cm (34,567) were included for the field-based AGB estimation (11,438.68 Mg). However, total AGB estimations (12,512.59 Mg) were made for all the individuals (297,777) with diameter $\geq$1 cm in the entire 50-ha plot. Results show that 11.6% of all the trees in the 50-ha plot contribute approximately 91% of total ACD, while vegetation with diameters <10 cm provide 9% of the total. We also found that ACD differ by up to 100% for a vertical variation of only 30 m. The variation in ACD (78–163 Mg·C·ha^{-1}) in a small elevation range suggests a strong influence of topographic position and confirms that topography should be taken into account in the forest inventory sampling design.

4. Discussion

Whether from a forest management or conservation perspective, it is important to establish a robust and confident methodology for estimating carbon stocks in tropical forest. The present study calibrated and evaluated ACD estimations by means of linear regression, considering the size and location of the plots in topographic positions used for fitting the general model.

We selected a regular grid of 1-ha subplots based on an underlying assumption that field plots (*ca.* 1 ha scale) are an unbiased sample of the landscape (*ca.* 10^2- to 10^4 -ha scale) [28,55], and previous findings of diminishing uncertainties between field-based and LiDAR-based estimates at this resolution [20]. The 1-ha grid scheme and the horizontal precision of the 50-ha plot corners (0.050 m) helped to reduce co-registration errors related to misalignment between field subplots and LiDAR data, as well as plot-edge effects. There is a tendency for errors to decrease in biomass estimates with increasing plot size, because large plots reduce the likelihood of plot-edge effects, which occur when the canopy of trees are found along the plot boundary [21]. Edge effects are likely more pronounced in less dense stands and where plot sizes are smaller [56]. The accuracy of plot-aggregate allometry used in this study appears to increase when averaging over more vegetation in larger plots since larger plots minimize the edge effects related to uncertainty in including or excluding a tree in the field survey. In the LiDAR calibration phase, use of small plots will always lead to inflated scatter and thus increased RMSE between LiDAR TCH (or any metric) and field-estimated ACD [17]. Although implementing larger plot sizes increase the costs and time needed for field sampling, large plots results in models with better performance, increase the accuracy of ACD predictions and reduce the variation in ALS-derived metrics [43].

The time lag between the forest census and LiDAR acquisition will also introduce errors in the final model. On the western 25 ha, Valencia *et al.* [51] examined aboveground biomass flux in different habitats and across diameter classes using data from two censuses separated by an average time interval of 6.3 years. They found that the forest lost small stems (4.6%), gained large trees (2.6%), and gained biomass (0.7%). The change in AGB stock was due entirely to this upward shift in size leading to more canopy trees and fewer saplings after just six year. Across habitats, the biggest increment in biomass was in the secondary forest patch (3.4% y^{-1}), whereas mature forest on ridges and valleys had small increases (0.10% and 0.09% y^{-1}, respectively). Relative to the difference between habitats, the 6 year change in AGB stock was almost trivial. The one exception was the increase in large trees in the secondary forest. The forest increased its standing biomass, but far less than the average reported for other Amazonian forest (*i.e.*, 0.30 *vs.* 0.98 Mg^{-1}·y^{-1}). Similarly, change in basal area in the three previous censuses (1997, 2002, 2009) was negligible and accounted for maximum 1% (32.94, 32.90, 33.25 $m^2 \cdot ha^{-1}$, respectively). This can change in abnormal time periods but is very unlikely that it changes in aseasonal forest like Yasuni. Even in extreme years, like 2010, such a change did not showed a huge impact on large trees, judging by a subsampled check in the forest dynamics we carried out in year 2012 (<2% change in basal area). From these results we infer that the change in ACD stocks due to the time lag between the field census and the LiDAR survey are minimal.

Field-based ACD estimations were calculated using an improved allometric model [30]. This pantropical model (Equation (1)) is used for estimating ACD in the absence of height measurements and incorporates wood density, trunk diameter, and a variable called E. Equation (1) shows that information on wood density (as inferred from the taxonomic determination of the tree), trunk diameter, and the variable E (as inferred from the geolocation of the plot) is sufficient to provide a robust ACD estimate [30]. However, to minimize bias, the development of locally derived diameter-height relationships is advised whenever possible. Our estimations showed that the ridge and valley of the Yasuni forest are remarkably different in ACD (Table 1); the difference is due almost entirely to a higher number of very large trees on the ridge, and to a lesser extent from higher density wood on the ridge. The results obtained confirm the estimates made by Valencia et al. [39], particularly the differences in AGB between valley and ridge.

To explore patterns of ACD distribution within the 50-ha plot, hot-spot analysis was performed. in which spatial patterns of high ACD (hot spots) and low ACD (cold spots) were identified. As expected, subplots with high ACD were located in the ridge while the lowest ACD were founded in valley containing swamps (Figure 2d). Spatial autocorrelation was calculated using Moran's index (<0.01), every 100 m to a distance of 500 m. ACD is less spatially clustered at 300 m (z-score = 3.479; p-value = 0.0005 and variance = 0). The compound topographic index (CTI), often referred to as the steady-state wetness index, is a quantification of catenary landscape position [57]. CTI is a useful indicator of ACD because it combines contextual and site information via the upslope catchment area and slope, respectively. Moore et al. [49] showed that the CTI is correlated with several soil attributes such as silt percentage, organic matter content, phosphorus and A horizon depth in the soil surface of a small toposequence. Kanagaraj et al. [58] found that CTI, slope and elevation were important drivers of species assemblage in the Barro Colorado Island plot. In our study, mean CTI at 1-ha subplot was correlated with field ACD estimations (r = −0.59); therefore, it was used as a variable together with mean elevation (r = 0.67) to allocate plots for the model calibration.

The high R^2 values reached can be attributed to the influence of 1-ha plot size and LiDAR density points (~20 m^{-2}) [24]. Although an analysis of LiDAR density and its influence on the estimation of ACD was not made in this study, a recent study by Leitold et al. [42] in tropical forest of Brazil, with complex topography (ranging from 100 m to 1100 m a.s.l.), reported errors of 80–125 $Mg \cdot ha^{-1}$ in predicted aboveground biomass for LiDAR return densities below 4 m^{-2}. The canopy heights calculated from reduced density LiDAR data declined as data density decreased due to a systematic effect of pulse density in the construction of digital terrain model (DTM) attributed to the algorithm used to classify ground points. The study requires some caution when using generalized ACD models based on a single LiDAR metric (e.g., TCH), as in our case study, especially at low LiDAR return densities. Yet increasing point density mitigates the problem of accurate canopy height and DTM generation. In contrast to the study made by Leitold et al. [42], Hansen et al. [43], showed that canopy metrics derived from sparse laser pulse density data can be used for ACD estimation in a tropical forest in Tanzania. Reducing the laser pulse density from 8 to 0.25 pulses m^{-2} increased the variation in the DTMs and canopy metrics. However, the replication effects, expressed by the reliability ratio, were not important at pulse densities of >0.5 pulses m^{-2}. In our study, using the four corners of the 50-ha plot, the mean difference in elevation between DTM and GNSS observations (−0.722 m) indicated a slight underestimate from LiDAR-derived terrain elevations in the study area. These results suggest that more research should be conducted to evaluate the influence of LiDAR point density on DTM generation and canopy metrics especially in tropical forests.

The calibration and validation accuracy of the model may depend on the number and size of the field plots available for analysis. For the calibration of the general approach used in this study, Asner et al. [17] reported an adj-R^2 = 0.86 and RMSE = 13.2 Mg·C·ha^{-1} using a network of 754 field inventory plots distributed across a wide range of tropical vegetation types, climates and successional states. Here, our local model considering topographic location yielded an adj-R^2 = 0.94 and RMSE = 5.81 Mg·C·ha^{-1} using 32 field plots of 1-ha for fitting the model and 18 field plots for validating the model. Given that all plots used for fitting and validating the local model were partitioned from a single large plot, our results should be interpreted with caution. It is important to recognize that having larger plots fully mapped and partitioned subplots, representing the host landscape, may lead to a more robust understanding of calibration uncertainties. We tested the general approach by acquiring the LiDAR model of top-of-canopy height, and combining them with spatially explicit estimates of basal area and wood density in the study area. However this approach allows for regional assessments of basal area and WD$_{BA}$ to replace exhaustive tree-specific measurements. In the simplest form, this approach used a single stocking coefficient (the ratio of BA to TCH) and a single wood density constant for each broad tropical region based on the literature. To apply the general model in a new region, exhaustive inventory would not be needed. Instead, spatially-explicit point-based estimates of BA (by relascope or prism method) and WD$_{BA}$ (by recording dominant species) could be collected within the coverage of a LiDAR TCH dataset. By regressing BA and WD$_{BA}$ onto TCH and substituting these regressions into the general model, regionally-tailored predictions could be generated.

5. Conclusions

A combination of LiDAR-derived topographic features (DTM and CTI), geo-statistical modeling techniques and plots tactically located and representative of the landscape, provide a consistent approach to calibrate a general LiDAR-based ACD model to a western Amazonian forest. In this study we assessed the errors in applying this calibration (i.e., the ability of LiDAR to predict the carbon density of a set of field inventory plots as assessed by the regression model), and not comparing to field-based methods. Fifty subplots of 1 ha were used to estimate and validate the general LiDAR-based ACD model, with uncertainty values of 5.81 Mg·C·ha^{-1} and a bias of 0.59 Mg·C·ha^{-1}, along with an adjusted R^2 of 0.94 indicates that the plot-aggregate allometric approach can be used to accurately estimate carbon stocks in the study area. The results showed that spatial stratification by topographic position may reduce bias in model calibration. The study also identified issues for further research related to the influence of pulse density and plot size to reliably estimate the metrics used to predict forest biomass in tropical forest.

Acknowledgments: Funding for this study was provided by the Secretariat for Higher Education, Science, Technology and Innovation of the Republic of Ecuador (SENESCYT). The Forest Dynamics Plot of Yasuní National Park has been made possible through the generous support of the US National Science Foundation, the Andrew W. Mellon Foundation, the Pontifical Catholic University of Ecuador, the Smithsonian Tropical Research Institute, and the University of Airbus of Denmark. The Yasuní Forest Dynamics Plot is part of the Center for Tropical Forest Science, a global network of large-scale demographic tree plots. The Principal Investigator of the Yasuni Forest Dynamics Plot is R. Valencia. We thank N. Vaughn of the Carnegie Airborne Observatory for assistance and advice with LiDAR data processing, and three anonymous reviewers for comments on a previous version of this manuscript. G.P. Asner was supported by the John D. and Catherine T. MacArthur Foundation. We gratefully acknowledge the support of Instituto Geografico Militar del Ecuador for LiDAR data collection. The Ecuadorian Ministerio del Ambiente granted permission for our work in Yasuní National Park.

Author Contributions: Patricio Xavier Molina, Mercedes Farjas Abadia and Gregory P. Asner conceived and designed the research. Juan Carlos Ojeda Manrique, Luis Alberto Sánchez Diez and Renato Valencia analyzed the data and revised the manuscript. Patricio Xavier Molina, Mercedes Farjas Abadía and Gregory P. Asner wrote the manuscript.

Conflicts of Interest: The authors declare no conflict of interest.

References

1. Angelsen, A.; Wertz-kanounnikoff, S.; Redd, W. *Moving Ahead with REDD Issues, Options and Implications*; Angelsen, A., Ed.; Center for International Forestry Research (CIFOR): Bogor, Indonesia, 2008.

2. Chave, J.; Andalo, C.; Brown, S.; Cairns, M.A.; Chambers, J.Q.; Eamus, D.; Fölster, H.; Fromard, F.; Higuchi, N.; Kira, T.; *et al.* Tree allometry and improved estimation of carbon stocks and balance in tropical forests. *Oecologia* **2005**, *145*, 87–99. [CrossRef] [PubMed]

3. Mitchard, E.T.A.; Feldpausch, T.R.; Brienen, R.J.W.; Lopez-Gonzalez, G.; Monteagudo, A.; Baker, T.R.; Lewis, S.L.; Lloyd, J.; Quesada, C.A.; Gloor, M.; *et al.* Markedly divergent estimates of Amazon forest carbon density from ground plots and satellites. *Glob. Ecol. Biogeogr.* **2014**, *23*, 935–946. [CrossRef] [PubMed]

4. Malhi, Y.; Wood, D.; Baker, T.R.; Wright, J.; Phillips, O.L.; Cochrane, T.; Meir, P.; Chave, J.; Almeida, S.; Arroyo, L.; *et al.* The regional variation of aboveground live biomass in old-growth Amazonian forests. *Glob. Chang. Biol.* **2006**, *12*, 1107–1138. [CrossRef]

5. Asner, G.P.; Powell, G.V.N.; Mascaro, J.; Knapp, D.E.; Clark, J.K.; Jacobson, J.; Kennedy-Bowdoin, T.; Balaji, A.; Paez-Acosta, G.; Victoria, E.; *et al.* High-resolution forest carbon stocks and emissions in the Amazon. *Proc. Natl. Acad. Sci. USA* **2010**, *107*, 16738–16742. [CrossRef] [PubMed]

6. Saatchi, S.; Houghton, R.A.; Dos Santos Alvalá, R.C.; Soares, J.V.; Yu, Y. Distribution of aboveground live biomass in the Amazon basin. *Glob. Chang. Biol.* **2007**, *13*, 816–837. [CrossRef]

7. Saatchi, S.S.; Harris, N.L.; Brown, S.; Lefsky, M.; Mitchard, E.T.; Salas, W.; Zutta, B.R.; Buermann, W.; Lewis, S.L.; Hagen, S.; *et al.* Benchmark map of forest carbon stocks in tropical regions across three continents. *Proc. Natl. Acad. Sci. USA* **2011**, *108*, 9899–9904. [CrossRef] [PubMed]

8. Lefsky, M.A.; Cohen, W.B.; Harding, D.J.; Parker, G.G.; Acker, S.A.; Thomas Gower, S. LiDAR remote sensing of above-ground biomass in three biomes. *Glob. Ecol. Biogeogr.* **2002**, *2*, 393–399. [CrossRef]

9. Baccini, A.; Asner, G.P. Improving pantropical forest carbon maps with airborne LiDAR sampling. *Carbon Manag.* **2013**, *4*, 591–600. [CrossRef]

10. Hyyppä, J.; Hyyppä, H.; Leckie, D.; Gougeon, F.; Yu, X.; Maltamo, M. Review of methods of small footprint airborne laser scanning for extracting forest inventory data in boreal forests. *Int. J. Remote Sens.* **2008**, *29*, 1339–1366. [CrossRef]

11. Petrokofsky, G.; Kanamaru, H.; Achard, F.; Goetz, S.J.; Joosten, H.; Holmgren, P.; Lehtonen, A.; Menton, M.C.; Pullin, A.S.; Wattenbach, M. Comparison of methods for measuring and assessing carbon stocks and carbon stock changes in terrestrial carbon pools. How do the accuracy and precision of current methods compare? A systematic review protocol. *Environ. Evid.* **2012**, *1*, 6. [CrossRef]

12. Mascaro, J.; Asner, G.P.; Knapp, D.E.; Kennedy-Bowdoin, T.; Martin, R.E.; Anderson, C.; Higgins, M.; Chadwick, K.D. A tale of two "Forests": Random Forest machine learning aids tropical Forest carbon mapping. *PLoS ONE* **2014**, *9*, 12–16. [CrossRef] [PubMed]

13. Asner, G.P.; Mascaro, J.; Anderson, C.; Knapp, D.E.; Martin, R.E.; Kennedy-Bowdoin, T.; van Breugel, M.; Davies, S.; Hall, J.S.; Muller-Landau, H.C.; *et al.* High-fidelity national carbon mapping for resource management and REDD+. *Carbon Balance Manag.* **2013**, *8*, 7. [CrossRef] [PubMed]

14. Asner, G.P.; Clark, J.K.; Mascaro, J.; Galindo García, G.A.; Chadwick, K.D.; Navarrete Encinales, D.A.; Paez-Acosta, G.; Cabrera Montenegro, E.; Kennedy-Bowdoin, T.; Duque, Á.; *et al.* High-resolution mapping of forest carbon stocks in the Colombian Amazon. *Biogeosciences* **2012**, *9*, 2683–2696. [CrossRef]

15. Taylor, P.; Asner, G.; Dahlin, K.; Anderson, C.; Knapp, D.; Martin, R.; Mascaro, J.; Chazdon, R.; Cole, R.; Wanek, W.; *et al.* Landscape-Scale controls on aboveground forest carbon Stocks on the Osa Peninsula, Costa Rica. *PLoS ONE* **2015**, *10*, 1–18. [CrossRef] [PubMed]

16. Asner, G.P.; Mascaro, J.; Muller-Landau, H.C.; Vieilledent, G.; Vaudry, R.; Rasamoelina, M.; Hall, J.S.; van Breugel, M. A universal airborne LiDAR approach for tropical forest carbon mapping. *Oecologia* **2012**, *168*, 1147–1160. [CrossRef] [PubMed]

17. Asner, G.P.; Mascaro, J. Mapping tropical forest carbon: Calibrating plot estimates to a simple LiDAR metric. *Remote Sens. Environ.* **2014**, *140*, 614–624. [CrossRef]

18. Martin, A.R.; Thomas, S.C. A reassessment of carbon content in tropical trees. *PLoS ONE* **2011**, *6*, e23533. [CrossRef] [PubMed]

19. Réjou-Méchain, M.; Muller-Landau, H.C.; Detto, M.; Thomas, S.C.; le Toan, T.; Saatchi, S.S.; Barreto-Silva, J.S.; Bourg, N.A.; Bunyavejchewin, S.; Butt, N.; *et al.* Local spatial structure of forest biomass and its consequences for remote sensing of carbon stocks. *Biogeosciences* **2014**, *11*, 6827–6840. [CrossRef]

20. Mascaro, J.; Detto, M.; Asner, G.P.; Muller-Landau, H.C. Evaluating uncertainty in mapping forest carbon with airborne LiDAR. *Remote Sens. Environ.* **2011**, *115*, 3770–3774. [CrossRef]

21. Hernández-Stefanoni, J.; Dupuy, J.; Johnson, K.; Birdsey, R.; Tun-Dzul, F.; Peduzzi, A.; Caamal-Sosa, J.; Sánchez-Santos, G.; López-Merlín, D. Improving species diversity and biomass estimates of tropical dry forests using airborne LiDAR. *Remote Sens.* **2014**, *6*, 4741–4763. [CrossRef]

22. Frazer, G.; Magnussen, S.; Wulder, M.A.; Niemann, K.O. Simulated impact of sample plot size and co-registration error on the accuracy and uncertainty of LiDAR-derived estimates of forest stand biomass. *Remote Sens. Environ.* **2011**, *115*, 636–649. [CrossRef]

23. Condit, R.; Ashton, P.S.; Baker, P.; Bunyavejchewin, S.; Gunatilleke, S.; Gunatilleke, N.; Hubbell, S.P.; Foster, R.B.; Itoh, A.; LaFrankie, J.V.; *et al.* Spatial patterns in the distribution of tropical tree species. *Science* **2000**, *288*, 1414–1418. [CrossRef] [PubMed]

24. Ruiz, L.A.; Hermosilla, T.; Mauro, F.; Godino, M. Analysis of the influence of plot size and LiDAR density on forest structure attribute estimates. *Forests* **2014**, *5*, 936–951. [CrossRef]

25. Zolkos, S.; Goetz, S.; Dubayah, R. A meta-analysis of terrestrial aboveground biomass estimation using LiDAR remote sensing. *Remote Sens. Environ.* **2013**, *128*, 289–298. [CrossRef]

26. Meyer, V.; Saatchi, S.S.; Chave, J.; Dalling, J.W.; Bohlman, S.; Fricker, G.A.; Robinson, C.; Neumann, M.; Hubbell, S. Detecting tropical forest biomass dynamics from repeated airborne LiDAR measurements. *Biogeosciences* **2013**, *10*, 5421–5438. [CrossRef]

27. Marvin, D.C.; Asner, G.P.; Knapp, D.E.; Anderson, C.B.; Martin, R.E.; Sinca, F.; Tupayachi, R. Amazonian landscapes and the bias in field studies of forest structure and biomass. *Proc. Natl. Acad. Sci. USA* **2014**, *111*, E5224–E5232. [CrossRef] [PubMed]

28. Chave, J.; Condit, R.; Aguilar, S.; Hernandez, A.; Lao, S.; Perez, R. Error propagation and scaling for tropical forest biomass estimates. *Philos. Trans. R. Soc. B Biol. Sci.* **2004**, *359*, 409–420. [CrossRef] [PubMed]

29. Ngomanda, A.; Engone Obiang, N.L.; Lebamba, J.; Moundounga Mavouroulou, Q.; Gomat, H.; Mankou, G.S.; Loumeto, J.; Midoko Iponga, D.; Kossi Ditsouga, F.; Zinga Koumba, R.; *et al.* Site-specific *versus* pantropical allometric equations: Which option to estimate the biomass of a moist central African forest? *For. Ecol. Manag.* **2014**, *312*, 1–9. [CrossRef]

30. Chave, J.; Réjou-Méchain, M.; Búrquez, A.; Chidumayo, E.; Colgan, M.S.; Delitti, W.B.C.; Duque, A.; Eid, T.; Fearnside, P.M.; Goodman, R.C.; *et al.* Improved allometric models to estimate the aboveground biomass of tropical trees. *Glob. Chang. Biol.* **2014**, 1–14. [CrossRef] [PubMed]

31. Feldpausch, T.R.; Lloyd, J.; Lewis, S.L.; Brienen, R.J.W.; Gloor, M.; Monteagudo Mendoza, A.; Lopez-Gonzalez, G.; Banin, L.; Abu Salim, K.; Affum-Baffoe, K.; *et al.* Tree height integrated into pan-tropical forest biomass estimates. *Biogeosciences* **2012**, *9*, 3381–3403. [CrossRef]

32. Picard, N.; Boyemba Bosela, F.; Rossi, V. Reducing the error in biomass estimates strongly depends on model selection. *Ann. For. Sci.* **2014**, 1–13. [CrossRef]

33. Bass, M.S.; Finer, M.; Jenkins, C.N.; Kreft, H.; Cisneros-Heredia, D.F.; McCracken, S.F.; Pitman, N.C.A.; English, P.H.; Swing, K.; Villa, G.; *et al.* Global conservation significance of Ecuador's Yasuní National Park. *PLoS ONE* **2010**, *5*, e8767. [CrossRef] [PubMed]

34. Valencia, R.; Foster, R.B.; Villa, G.; Condit, R.; Svenning, J.-C.; Hernández, C.; Romoleroux, K.; Losos, E.; Magård, E.; Balslev, H. Tree species distributions and local habitat variation in the Amazon: Large forest plot in eastern Ecuador. *J. Ecol.* **2004**, *92*, 214–229. [CrossRef]

35. Anderson-Teixeira, K.J.; Davies, S.J.; Bennett, A.C.; Gonzalez-Akre, E.B.; Muller-Landau, H.C.; Joseph Wright, S.; Abu Salim, K.; Almeyda Zambrano, A.M.; Alonso, A.; Baltzer, J.L.; *et al.* CTFS-ForestGEO: A worldwide network monitoring forests in an era of global change. *Glob. Chang. Biol.* **2015**, *21*, 528–549. [CrossRef] [PubMed]

36. Valencia, R.; Condit, R.; Foster, R.B.; Romoleroux, K.; Villa Munoz, G.; Svenning, J.; Magård, E.; Bass, M.S.; Losos, E.C.; Balslev, H. Yasuní Forest Dynamics Plot, Ecuador. In *Tropical Forest Diversity and Dynamism: Findings from a Large-Scale Plot Network*; Losos, E.C., Leigh, E.G., Eds.; University of Chicago Press: Chicago, IL, USA, 2004; pp. 609–628.

37. Condit, R. *Tropical Forest Census Plots*; Springer-Verlag: Berlin, Germany, 1998.

38. Romero-Saltos, H.; Valencia, R.; Macía, M.J. Patrones de distribución y rareza de plantas leñosas en el Parque Nacional Yasuní y la Reserva Étnica Huaorani , Amazonía ecuatoriana. In *Evaluación de Recursos Vegetales no Maderables en la Amazonia Noroccidental*; Duivenvoorden, J.F., Balslev, H., Cavelier, J., Grandez, C., Tuomisto, H., Valencia, R., Eds.; Universiteit van Amsterdam: Amsterdam, The Netherlands, 2001; pp. 131–162.

39. Romero-Saltos, H.; Hernández, C.; Valencia, R. Diversidad y dinámica de árboles en una parcela de gran escala. In *Arboles Emblemáticos de Yasuní, Ecuador*; Publicaciones del Herbario QCA, Escuela de Ciencias Biológicas, Pontificia Universidad Católica del Ecuador: Quito, Ecuador, 2014; pp. 14–22.

40. Valbuena, R.; Mauro, F.; Rodriguez-Solano, R.; Manzanera, J.A. Accuracy and precision of GPS receivers under forest canopies in a mountainous environment. *Spanish J. Agric. Res.* **2010**, *8*, 1047–1057. [CrossRef]

41. Sistema de Referencia Geocéntrico para las Américas. Available online: http://www.sirgas.org/index.php (accessed on 19 September 2015).

42. Leitold, V.; Keller, M.; Morton, D.C.; Cook, B.D.; Shimabukuro, Y.E. Airborne LiDAR-based estimates of tropical forest structure in complex terrain: Opportunities and trade-offs for REDD+. *Carbon Balance Manag.* **2015**, *10*. [CrossRef] [PubMed]

43. Hansen, E.; Gobakken, T.; Næsset, E. Effects of pulse density on digital terrain models and canopy metrics using airborne laser scanning in a tropical rainforest. *Remote Sens.* **2015**, *7*, 8453–8468. [CrossRef]

44. Kennel, P.; Tramon, M.; Barbier, N.; Vincent, G. Canopy height model characteristics derived from airborne laser scanning and its effectiveness in discriminating various tropical moist forest types. *Int. J. Remote Sens.* **2013**, *34*, 8917–8935. [CrossRef]

45. Meng, X.; Currit, N.; Zhao, K. Ground filtering algorithms for airborne LiDAR data: A review of critical issues. *Remote Sens.* **2010**, *2*, 833–860. [CrossRef]

46. Rapidlasso GmbH. LAStools Rapid LiDAR Processing. Available online: http://rapidlasso.com/ (accessed on 22 September 2015).

47. Hyyppä, J.; Hyyppä, H.; Litkey, P.; Yu, X.; Haggrén, H.; Rönnholm, P.; Pyysalo, U.; Pitkänen, J.; Maltamo, M. Algorithms and methods of airborne laser-scanning for forest measurements. *Int. Arch. Photogramm. Remote Sens. Spat. Inf. Sci.* **2004**, *36*, 82–89.

48. Evans, J.; Oakleaf, J.; Cushman, S.; Theobald, D. An ArcGIS Toolbox for surface Gradient and Geomorphometric Modelling, version 2.0-0. Available online: http://evansmurphy.wix.com/evansspatial (accessed on 2 March 2015).

49. Moore, I.; Gessler, P.; Nielsen, G.A.; Peterson, G.A. Soil attribute prediction using terrain analysis. *Soil Sci. Soc. Am. J.* **1993**, *57*, 443–452.

50. Rampi, L.P.; Knight, J.F.; Pelletier, K.C. Wetland mapping in the Upper Midwest United States. *Photogramm. Eng. Remote Sens.* **2014**, *80*, 439–448. [CrossRef]

51. Valencia, R.; Condit, R.; Muller-Landau, H.C.; Hernandez, C.; Navarrete, H. Dissecting biomass dynamics in a large Amazonian forest plot. *J. Trop. Ecol.* **2009**, *25*, 473. [CrossRef]

52. Fox, J. The R commander: A basic-statistics graphical user interface to R. *J. Stat. Softw.* **2005**, *14*, 1–42. [CrossRef]

53. Peña Sánchez de Rivera, D. *Regresión y Diseño de Experimentos*; Alianza Editorial: Madrid, Spain, 2002; pp. 459–526.

54. Baskerville, G.L. Use of logarithmic regression in the estimation of plant biomass. *Can. J. For. Res.* **1972**, 49–53. [CrossRef]

55. Muller-Landau, H.; Detto, M.; Chisholm, R.; Hubbell, S.; Condit, R. Detecting and projecting changes in forest biomass from plot data. In *Forests and Global Change*; Coomes, D., Burslem, D., Simonsen, W., Eds.; Cambridge University Press: Cambridge, UK, 2014; pp. 381–416.

56. Andersen, H.-E.; McGaughey, R.J.; Reutebuch, S.E. Estimating forest canopy fuel parameters using LiDAR data. *Remote Sens. Environ.* **2005**, *94*, 441–449. [CrossRef]

57. Gessler, P.E.; Moore, I.D.; McKenzie, N.J.; Ryan, P.J. Soil-landscape modelling and spatial prediction of soil attributes. *Int. J. Geogr. Inf. Syst.* **1995**, *9*, 421–432. [CrossRef]
58. Kanagaraj, R.; Wiegand, T.; Comita, L.S.; Huth, A. Tropical tree species assemblages in topographical habitats change in time and with life stage. *J. Ecol.* **2011**, *99*, 1441–1452. [CrossRef]

Multispectral Radiometric Analysis of Façades to Detect Pathologies from Active and Passive Remote Sensing

Susana Del Pozo [1,*], Jesús Herrero-Pascual [1], Beatriz Felipe-García [2], David Hernández-López [2], Pablo Rodríguez-Gonzálvez [1] and Diego González-Aguilera [1]

Academic Editors: Fabio Remondino, Norman Kerle and Prasad S. Thenkabail

[1] Department of Cartographic and Land Engineering, University of Salamanca, Hornos Caleros, 05003 Ávila, Spain; sabap@usal.es (J.H.-P.); pablorgsf@usal.es (P.R.-G.); daguilera@usal.es (D.G.-A.)
[2] Institute for Regional Development (IDR), Albacete, University of Castilla La Mancha, 02071 Albacete, Spain; bfelipe@jccm.es (B.F.-G.); david.hernandez@uclm.es (D.H.-L.)
* Correspondence: s.p.aguilera@usal.es

Abstract: This paper presents a radiometric study to recognize pathologies in façades of historical buildings by using two different remote sensing technologies covering part of the visible and very near infrared spectrum (530–905 nm). Building materials deteriorate over the years due to different extrinsic and intrinsic agents, so assessing these affections in a non-invasive way is crucial to help preserve them since in many cases they are valuable and some have been declared monuments of cultural interest. For the investigation, passive and active remote acquisition systems were applied operating at different wavelengths. A 6-band Mini-MCA multispectral camera (530–801 nm) and a FARO Focus3D terrestrial laser scanner (905 nm) were used with the dual purpose of detecting different materials and damages on building façades as well as determining which acquisition system and spectral range is more suitable for this kind of studies. The laser scan points were used as base to create orthoimages, the input of the two different classification processes performed. The set of all orthoimages from both sensors was classified under supervision. Furthermore, orthoimages from each individual sensor were automatically classified to compare results from each sensor with the reference supervised classification. Higher overall accuracy with the FARO Focus3D, 74.39%, was obtained with respect to the Mini MCA6, 66.04%. Finally, after applying the radiometric calibration, a minimum improvement of 24% in the image classification results was obtained in terms of overall accuracy.

Keywords: cultural heritage; multispectral camera; laser scanning; radiometric calibration; remote sensing; close range photogrammetry; multispectral classification

---

1. Introduction

Historical buildings and monuments are valuable constructions for the area where they are placed. The degradation of their construction materials is caused mainly by environmental factors such as pollution and meteorological conditions. Specifically, the presence of water plays an important role in stone deterioration processes [1]. It accelerates the weathering processes contributing to dissolution and frost/thaw cycles among others [2] allowing the formation of black crust on the rock surface resulting in mechanical and chemical degradations of stones. For that reason the use of non-contact and non-destructive technologies to study stone damages is important for the preservation of buildings and for the choice of the best technique for restoration [3,4].

Terrestrial laser scanners and multispectral digital cameras are two different technologies that are suitable for these studies. They are non-destructive and non-invasive sensors that allow researchers to acquire massive geometric and radiometric information across the building with high accuracy and in a short acquisition time. The geometrical information provided by laser scanner technology has been successfully applied in a large number of fields such as archaeology [5], civil engineering [6], geology [7] and geomorphological analysis [8]. On the other hand, radiometric information, provided by the laser intensity data and the multispectral digital cameras, is used less frequently. Even so, its high potential for classification tasks and recognition of different materials has been demonstrated [9]. Nowadays, in the literature, one can find works related to this issue ranging from methodologies of radiometric calibration [10] to corrections of intensity values [9,11] including applications of the intensity data [12]. Spectral classification methods are based on the properties of the reflected radiation from each surface and the fact that each specific material has wavelength dependent reflection characteristics. There are many classification methods, which vary in complexity. These methods include hard and soft classifiers, parametric and non-parametric methods and supervised and unsupervised techniques [13]. There are several works related to the application of these techniques to the identification of damage on building surfaces [14–18].

The main objective of this paper is the classification and mapping of pathologies and materials of a historical building façade from reflectance values at different wavelengths by combining intensity calibrated data from a FARO Focus3D laser scanner and calibrated images from a 6-band Mini-MCA multispectral camera. Additional goals were evaluating the degree of automation in the pathology detection process of façades. To achieve these objectives, the paper is divided into the following sections: Section 2 gives the details and specifications of the equipment employed and thoroughly describes the methods employed in the workflow methodology. Section 3 shows the classification maps and accuracy results for both unsupervised and supervised classifications, closing with Section 4 which summarizes the main conclusions and findings drawn from the study.

2. Material and Methods

The methodology developed to reach the objectives of the paper consists of three main stages: the data acquisition, the pre-processing and the processing of data as is outlined in Figure 1. For the data acquisition, two sensors with different operating principles were implemented: a passive multispectral camera and an active terrestrial laser scanner. The pre-processing step involved data filtering and several corrections applied to the spectral information to finally obtain data in reflectance values. During the last step and taking advantage of the metrics from the scan points, reflectance orthoimages were generated for both the multispectral images and the laser intensity. These orthoimages were the input for two different classifications processes: a clustering classification with data from each sensor and a supervised classification with the set of all data from both sensors.

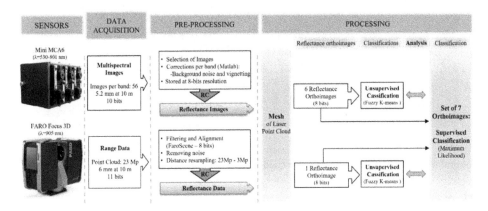

Figure 1. The workflow of the methodology presented. Acronyms: RC = Radiometric calibration and Mp = Millions of points.

2.1. Equipment

For the documentation of the façade, the following equipment was used: two radiometrically calibrated sensors with different characteristics and data acquisition principles, a passive multispectral camera and an active terrestrial laser scanner. Figure 1 shows the main characteristics of them and the different stages of the workflow followed in this research.

For the multispectral data acquisition, a calibrated lightweight Multiple Camera Array (MCA-Tetracam) was employed. This low-cost sensor allows versatility in data acquisition; however it requires the radiometric and geometric corrections to ensure the quality of the results [19]. It includes a total of 6 individual sensors with filters for the visible and near infrared spectrum data acquisition. More specifically, the individual bands of 530, 672, 700, 742, 778 and 801 nm were used. The longest wavelength was chosen taking into account that the multispectral sensor is not externally cooled. In spite of its 1280×1024 pixels of image resolution, the camera has a radiometric resolution of 10 bits. The focal length of 9.6 mm and the pixel size of 5.2 μm yield a façade sample distance (FSD) of 5.4 mm for a distance of 10 m, which should be taken into account for the pathology detection performance in small elements. The main limitation of this camera is the field of view ($38° \times 31°$), so several captures were needed to keep the FSD.

The FARO Focus3D is a phase shift continuous wave terrestrial laser scanner (TLS) operating at a wavelength of 905 nm. It is not common to use this kind of sensor to perform radiometric studies but it guarantees a comprehensive data acquisition whose results are not influenced by changes in light. This device measures distances in a range of 0.60–120 m with a point measurement rate of 976,000 points per second. It has an accuracy of 0.015° in normal lighting and reflectivity conditions and a beam divergence of 0.19 mrad, equivalent to 19 mm per 100 m range. The field of view covers 320° vertically and 360° horizontally with a 0.009° of angular resolution and the returning intensity is recorded at 11 bits. This laser scanner includes, in addition, a double compensator in the horizontal and vertical axis that can be used as constraint for the scan alignment.

Additionally, a high resolution spectroradiometer (ASD FieldSpec3) (Figure 2) was used as a remote detector of radiant intensity from the visible to the shortwave infrared ranges (350 to 2500 nm with a maximum spectral resolution of 3 nm and ±1 nm wavelength accuracy) to validate the spectral results of the study [20]. Equipped with optical fiber cables, it measured reflectances from the different materials and covers of the façade with a 25° field of view. Measures were made by positioning the spectroradiometer gun (Figure 2a) as orthogonal as possible and at a distance of approximately 10 cm from the sample, trying to cover a relatively homogeneous area of the material.

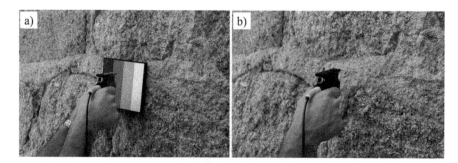

Figure 2. ASD FieldSpec3 spectroradiometer collecting spectral radiation reflected from (**a**) the Spectralon target and (**b**) mortar between contiguous stones of the examined façade.

2.2. Data Acquisition

Since each material has a unique reflectance behavior depending of the wavelength, the presence of pathologies on façades, such as moisture, moss or efflorescence, is likely to be successfully detected by analyzing the reflected visible and very near infrared radiation from the façades in reflectance values instead of digital levels (output digital format of the device). That is why these two sensors were radiometrically pre-calibrated and used to obtain orthoimages with surface reflectance values instead of digital levels. Since reflectance, for the specific case of a passive sensor, is a function of the solar incident radiance, a standard calibrated reflection target (Spectralon, Labsphere) was required and placed on the façade (Figure 2a), thus it appeared in every multispectral image to be able to calculate the solar irradiance (E) of each capture moment.

Illumination is a crucial parameter for data acquisition with passive sensor, particularly when several shot positions are required to cover the object of study. For that reason and to ensure the greatest resolution, taking fewer photos as far as possible was prioritized in this study. A total of 56 captures were collected with a FSD of 5.4 mm for the worst case so that the standard calibrated reflection target appeared in all of them.

On the other hand, the laser scanner data acquisition was designed so that the effect of the laser beam incidence angle [21,22] was minimized. Intensity data at 11-bit resolution was collected at an average distance of 10 m through three scans with scan area restrictions. Thus, 7 m of façade were covered for each of the scans assuming a maximum incidence angle error of 5.6% regarding the maximum oblique angle of incidence (19.29°). In addition, scanning positions were selected according to the different technical specifications of the scanner for an spatial resolution of 6 mm at the working distance. The laser network was adapted and filtered due to the presence of obstacles that hinder a single station data acquisition.

2.3. Pre-Processing

Before the reflectance orthoimage generation some corrections to raw data were applied to avoid error propagation in the radiometric calibration process. In this section, these radiometric corrections and the final radiometric calibration were described as well as the orthoimages generation process. Finally, the orthoimages were classified to obtain maps of different pathologies and building materials.

2.3.1. Multispectral Images Corrections

Low-cost sensors, such as the Mini MCA6, are more likely to be affected by different noise sources so that the actual value of radiation collected by them is altered (Equation (1)) [23]. Specifically, the Mini MCA6 was affected by two different sources errors: a background noise and a vignetting effect [20]. Both errors were studied under precise laboratory controlled conditions for each wavelength band.

The background noise is a systematic error caused by the sensor electronics of the camera. It was analyzed in a completely dark room in the absence of light determining the noise per band depending on the exposure time. For this study, the maximum background error was for the 801-nm band and involved a 1.07% increment of the actual digital level value. Regarding the vignetting effect [24], the radial attenuation of the brightness was studied taking images of a white pattern with uniform lighting conditions. Digital levels of each multispectral image were corrected for these two effects through a script developed in Matlab to improve the data quality before the radiometric calibration.

$$DL_{raw} = DL_{radiance} + (DL_{bn} + DL_v) \tag{1}$$

where DL_{raw} are the digital levels of the raw images, $DL_{radiance}$ are the digital levels from the radiance component, DL_{bn} are the digital levels from background noise and DL_v are the digital levels from the vignetting component.

2.3.2. Filtering and Alignment of the Point Clouds

The raw laser scanner data were filtered and segmented in order to remove those points that were not part of the object of study (adjacent building, artificial elements, trees, *etc.*). The individual point cloud alignment was done by a solid rigid transformation by the use of external artificial targets (spheres). The spheres were stationed in tripods at the plumb-line plane surveyed by the global navigation satellite system (GNSS). The laser local coordinate system could be transformed to a global coordinate system (UTM30N in ETRS89), allowing the geo-referencing of the subsequent classification for a global analysis and interpretation. This proposed workflow allowed a final relative precision of the coordinates of the artificial targets of 0.01 m and an absolute error of 0.03 m after post-processing. As a result, a unique point cloud in a local coordinate system with 11 mm precision (due to the error propagation of inherent error sources of laser scanner [25] and the error associated to the definition of the coordinate system) was generated.

2.3.3. Radiometric Calibrations

To perform the radiometric calibration of both sensors, auxiliary equipment such as lambertian surfaces with known spectral behavior (Spectralon) and/or a spectroradiometer are needed to solve the calibration. Thus, after the calibration process images values, in the case of the camera, and points' intensities, in the case of the laser scanner, correspond to the radiation emitted by the surface expressed in radiance or reflectance. The Mini MCA6 multispectral camera was calibrated in a previous field campaign [20] through *in situ* spectroradiometer measurements of artificial surfaces, with known and unknown reflectance behavior (Spectralon and polyvinyl chloride vinyl sheets respectively). Regarding the radiometric calibration of the TLS, it was carried out in laboratory by using a Spectralon and in absence of light.

The multispectral camera was calibrated by the radiance-based vicarious method [26–28], being the transformation equation from raw images into images with reflectances values Equation (2):

$$\rho_{MCA} = \frac{c0_\lambda + c1_\lambda \cdot DL/Fv_\lambda}{E_\lambda} \cdot \pi \tag{2}$$

where $c0_\lambda$ and $c1_\lambda$, offset and gain, are the calibration coefficients of each camera band, Fv_λ the shutter opening time factor and E_λ the solar irradiance at the ground level. Table 1 summarizes the multispectral camera calibration coefficients and the R^2 determination coefficient achieved per band.

Table 1. Calibration coefficients of the Mini MCA6 per band.

Bands	$c0_\lambda$	$c1_\lambda$	R^2
530 nm	0.000264	0.057718	0.9816
672 nm	−0.000795	0.050005	0.9823
700 nm	−0.000861	0.041353	0.9820
742 nm	−0.001205	0.074335	0.9843
778 nm	−0.001510	0.047292	0.9846
801 nm	−0.000834	0.047656	0.9827

In order to obtain reflectance values directly from laser data, a reflectance-based radiometric calibration [28] consisting of analyzing the distance-behavior of the intensity data (Figure 3) was performed (Equation (3)).

$$\rho_{FARO} = e^{a \cdot d} \cdot b \cdot d^2 \cdot e^{c1_F \cdot DL_F} \tag{3}$$

where a and b were the empirical coefficients related to the signal attenuation and internal TLS conversion from the received power to the final digital levels, d the distance between the laser scanner and the object, $c1_F$ the gain of the TLS and DL_F the raw intensity data in digital levels (11 bits). Please note that the empirical coefficients were obtained by a laboratory study, since the TLS internal electronics and intermediate signal processing is not disclosed.

In this case, a laboratory experiment from 5 to 36 m at one-meter intervals provided enough information to study the FARO Focus3D internal behavior (Figure 3). It was conducted in low-light conditions at a controlled temperature of 20 °C to model and simulate the system behavior. By positioning a Spectralon at each distance increment, intensity data were acquired at a quarter of the maximum resolution of the laser scanner (6 mm). The calibrated surface (Figure 2a) consists of four panels of 12%, 25%, 50% and 99% reflectance and it was assembled on a stable tripod to ensure its verticality. The raw intensity data from each reflectance panel were obtained by averaging the intensity values of the points belonging to each panel. The mean intensity value was plotted per panel and distance resulting in the Figure 4.

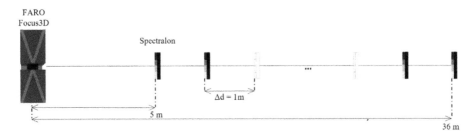

Figure 3. Sketch of the test performed to analyze the internal radiometric behavior of the FARO Focus3D.

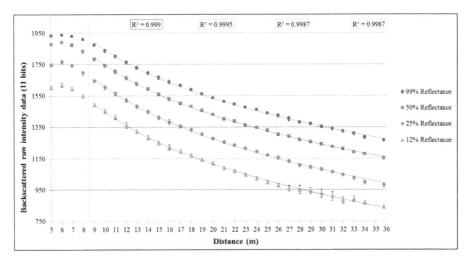

Figure 4. FARO Focus3D backscattered intensity behavior for the measurements of the four Spectralon panels at 1 m distance increments related to the signal attenuation (Equation (3)).

Figure 4 shows the signal attenuation of the FARO Focus 3D with distance as well as the logarithmic model that the measurements follow for distances up to 9 m. This particular behavior was noted in previous research works with similar sensors [29] and it is explained by the lidar equation [30]. By knowing the calibrated reflectance values of each Spectralon panel for 905 nm, the wavelength of the laser scanner, field measurements could be related with these reflectance values at each studied distance. Being 0.992, 0.560, 0.287 and 0.139 the normalized (0–1) reflectance values for the panel of 99%, 50%, 25% and 12% of reflectance respectively. Figure 5 shows how these values relate at a 10 m distance, and follow an exponential relationship which is shown in Equation (3). This distance was chosen as a threshold since for lower distances the calibration model changes due to the internal measurement system, involving alternative mathematical models.

As Figure 4 shows, the greater the distance the greater the intensity errors in the measurements. This behavior is related to the decrease of the received power due to the distance attenuation and signal scattering. Since the effective range of the employed TLS is higher than the studied distance, this error only appears significantly in the lower reflectance surface (12% panel).

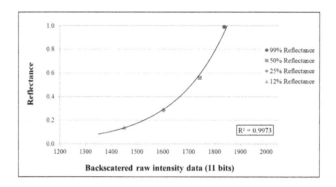

Figure 5. Relationship between TLS raw intensity data and reflectance for each spectralon panel at 10 m distance.

Based on the empirical study of the laser response, the attenuation of the signal with the distance (Figure 4) and the logarithmic behavior of the sensor [29], the relationship between digital levels and reflectances was finally approximated according to the Equation (4).

$$\rho_{FARO} = e^{0.214 \cdot d} \cdot 3.907 \cdot 10^{-7} d^2 \cdot e^{0.005415 \cdot DL_F} \tag{4}$$

This empirical equation can be applied only to objects at a distance over than 8 m since as can be shown in Figure 4, the FARO Focus3D has a completely different behavior for shorter distances.

2.4. Processing

In this subsection reflectance products are joined to achieve the orthoimages at each wavelength and they are finally classified to obtain maps of different building materials and pathologies.

2.4.1. Orthoimages Generation

Once the final point cloud was filtered, aligned and calibrated, a triangulation was applied to create the digital façade model (DFM). This step was required in order to generate continue 2D products (in the form of true orthoimages) and carry out the pathology detection by the classification process. For the DSM generation the incremental triangulation Delaunay algorithm was applied [31]. The output was refined to avoid artifact, meshing gaps, and other errors [32].

Orthoimages are highly demanded products that offer many benefits: metric accuracy and radiometric information useful to analyze different information quantitatively and qualitatively.

For the orthoimage generation, it was necessary to know the external orientation of the images with respect to the coordinate system of the laser point cloud model. For that purpose an average of 20 corresponding points between the point cloud and images were manually established. The image projection was characterized by a rigid transformation (rotation and translation) together with the internal camera parameters.

Orthoimages were generated based on the anchor point method [33]. This method consists of applying an affine transformation to each one of the planes formed by the optimized triangular mesh, which was obtained from the point cloud determined by the laser. Through the collinearity condition [34], the pixel coordinates of the vertices of the mesh were calculated, and the mathematical model of the affine transformation directly relates the pixel coordinates of the registered image and of the orthoimage.

2.4.2. Orthoimages Classifications

In order to categorize the orthoimages in different informational classes a previous automatic unsupervised classification and a posterior supervised classification were performed. The unsupervised classification was based on the Fuzzy K-means clustering algorithm where each observation can concurrently belong to multiple clusters [35]. For a set of n multidimensional pixels, the automatic management in l clusters iteratively minimizes the Equation (5) [36]:

$$J_m = \sum_{i=1}^{n} \sum_{l=1}^{\lambda} u^m_{i,l} \| x_i - c_l \|^2 ; \ 1 \leqslant m < \infty \tag{5}$$

where m represents any real number greater than 1, x_i the i-th of d-dimensional measured data, u_{il} the degree of membership of x_i in the cluster l, c_l the d-dimensional center of the cluster and $\| ** \| =$ Euclidean norm expressing the similarity between any measured data and the center.

Fuzzy partitioning is carried out through an iterative optimization of the objective function shown above, with the update of membership and the cluster centers by Equation (6).

$$u_{il} = \frac{1}{\sum_{k=1}^{c} \left[\frac{\| x_i - c_l \|}{\| x_i - c_k \|} \right]^{\frac{2}{m-1}}} ; \ c_l = \frac{\sum_{i=1}^{n} u_{il}^m \cdot x_i}{\sum_{i=1}^{n} u_{il}^m} \tag{6}$$

This iteration will stop when $\max_{il} \left\{ \left| u_{il}^{(k+1)} - u_{il}^{(k)} \right| \right\} < \varepsilon$, where ε is the stop criterion between 0 and 1 and k represents the iteration steps.

After this classification, a first approach of the spectral classes and different construction materials was obtained. With a subsequently supervised classification and applying the expert knowledge of some classes, the final results improved. Furthermore, this supervised classification will serve as reference to discuss which sensor is the ideal one for detecting materials and pathologies in façades.

In this case, a maximum likelihood (ML) classification algorithm [37] was applied. The ML classifier quantitatively evaluates both the variance and covariance of the category spectral response patterns when classifying an unknown pixel. The resulting bell-shaped surfaces are called probability functions, and if the prior distributions of this function are not known, then it is possible to assume that all classes are equally probable. As a consequence, we can drop the probability in the computation of the discriminant function $F(g)$ (Equation (7)), and there is one such function for each spectral category [38].

$$F(g) = -\ln |\Sigma_p| - (g - \mu_p)^T \Sigma_p^{-1} (g - \mu_p) \tag{7}$$

where p is the p-th cluster, Σ_p is the variance-covariance matrix and μ_p represents the class mean vector and g the observed pixel.

3. Experimental Results

The study area was the Shrine of San Segundo declared World Cultural Heritage in 1923 [39] (Figure 6). This Romanesque shrine is located in the west of the city of Ávila (Spain) and was built in the 12th century with unaltered grey granite plinths and walls with the alternation of granite blocks with different alteration degrees. The unaltered granite is mainly present in the blocks of low areas because of its high compressive strength and resistance to water absorption.

The field work was carried out on 27 July 2012 around the southern façade of the church (Figure 6), the most interesting façade from a historical point of view because it preserves the Romanesque main front. The five archivolts and capitals are decorated with plant and animal motifs. A total of 3 stations for the case of laser scanner were performed to cover the façade at a distance of 10 m (see Figure 6 right). The resolution of the data capture of the FARO Focus3D was a quarter of the full resolution provided by the manufacturer, 6 mm at 10 m. Moreover, the façade was photographed at the same distance with the Mini MCA6 multispectral camera with a FSD of 5.4 mm. A selection of 9 multispectral images of the 56 (7 per station) were used for the orthoimages generation. This selection was related with the most suitable images regarding the area of study and the optimal sharpness and quality of the set of images. The total volume of information generated amounted to 10.7 GB, where the great part was due to the meshes and orthoimages generation projects. Figure 7 shows the set of the 7 final orthoimages with a 6 mm FSD.

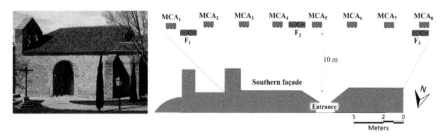

Figure 6. South façade of the Church of San Segundo in Ávila (Spain) (**left**) and a sketch of the acquisition setup with the different sensor's stations (MCA6-multispectral camera, FARO Focus3D) (**right**).

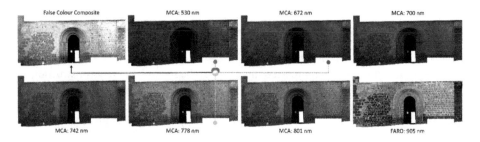

Figure 7. Set of 7 orthoimages of the façade in reflectance values from the two analyzed sensors (MCA6 multispectral camera and FARO Focus3D) and a false colorcolor composite orthoimage.

3.1. Reflectance Orthoimages

In order to compare the discrimination capability of both technologies to distinguish building materials and pathologies a first unsupervised classification of the orthoimages belonging to each sensor was performed (Figures 8 and 9). A final supervised classification with the complete set of 7 orthoimages was carried out. For each informational class manually representative areas distributed throughout the façade (between 5 and 10 polygons per class) were selected. This last classification serves as a reference with which to compare each individual unsupervised classification. The steps followed by the workflow are shown in Figure 1.

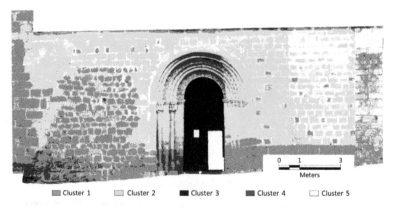

Figure 8. Mini MCA6 map for the 5-clusters unsupervised classification.

Cluster 1   Cluster 2   Cluster 3   Cluster 4   Cluster 5

Figure 9. FARO Focus3D map for the 5-clusters unsupervised classification.

3.2. Orthoimages Classifications

Ten predefined clusters were used in each case for the unsupervised classification algorithm. In all of them, the resulting map showed the existence of affected areas. Post-analysis reduced the number of clusters. The number of clusters decreased from 10 (initial clusters) to 5 thematic classes with real meaning: (1) unaltered granite; (2) altered granite; (3) wood (door of the church); (4) areas with moisture evidences (caused by capillarity or filtration water) and (5) mortar between blocks.

It is noteworthy that results from Mini MCA6 are not fully satisfactory due to large variability in lighting conditions during the data acquisition. As mentioned at the beginning of Section 3, the fieldwork took place on 27 July 2012, with a 6-h total acquisition time. Although radiometric calibration reduces the effects of the lighting variability between different data acquisition time, passive sensors are really sensitive to shady areas. These areas could be seen in Figure 7, specifically in the orthoimages from the Mini MCA6, and also in the classification results of the entrance area in Figure 8 (blue color). However, this is not the case for the active sensor, FARO Focus3D, where the continuity of materials and pathologies is a remarkable aspect.

Comparing the results with a visual inspection, results correspond quite well to reality for both types of existing granites (unaltered and altered) and wood by three well differentiated clusters in all classification maps (Figures 8 and 9). Regarding pathologies detection, it was not possible to draw final conclusions with these first unsupervised classifications. However, this process served to perform a better defined supervised classification.

With the aim of having a reference with which to compare both unsupervised classification maps, a supervised classification of the full set of 7 orthoimages in reflectance values was performed (Figure 10) taking into account the two existing variants of granite, their pathologies derived primarily from moisture and the other informational classes.

The best overall accuracy for the Fuzzy K-means unsupervised classifications was 74.39%, achieved for the FARO Focus3D map in contrast with the 66.04% accuracy for the Mini MCA6 map. This indicates that the best correlation between the number of pixels correctly classified and the total number of pixels occurred for this near infrared active sensor.

Table 2 contrasts the results of the supervised classification (based on training areas) with the unsupervised classification for each sensor. The table shows the sum of pixels belonging to each class for each of the classifications performed. The count is expressed as a percentage of the total number of classified pixels (1,154,932 without taking the background class into account).

▨ Unaltered Granite   ▢ Altered Granite   ■ Wood   ■ Moisture   ▢ Mortar

Figure 10. Multisensory map for the 5 informational classes supervised classification.

Table 2. Pixels computation belonging to each thematic class.

Class	Reference Map	Multispectral Map	Laser Map
Unaltered granite	30.04%	28.53%	27.33%
Altered granite	42.60%	48.06%	47.27%
Wood	5.35%	5.67%	5.82%
Moisture	1.88%	4.74%	1.31%
Mortar	20.13%	13.00%	18.27%

In a quantitative analysis for the estimation of the two types of granite and wood, results of both sensors are quite similar and really close to the reference map while intensity data from laser scanner are the closest to the reference map results for the estimation of moisture and mortar. Results show higher pixels classified as moisture in the case of multispectral map (2.86% higher with respect to the reference map) and few pixels classified as mortar (7.13% lower than the reference map) due mainly to the altered granite count (whose spectral response has the greatest similarity). Results from the laser sensor are quite similar, greater amount of altered granite by reducing the unaltered granite and mortar detected classes. Note that the best results for moisture detection are achieved with the FARO Focus3D, since humidity has a major interference with this wavelength [40]. Since the pathological classes (moisture and altered granite) are better recognized by the laser scanner and it is the most comprehensive sensor with results closer to the reference, it can be concluded that the active sensor has proven to be the best option to study and detect pathologies and different construction materials for studies with high variability in light conditions where passive sensors are greatly affected.

To evaluate the separability between classes the transformed divergence indicator [41], ranging from 0 to 2, was used as the most widely used quantitative estimator for this purpose [42]. Table 3 shows the separability between the final 5 classes.

Table 3. Transformed divergence for the supervised classification.

	Unaltered Granite	Altered Granite	Wood	Moisture
Altered granite	1.87	-	-	-
Wood	2.00	2.00	-	-
Moisture	1.99	1.99	2.00	-
Mortar	1.98	1.42	2.00	2.00

In general, a high separability was achieved for all 5 classes, highlighting the good separability between the spectral signatures of the two granite types. The worst results were for the mortar and the altered granite classes. This fact is explained by two reasons: on the one hand, the façade sample distance (FSD) of the orthoimages (11 mm in the worst case) was not enough to detect façade areas with smaller thickness of mortar; and on the other, altered granite class presented the closest spectral behavior regarding mortar. With respect to the moisture of the façade, it appeared in lower areas of the shrine (capillarity rising damp) and in the buttress, acting as a filter system for the water from the roof (filtration moisture). These areas are built with unaltered granite blocks since lower areas need to support the loads of the whole building (also in buttress). The radiometric misunderstanding between moisture and unaltered granite did not occur in the case of altered granite since the latter is part of the center of the façade, a low humidity area.

3.3. Accuracy Assessment

In order to assess the accuracy of the unsupervised classifications, the supervised classification approach based on maximum likelihood algorithm served as reference. Five classes and the seven bands available were considered in the classification process. Accuracy results for the case of the Mini MCA6 multispectral camera and the FARO Focus3D laser scanner were 66.04% and 74.39% respectively as mentioned above, and according to the Cohen's Kappa coefficient [43] the level of agreement was 0.50 and 0.621 respectively (excluding the null class).

Furthermore, as mentioned in Section 2.1, an ASD FieldSpec3 spectroradiometer was used to measure several samples of granite for a parallel study. Those measures, in this study, have been used as reference and as a complement to the above analysis to compare the spectral signatures of these construction materials with the discrete reflectance results obtained from the Mini MCA6 and the FARO Focus3D (Figure 11). The spectral signatures and deviations of the two types of granite present in the façade are plotted for the wavelength range covered by both sensors (530–905 nm).

In Figure 11, the graph continuous lines show at any wavelength the mean value of the reflectances of unaltered and altered granite samples distributed along the façade and measured with the spectroradiometer (a total of 6 and 7 samples of granite, respectively). On the other hand, the colored areas represent the standard deviation of that spectroradiometer measurements. Regarding the discrete values of reflectance achieved with the sensors (discrete points) they result from the mean reflectance value of the "unaltered granite" and "altered granite" classes for each sensor's wavelength of the supervised classification map. The "mortar" class was not finally evaluated due to its variability in thickness along the façade and due to the fact that the FSD achieved was in many areas greater than its thickness.

It should be mentioned that a great fit of the reflectance values from both sensors (discrete points) was achieved for both granite real spectral behaviors (spectroradiometer measurements) with admissible standard deviations associated (lower than those associated with spectroradiometer measurements). For both evaluated materials, the mean error was 0.007 (in the range 0–1), being the maximum 0.049 (in the range 0–1), which is better than the expected error for this vicarious calibration technique (around 5%).

The confusion matrices for the assessment of both sensors are shown in Tables 4 and 5 where the main diagonal indicates the percentage of pixels that have been correctly classify and the off-diagonal values represent misclassification. The producer and user accuracies as well as the overall accuravy are given. Regarding the moisture class, a significant performance improvement of the classifier is observed for this class for the operating wavelength of the FARO Focus3D. In the case of the mortar class, the Mini MCA6 do not bring good results mainly due to the errors produced during the 6-band registration process. Finally, we mention that in the case of the unaltered and altered granites, little variations were observed between both sensors.

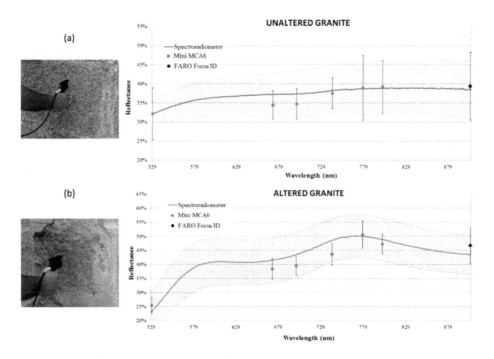

Figure 11. Spectral signatures of the two different types of granites, (**a**) unaltered and (**b**) altered, measured with the ASD spectroradiometer for the wavelength interval covered by the sensors used (Mini MCA6 and FARO Focus3D) where points are obtained from the orthoimages in reflectance values.

Table 4. Confusion matrix of the Mini MCA6 unsupervised classification.

	Moisture	Mortar	Altered Granite	Unaltered Granite	Wood	User Accuracy
Moisture	40.14%	25.40%	0.54%	33.92%	0.01%	40.14%
Mortar	0.06%	39.11%	59.53%	1.31%	0.00%	39.11%
Altered granite	3.79%	7.69%	73.86%	14.61%	0.05%	73.86%
Unaltered granite	6.56%	5.30%	16.39%	70.75%	1.01%	70.75%
Wood	0.00%	0.00%	0.00%	0.00%	100.00%	100.00%
Producer accuracy	21.10%	57.71%	66.00%	74.28%	94.26%	
					Overall accuracy:	**66.04%**

Table 5. Confusion matrix of the FARO Focus3D unsupervised classification.

	Moisture	Mortar	Altered Granite	Unaltered granite	Wood	User Accuracy
Moisture	52.46%	1.04%	23.17%	19.81%	3.52%	52.46%
Mortar	0.00%	60.46%	37.36%	2.18%	0.00%	60.46%
Altered granite	0.05%	12.81%	78.29%	8.64%	0.20%	78.29%
Unaltered granite	0.29%	1.49%	21.70%	75.47%	1.04%	75.47%
Wood	0.48%	0.00%	0.00%	0.00%	99.52%	99.52%
Producer accuracy	89.64%	68.32%	68.68%	83.62%	91.81%	
					Overall accuracy:	**74.39%**

To conclude, it should be highlight that the improvement in both the overall accuracy and the Kappa coefficient is significant in the case of working with radiometrically calibrated sensors as opposed to the use uncalibrated ones [17]. The results for the Mini MCA6 have experienced a 24% improvement in terms of overall accuracy and 23% regarding the Kappa coefficient. Furthermore, the improvement from the use of the calibrated FARO Focus3D was of 29% and 35% regarding the overall accuracy and Kappa coefficient, respectively. Results worsen in the case of the Mini MCA6 due to two factors; the first is that the camera is a passive sensor, so it is sensitive to changes in light conditions and shadow areas during data acquisition. The second error factor is caused by the slave image registration process, as has been mentioned above, due to the errors in the determination of baselines, angular misalignments and the internal parameters of the camera. Any error in those parameters is propagated into the final multispectral orthoimage, being worsened for higher spatial resolutions, where the geometric pixel footprint in the object may differ depending on the wavelength.

4. Conclusions

The work presented in this paper shows a comparison of the classification results from the use of different radiometrically calibrated sensors to detect pathologies in materials of historical buildings façades. By combining the use of two different data acquisition techniques (active and passive), two sensors were examined: a multispectral camera and a 3D laser scanner. The results show the different radiometric responses of the ashlars of a church with different damages levels (mainly moisture). The classification algorithms used for the classification processes were the Fuzzy K-means and the maximum likelihood classification algorithms.

A complete description of the workflow followed is outlined describing the data acquisition, pre-processing (including sensors radiometric calibrations), orthoimages generation and the application of two classification algorithms to assess the final results. Our results show that the most comprehensive sensor for which the best results were obtained is the FARO Focus3D. This is possibly due to the advantage of working in an active way with no need of external radiation. As a result, classification maps were not affected by different lighting conditions during data acquisition. Furthermore, geometric models of the study object can be derived thanks to its data capture. With these models, physical pathologies (such as fissures, desquamations, *etc.*) could be analyzed and both these damages and chemical pathologies could be quantified. However, for the challenge of the registration of 6 wavelength bands, the results from the Mini MCA6 were quite good. Considering all those issues and with the experience of working with these sensors in previous studies, it is concluded that the radiometric calibration of the sensors is crucial since it contributes to improving the accuracy of the outcomes (a 35% Kappa coefficient improvement in the case of the FARO Focus3D). Thus, a sensor combination with laser scanning as a primary choice is the best solution for pathology detection and quantification. By adding the intensity information to visible or multispectral information, results of classification improve in a quantitative and a qualitative way.

In future work, the use of a hyperspectral camera or another laser scanner operating in the shortwave infrared as a complement of the sensors proposed will improve the pathologies detection and the overall accuracy results since the spectral resolution of the study would be increased. In addition, and for non-carved historical buildings, the roughness of the façade would be calculated from the scan points in order to have additional data of the materials so it can help in the discrimination process. Finally, and regarding the data acquisition of passive sensors, constant favorable climatic conditions will be planned so that the accuracy of its classification results may be significantly improved.

Acknowledgments: This work has been partially supported by the Spanish Ministry of Economy and Competitiveness through the ENERBIUS project "Integrated system for the energy optimization of buildings" (ENE2013-48015-C3-3-R).

Author Contributions: All authors conceived, designed and performed the experimental campaign. Susana del Pozo and Pablo Rodriguez-Gonzálvez implemented the methodology and analyzed the results. Susana del Pozo wrote the manuscript and all authors read and approved the final version.

Conflicts of Interest: The authors declare no conflict of interest.

References

1. Marszalek, M. Deterioration of stone in some monuments exposed to air pollution: A Cracow case study. In *Air Pollution and Cultural heritage*; Taylor and Francis: London, UK, 2004; pp. 151–154.

2. Corvo, F.; Reyes, J.; Valdes, C.; Villaseñor, F.; Cuesta, O.; Aguilar, D.; Quintana, P. Influence of air pollution and humidity on limestone materials degradation in historical buildings located in cities under tropical coastal climates. *Water Air Soil Pollut.* **2010**, *205*, 359–375. [CrossRef]

3. Fort, R.; de Azcona, M.L.; Mingarro, F. Assessment of protective treatment based on their chromatic evolution: Limestone and granite in the Royal Palace of Madrid, Spain. In *Protection and Conservation of the Cultural Heritage in the Mediterranean Cities*; Galan, E., Zezza, F., Eds.; CRC Press/Balkema: Sevilla, Spain, 2002; pp. 437–441.

4. Weritz, F.; Kruschwitz, S.; Maierhofer, C.; Wendrich, A. Assessment of moisture and salt contents in brick masonry with microwave transmission, spectral-induced polarization, and laser-induced breakdown spectroscopy. *Int. J. Archit. Herit.* **2009**, *3*, 126–144. [CrossRef]

5. Lambers, K.; Eisenbeiss, H.; Sauerbier, M.; Kupferschmidt, D.; Gaisecker, T.; Sotoodeh, S.; Hanusch, T. Combining photogrammetry and laser scanning for the recording and modelling of the late intermediate period site of pinchango alto, Palpa, Peru. *J. Archaeol. Sci.* **2007**, *34*, 1702–1712. [CrossRef]

6. González-Aguilera, D.; Gómez-Lahoz, J.; Sánchez, J. A new approach for structural monitoring of large dams with a three-dimensional laser scanner. *Sensors* **2008**, *8*, 5866–5883. [CrossRef]

7. Buckley, S.J.; Howell, J.; Enge, H.; Kurz, T. Terrestrial laser scanning in geology: Data acquisition, processing and accuracy considerations. *J. Geol. Soc.* **2008**, *165*, 625–638. [CrossRef]

8. Armesto, J.; Ordóñez, C.; Alejano, L.; Arias, P. Terrestrial laser scanning used to determine the geometry of a granite boulder for stability analysis purposes. *Geomorphology* **2009**, *106*, 271–277. [CrossRef]

9. Höfle, B.; Pfeifer, N. Correction of laser scanning intensity data: Data and model-driven approaches. *ISPRS J. Photogramm. Remote Sens.* **2007**, *62*, 415–433. [CrossRef]

10. Kaasalainen, S.; Kukko, A.; Lindroos, T.; Litkey, P.; Kaartinen, H.; Hyyppä, J.; Ahokas, E. Brightness measurements and calibration with airborne and terrestrial laser scanners. *IEEE Trans. Geosci. Remote Sens.* **2008**, *46*, 528–534. [CrossRef]

11. Franceschi, M.; Teza, G.; Preto, N.; Pesci, A.; Galgaro, A.; Girardi, S. Discrimination between marls and limestones using intensity data from terrestrial laser scanner. *ISPRS J. Photogramm. Remote Sens.* **2009**, *64*, 522–528. [CrossRef]

12. Lichti, D.D. Spectral filtering and classification of terrestrial laser scanner point clouds. *Photogramm. Record* **2005**, *20*, 218–240. [CrossRef]

13. Mather, P.; Tso, B. *Classification Methods for Remotely Sensed Data*, 2nd ed.; CRC Press: Boca Raton, FL, USA, 2009.

14. Lerma, J.L. Multiband *versus* multispectral supervised classification of architectural images. *Photogramm. Record* **2001**, *17*, 89–101. [CrossRef]

15. Lerma, J.L. Automatic plotting of architectural facades with multispectral images. *J. Surv. Eng.* **2005**, *131*, 73–77. [CrossRef]

16. Ruiz, L.; Lerma, J.; Gimeno, J. Application of computer vision techniques to support in the restoration of historical buildings. *ISPRS Int. Arch. Photogramm. Remote Sens. Spatl. Inf. Sci.* **2002**, *34*, 227–230.

17. Hemmleb, M.; Weritz, F.; Schiemenz, A.; Grote, A.; Maierhofer, C. Multi-spectral data acquisition and processing techniques for damage detection on building surfaces. In Proceedings of the ISPRS Commission V Symposium "Image Engineering and Vision Metrology", Dresden, Germany, 25–27 September 2006.

18. Del Pozo, S.; Herrero-Pascual, J.; Felipe-García, B.; Hernández-López, D.; Rodríguez-Gonzálvez, P.; González-Aguilera, D. Multi-sensor radiometric study to detect pathologies in historical buildings. *ISPRS Int. Arch. Photogramm. Remote Sens. Spat. Inf. Sci.* **2015**, *XL-5/W4*, 193–200. [CrossRef]

19. López, D.H.; García, B.F.; Piqueras, J.G.; Alcázar, G.V. An approach to the radiometric aerotriangulation of photogrammetric images. *ISPRS J. Photogramm. Remote Sens.* **2011**, *66*, 883–893. [CrossRef]

20. Del Pozo, S.; Rodríguez-Gonzálvez, P.; Hernández-López, D.; Felipe-García, B. Vicarious radiometric calibration of a multispectral camera on board an unmanned aerial system. *Remote Sens.* **2014**, *6*, 1918–1937. [CrossRef]

21. Kaasalainen, S.; Ahokas, E.; Hyyppä, J.; Suomalainen, J. Study of surface brightness from backscattered laser intensity: Calibration of laser data. *IEEE Geosci. Remote Sens. Lett.* **2005**, *2*, 255–259. [CrossRef]

22. Soudarissanane, S.; Lindenbergh, R.; Menenti, M.; Teunissen, P. Scanning geometry: Influencing factor on the quality of terrestrial laser scanning points. *ISPRS J. Photogramm. Remote Sens.* **2011**, *66*, 389–399. [CrossRef]

23. Al-amri, S.S.; Kalyankar, N.V.; Khamitkar, S.D. A comparative study of removal noise from remote sensing image. *Int. J. Comput. Sci. Issue* **2010**, *7*, 32–36.

24. Zheng, Y.; Lin, S.; Kambhamettu, C.; Yu, J.; Kang, S.B. Single-image vignetting correction. *IEEE Trans. Pattern Anal. Mach. Intell.* **2009**, *31*, 2243–2256. [CrossRef] [PubMed]

25. Reshetyuk, Y. *Self-Calibration and Direct Georeferencing in Terrestrial Laser Scanning. Doctoral Thesis in Infrastructure, Geodesy*; Royal Institute of Technology (KTH): Stockholm, Suecia, 2009.

26. Honkavaara, E.; Arbiol, R.; Markelin, L.; Martinez, L.; Cramer, M.; Bovet, S.; Chandelier, L.; Ilves, R.; Klonus, S.; Marshal, P. Digital airborne photogrammetry—A new tool for quantitative remote sensing? A state-of-the-art review on radiometric aspects of digital photogrammetric images. *Remote Sens.* **2009**, *1*, 577–605. [CrossRef]

27. Biggar, S.; Slater, P.; Gellman, D. Uncertainties in the in-flight calibration of sensors with reference to measured ground sites in the 0.4–1.1 μm range. *Remote Sens. Environ.* **1994**, *48*, 245–252. [CrossRef]

28. Dinguirard, M.; Slater, P.N. Calibration of space-multispectral imaging sensors: A review. *Remote Sens. Environ.* **1999**, *68*, 194–205. [CrossRef]

29. Kaasalainen, S.; Krooks, A.; Kukko, A.; Kaartinen, H. Radiometric calibration of terrestrial laser scanners with external reference targets. *Remote Sens.* **2009**, *1*, 144–158. [CrossRef]

30. Jelalian, A.V. *Laser Radar Systems*; Artech House: New York, NY, USA, 1992.

31. Bourke, P. An algorithm for interpolating irregularly-spaced data with applications in terrain modelling. In Proceedingsof the Pan Pacific Computer Conference, Beijing, China, 1 January 1989.

32. Attene, M. A lightweight approach to repairing digitized polygon meshes. *Vis. Comput.* **2010**, *26*, 1393–1406. [CrossRef]

33. Kraus, K. *Photogrammetry: Geometry from Images and Laser Scans*, 2nd ed.; Walter de Gruyter: Berlin, Germany, 2007.

34. Albertz, J.; Kreiling, W. *Photogrammetrisches Taschenbuch*; Herbert Wichmann Verlag: Berlin, Germany, 1989.

35. Bezdek, J.C. *Pattern Recognition with Fuzzy Objective Function Algorithms*; Plenum Press: New York, NY, USA, 1981.

36. Kannan, S.; Sathya, A.; Ramathilagam, S.; Pandiyarajan, R. New robust fuzzy C-Means based gaussian function in classifying brain tissue regions. In *Contemporary Computing*; Ranka, S., Aluru, S., Buyya, R., Chung, Y.-C., Dua, C., Grama, A., Gupta, S.K.S., Kumar, R., Phoha, V.V., Eds.; Springer: Berlin, Germany, 2009; pp. 158–169.

37. Richards, J.A. *Remote Sensing Digital Image Analysis*; Springer: Berlin, Germany, 1999; Volume 3.

38. Lillesand, T.; Kiefer, R.W.; Chipman, J. *Remote Sensing and Image Interpretation*, 7th ed.; John Wiley & Sons: Hoboken, NJ, USA, 2015.

39. García, F.A.F. *La Invención de la Iglesia de san Segundo. Cofrades y Frailes Abulenses en los Siglos xvi y xvii*; Institución Gran Duque de Alba: Ávila, Spain, 2006.

40. Rantanen, J.; Antikainen, O.; Mannermaa, J.-P.; Yliruusi, J. Use of the near-infrared reflectance method for measurement of moisture content during granulation. *Pharm. Dev. Technol.* **2000**, *5*, 209–217. [CrossRef] [PubMed]

41. Davis, S.M.; Landgrebe, D.A.; Phillips, T.L.; Swain, P.H.; Hoffer, R.M.; Lindenlaub, J.C.; Silva, L.F. *Remote Sensing: The Quantitative Approach*; McGraw-Hill International Book Co.: New York, NY, USA, 1978.

42. Tolpekin, V.; Stein, A. Quantification of the effects of land-cover-class spectral separability on the accuracy of markov-random-field-based superresolution mapping. *IEEE Trans. Geosci. Remote Sens.* **2009**, *47*, 3283–3297. [CrossRef]

43. Cohen, J. Weighted kappa: Nominal scale agreement provision for scaled disagreement or partial credit. *Psychol. Bull.* **1968**, *70*, 213–220. [CrossRef] [PubMed]

Modeling and Mapping Agroforestry Aboveground Biomass in the Brazilian Amazon Using Airborne Lidar Data

Qi Chen [1,2], **Dengsheng Lu** [1,3,*], **Michael Keller** [4,5], **Maiza Nara dos-Santos** [4], **Edson Luis Bolfe** [4], **Yunyun Feng** [1] and **Changwei Wang** [2]

Academic Editors: Guomo Zhou, Conghe Song, Guangxing Wang, Nicolas Baghdadi and Prasad S. Thenkabail

[1] Key Laboratory of Carbon Cycling in Forest Ecosystems and Carbon Sequestration of Zhejiang Province, School of Environmental & Resource Sciences, Zhejiang A&F University, Lin An 311300, China; qichen@hawaii.edu (Q.C.); fengyyfm@163.com (Y.F.)
[2] Department of Geography, University of Hawaii at Manoa, Honolulu, HI 96822, USA; changwei_wang@scau.edu.cn
[3] Center for Global Change and Earth Observations, Michigan State University, East Lansing, MI 48823, USA
[4] Brazilian Agricultural Research Corporation—Embrapa, Campinas, SP 13070-115, Brazil; mkeller.co2@gmail.com (M.K.); maizanara@gmail.com (M.N.-S.); edson.bolfe@embrapa.br (E.L.B.)
[5] USDA Forest Service, International Institute of Tropical Forestry, San Juan, PR 00926, USA
* Correspondence: luds@zafu.edu.cn

Abstract: Agroforestry has large potential for carbon (C) sequestration while providing many economical, social, and ecological benefits via its diversified products. Airborne lidar is considered as the most accurate technology for mapping aboveground biomass (AGB) over landscape levels. However, little research in the past has been done to study AGB of agroforestry systems using airborne lidar data. Focusing on an agroforestry system in the Brazilian Amazon, this study first predicted plot-level AGB using fixed-effects regression models that assumed the regression coefficients to be constants. The model prediction errors were then analyzed from the perspectives of tree DBH (diameter at breast height)—height relationships and plot-level wood density, which suggested the need for stratifying agroforestry fields to improve plot-level AGB modeling. We separated teak plantations from other agroforestry types and predicted AGB using mixed-effects models that can incorporate the variation of AGB-height relationship across agroforestry types. We found that, at the plot scale, mixed-effects models led to better model prediction performance (based on leave-one-out cross-validation) than the fixed-effects models, with the coefficient of determination (R^2) increasing from 0.38 to 0.64. At the landscape level, the difference between AGB densities from the two types of models was ~10% on average and up to ~30% at the pixel level. This study suggested the importance of stratification based on tree AGB allometry and the utility of mixed-effects models in modeling and mapping AGB of agroforestry systems.

Keywords: agroforestry; aboveground biomass; lidar; mixed-effects models; allometry; wood density

1. Introduction

Agroforestry denotes land use systems where woody perennials (trees, shrubs, palms, *etc.*) are cultivated on the same land units as agricultural crops and/or animals [1,2]. In this study, we use agroforestry as a general term to refer to the land use system of cultivating woody perennials, either polyculture or monoculture (*i.e.*, plantation) on agricultural land, regardless the current existence of crops or animals. Over 10 million km^2 of agricultural lands have greater than 10% tree cover [3]. Through the provision of diversified products, agroforestry has been advocated and practiced by many countries to offer a wide range of economic, social, and ecological benefits: (1) increasing the per capita farm income by planning high-value tree products [4]; (2) improving soil fertility and land productivity [5]; (3) increasing household resilience [4]; (4) mitigating the impacts of climate variability and change [6,7]; (5) conserving biodiversity [8–10]; and (6) improving air and water quality [11–13].

Agroforestry offers high potential for carbon (C) sequestration [14] not only because the carbon density of agroforestry is usually higher than annual crops or pasture [15–17] but also because the trees produce fuelwood and timbers that otherwise would be harvested from natural forests [1]. The fine and coarse wood debris of plants are also stored in the soil C pool for long periods [18]. Thus, the role of agroforestry for C sequestration is increasingly recognized by the IPCC (Intergovernmental Panel on Climate Change) [19]. However, for agroforestry to be successful as a strategy for C sequestration, it is necessary to have sound detection and monitoring systems [20].

Remote sensing provides an effective way to estimate and monitor the biomass and C stock of different vegetation types [21] including agroforestry systems [22–25]. In particular, airborne lidar (Light Detection and Ranging) has emerged in the 21 century as the most accurate technology to quantify forest aboveground biomass (AGB) at the landscape level [21,26]. Lidar is especially powerful over forests of high biomass where passive optical imaging or radar sensors have saturation problems. One of the main advantages of lidar is that it can penetrate through the small canopy gaps for detecting vertical structure and extracting ground elevation. The height information derived from lidar is strongly related to biomass for most tree species. A large number of studies have reported the use of airborne lidar for mapping AGB in boreal (e.g., [27–29]), temperate (e.g., [30,31]), and tropical (e.g., [32–36]) forests.

Compared to the large body of literature of lidar remote sensing of carbon and biomass for forests, the use of this frontier technology for quantifying agroforestry AGB is very limited. One of the best examples of agroforestry of relative economic success of colonization in the Brazilian Amazon occurred in the Tomé-Açu [37]. Focusing on an agroforestry system in Tomé-Açu, this study aims to (1) predict plot-level agroforestry AGB using regular fixed-effects regression models that treat the regression coefficients as constants; (2) identify the causes of AGB prediction errors when such models are used; (3) propose new ways to stratify vegetation types and apply mixed-effects models to reduce the AGB prediction errors; and (4) discuss the challenges and future directions for modeling and mapping agroforestry AGB.

2. Study Area

This study was conducted in the Quatro Bocas district in the municipality of Tomé-Açu (2°28′S and 48°20′W), located in the northeastern region of Pará state in Brazil (Figure 1). The study area is 2 km by 5 km. Topography in the region is characterized by low flat plateaus, terraces, and lowlands with altitudes varying from 14 to 96 m. The soils are classified as Ferralsols, Plintosols, and Fluvisols. The average annual rainfall is 2300 mm [38]. The Tomé-Açu region has a humid mesothermal climate—Ami according to the Köppen classification—with high average annual temperatures (26 °C) and relative air humidity rate of about 85%. The original vegetation is lowland dense ombrophilous forest, which has been intensely altered. The landscape mosaic is dominated by pasture, agricultural fields, and secondary forests. Forest remnants are observed especially at the margins of streams.

Tomé-Açu started its agricultural development in the 1920′s, with the beginning of the Japanese immigration to the region. The immigrants implanted horticulture and, later, black pepper (*Piper nigrum* L.). They were provided with lands by the Brazilian government, which made technological development possible and turned Pará into the greatest black pepper producer in Brazil. With the decay of the black-pepper cycle from the 1970′s on, caused by fusarium blight, the farmers looked for new production alternatives. According to Homma [39], the way out of this ecological crisis for the immigrants was to diversify their activities, with emphasis in fruit crops, especially papaya, melon, acerola, orange, dende (oil palm), cupuacu, passion fruit, and other native and exotic fruit trees and vines that initiated a new economic cycle for the region. The current agroforestry systems

(1 to 34 years old) have a great variety of fruit and timber tree species. Homma [40] pointed out that the success of the region's agricultural development resulted from the Japanese-Brazilian farmers' innovative thinking, their holistic view of future markets, and their social-minded spirit, which made possible the creation of the Cooperativa Agrícola Mista de Tomé-Açu (Camta) in 1931, whose intention was to sell vegetables and nowadays commercializes the agroforestry products (fruit, pulp, juice, and oil) in various countries.

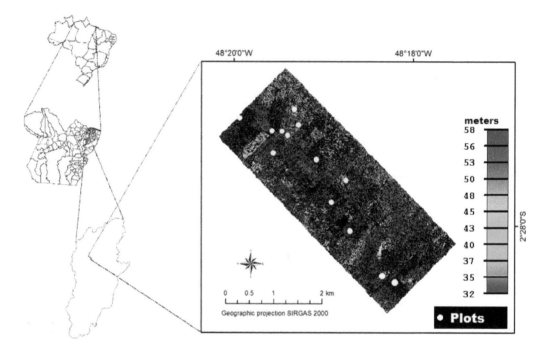

Figure 1. The study area—Tomé-Açu at Pará state in Brazil (Data source: Lidar data were acquired in 2013 and 13 plots were measured in 2014) (Note: colors from red to blue represent the elevation of the laser points).

3. Methods

3.1. Field Data Collection and AGB Estimation

Field inventory followed methods consistent with standardized protocols for tropical forests (e.g., [41]). Thirteen field plots of 30 m × 30 m were established in agroforestry regions using a differential GNSS (Trimble GeoXH-6000) during August and October 2014. The plot locations were selected to cover agroforestry fields with different ages and floristic compositions [22]. Diameter at breast height (DBH) was measured with metric tapes. Tree height was measured with tape and clinometer with the estimated accuracy of 10% of the tree height [42,43].

Over the 13 plots, the field crew collected measurements for a total of 1173 live trees to a minimum DBH of 5 cm from 22 species (Table 1). Four out of 13 plots grew monocultures including teak (plots 9 and 10), dende (plot 11), and cupuacu (plot 15), while the rest of the plots were mixed with two to eight species (Table 2).

Table 1. Tree species of the field plots and their wood densities.

Common Name	Scientific Name	Tree Count	ρ (g/cm^3)	Type
Acai	*Euterpe oleracea*	287	0.41	Palm
Ameixa (Tallow plum)	*Ximenia americana*	108	0.64	Other
Andiroba	*Carapa guianensis*	7	0.57	Other
Cacau (Cocoa)	*Theobroma glaucum*	408	0.53	Other
Castanha dopara (Brazil nut)	*Bertholletia excelsa*	4	0.64	Other
Cupuacu	*Theobroma grandiflorum*	72	0.53	Other
Dende (American oil palm)	*Elaeis oleifera*	13	0.41	Palm
Embauba	*Cecropia ficifolia*	5	0.27	Other
Freijo cinza	*Cordia goeldiana*	1	0.50	Other
Goiaba (Guava)	*Myrciaria floribunda*	6	0.77	Other
Guariuba	*Clarisia racemosa*	5	0.59	Other
Ipe	*Tabebuia chrysotricha*	34	0.64	Other
Ipe rosa	*Tabebuia roseo-alba*	2	0.52	Other
Mogno (Mahogany)	*Swietenia macrophylla*	11	0.51	Other
Molongo	*Ambelania acida*	2	0.52	Other
Murta	*Strychnos subcordata*	1	0.54	Other
Paliteira	*Clitoria fairchildiana*	2	0.64	Other
Parica	*Schizolobium amazonicum*	2	0.49	Other
Pelo de Cutia	*Banara guianensis*	24	0.61	Other
Seringa (Rubber)	*Hevea brasiliensis*	35	0.49	Other
Tamanqueira	*Zanthoxylum rhoifolium*	1	0.49	Other
Teca (Teak)	*Tectona grandis*	143	0.64	Other

Note: ρ is species-specific wood density (g/cm^3).

Table 2. Tree species of individual field plots and plot attributes including AGB, height (H), basal area (BA), and wood density (ρ_{plot}).

Plot ID	Species (Tree Count)	Type	AGB (Mg/ha)	H (m)	BA (m^2/ha)	ρ_{plot} (g/cm^3)
1	Acai (90), Cacau (95)	Polyculture	46.0	6.8	18.9	0.44
3	Cacau (59), Ipe (29), Parica (2)	Polyculture	78.0	8.6	17.8	0.58
5	Cacau (50), Seringa (35), other (2)	Polyculture	107.7	9.6	21.1	0.49
7	Cacau (51), Andiroba (7), Ipe Rosa (2), Molongo (2), Paliteira (2), Cupuacu (1)	Polyculture	159.4	7.4	23.3	0.56
8	Acai (69), Cacau (63), Cupuacu (25)	Polyculture	124.3	8.0	25.6	0.57
9	Teca (30)	Monoculture	178.4	18.9	21.1	0.64
10	Teca (113)	Monoculture	255.5	17.7	31.1	0.64
11	Dende (13)	Monoculture	219.8	11.2	103.3	0.41
12	Cacau (45), IPE (3)	Polyculture	13.1	3.9	6.7	0.52
13	Acai (70), Pelo de Cutia (23), Ameixa (14), Guariuba (5), Goiaba (2), other (2)	Polyculture	41.7	7.5	11.1	0.56
14	Ameixa (94), Acai (54), Embauba (4), Goiaba (4), other(1)	Polyculture	105.1	9.2	22.2	0.61
15	Cupuacu (46)	Monoculture	28.1	4.5	15.6	0.53
16	Cacau (45), Mogno (5), Ipe (2)	Polyculture	10.6	3.5	5.6	0.53

Note: Plot-level wood density (ρ_{plot}) was calculated by weighting individual tree wood density with their size (basal area ∗ tree height). H is the mean of individual tree heights.

Based on wood density, DBH, and tree height, the oven-dry aboveground biomass ($\hat{B}_{tree}$) of individual trees was estimated using the following allometric models:

For palm trees (Acai and American oil palm),

$$\hat{B}_{tree} \ (kg) = 0.001 * \left(\exp \left(0.9285 * \ln \left(DBH^2 \right) + 5.7236 \right) * 1.05001 \right) \tag{1}$$

For other trees:

$$\hat{B}_{tree} \ (kg) = 0.0704 * \left(\rho * DBH^2 * H \right)^{0.9701} \tag{2}$$

where DBH is in cm, ρ is wood density in g/cm^3, H is total plant height in meter. Model 1 was developed from stemmed palms in Amazon [44]. Model 2 was a model derived from the pan-tropical destructive tree AGB database [45]. The model was fitted using a generalized nonlinear least squares method [31] instead of log-transformation-based approaches [45]. Species-specific wood density was based on the Global Wood Density Database [46,47] for all species except American oil palm, for which we found more recent estimate from Fathi [48]. The plot-level AGB density $\hat{B}_{plot}$ was calculated by summing the individual tree AGB at each plot and then divided by the plot area (900 m^2). The AGB density of the plots is within the range of 10.6 to 255.5 Mg/ha (Table 2).

3.2. Airborne Lidar Data Acquisition and Processing

Airborne lidar was collected by GEOID Laser Mapping on 2 September 2013 using an Optech ALTM Orion M200 sensor with an integrated GPS, IMU system at an average height of 853 m above ground and a scan angle of 11°. Global positioning system data were collected at a ground station simultaneously with the flight at a ground survey location to permit post-processing for estimated horizontal and vertical accuracies (1 σ) of ±0.3 m and 0.15 m, respectively. The average point density is 24 pt/m^2.

The airborne lidar data were processed using the Toolbox for Lidar Data Filtering and Forest Studies (Tiffs) [49]. The lidar point cloud was classified into ground returns and non-ground returns by the data provider. These ground returns were interpolated into a 1-m Digital Terrain Model (DTM). The relative height of every laser point was calculated as the difference between its Z coordinate and the underlying DTM elevation. Based on the relative height of every laser point, statistics such as mean, standard deviation, skewness, kurtosis, quadratic mean height, percentile height (10th, 20th, ..., 90th), and the proportion of points within different bins (0–5 m, 5–10 m, ... , 45–50 m) for each plot were extracted. Note that we calculated the lidar metrics based on all returns, including those from ground, to incorporate both horizontal and vertical canopy structure information [20,21]. To map AGB over the landscape, we also generated lidar metric rasters at 30 m resolution, the same dimension as the field plots.

3.3. Lidar-Based AGB Modeling and Mapping

We developed two types of models for biomass estimation. First, we used a fixed-effects model to estimate biomass for all plots. This was developed with multiplicative power models [36,50] at the plot level for biomass estimation:

$$\hat{B}_{plot} = f_{plot} \left(\varphi, z \right) = \varphi_0 z_1^{\varphi_1} z_2^{\varphi_2} \dots z_m^{\varphi_m} \tag{3}$$

where z_1, z_2, and z_m are different lidar metrics, $\varphi_0, \varphi_1, \varphi_2, \dots, \varphi_m = \pi r^2$ are model parameters. To select the most relevant lidar metrics for biomass prediction, we used the same approach as [36] with feature selection using log-scale forward stepwise regression and then iteratively fitted nonlinear multiplicative power models by keeping the statistically significant variables. We then examined the residual errors of the above model from the perspectives of both tree DBH-H relationship and wood density at the plot scale. We calculated the plot-level average wood density by weighting each tree's wood density with a proxy of tree stem volume:

$$\rho_{plot} = \frac{\sum_i \left(DBH_i^2 * H_i\right) \rho_i}{\sum_i \left(DBH_i^2 * H_i\right)} \tag{4}$$

where i is the subscript for a tree in a plot and $DBH^2 \times H$ is the proxy of tree stem volume.

To test whether AGB prediction can be improved by stratification, we divided the plots into n different vegetation groups and estimated AGB using nonlinear mixed-effects models. The mixed-effects models assume that the parameters of lidar-based AGB models vary by vegetation groups. Each model parameter is considered as a combination of fixed and random effects:

$$\varphi_k = \beta + b_k \tag{5}$$

where φ_k is a parameter for group k (k $\subset$ [1,n]), β is the fixed effect or population-level constant, b_k is the random effect that varies across different vegetation groups. The random effects can be modeled as Gaussian variables with mean of zeros and covariance matrix ψ.

We used the nonlinear power model as the starting point for developing the mixed-effects models. In other words, we assumed the mixed-effects models have the same predictor(s) and model form as the fixed-effects models, but their parameters were modeled as random variables. We began our fitting of the mixed-effects models by assuming the random effects covariance matrix ψ to be fully unconstructed. If the estimate of any random effects was statistically insignificant from zero across all different vegetation types, the random effect was dropped from the model.

The RMSE (Root Mean Squared Error) and AIC (Akaike Information Criterion) were calculated to compare the performance of different models [51]. Pesudo-R^2 was denoted and calculated as:

$$R^2 = 1 - \frac{\sum_i \left(B_{plot,i} - \hat{B}_{plot,i}\right)^2}{\sum_i \left(B_{plot,i} - \overline{B}_{plot}\right)^2} \tag{6}$$

where i is the subscript of individual field plot, $B_{plot,i}$ is the plot-level AGB estimated from field data, $\overline{B}_{plot}$ is the mean of all plots' AGB field estimates, $\hat{B}_{plot,i}$ is the predicted plot AGB based on fixed-effects or mixed-effects models. We calculated R^2 and RMSE using both re-substitution and cross-validation methods. We mapped the AGB density by applying the plot-level models using the 30-m airborne lidar metrics at the landscape level.

3.4. Mapping Agroforest Distribution with Visual Interpretation

The results of the mixed-effects models suggested the need to classify the study area into forests, American oil palm, teak, and other agroforests for predicting AGB over the landscape. We classified the landscape via visual interpretation of airborne lidar data in 2D and 3D GIS environment with the aid of Google Earth imagery acquired in 2010. The visual interpretation was mainly based on information such as tree shape, plantation pattern, and color derived from airborne lidar point cloud and high spatial resolution Google Earth imagery. A Canopy Height Model (CHM) of 1-m spatial resolution was generated and used as a backdrop image for delineating the boundary of vegetation patches and roads. Visual interpretation instead of automatic classification was chosen to ensure the highest classification accuracy (without the salt-and-pepper noise in pixel-based classification or boundary inaccuracy in object-based classification), which can help us to conduct a focused analysis of the importance of stratification on landscape-level AGB mapping.

Some remnant forests also exist in the study area. Because the focus of this study is agroforestry, the areas of forests were identified and masked out via visual interpretation. Often, the differences between agroforests and forests were sharp in the lidar image (see Figure 2a,b). However, for some fields that have large trees in the overstory and were only marginally managed with few trees planted underneath, the distinction between agroforests and forests was not so obvious, both conceptually and visually (see Figure 2c,d). Our strategy was to classify a field into agroforestry only if we saw clear evidence of planting activities (e.g., rows of trees). This might lead to large omission errors but will have minimal commission errors in the identification of agroforests. Such a strategy can be justified because our focus in this study is the AGB of agroforestry. Within the agroforestry areas, the American oil palm and teak can be easily identified with visual interpretation due to their unique characteristics (Figure 3) and the small size of our study area.

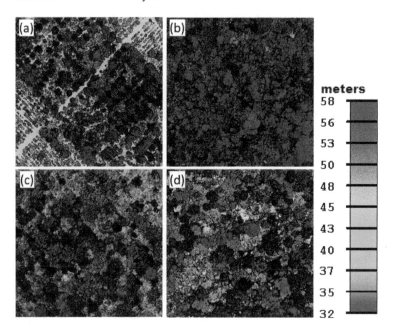

Figure 2. Examples (**a–d**) of agroforestry (**left**) and forests (**right**) fields shown in airborne lidar. Color (from red to blue) represents the elevation of the laser points.

Figure 3. Teak plantations shown in the field (**a**); aerial photo (**b**); and lidar data (**c**); American oil palm sample tree (**d**); and plantation shown in aerial photos (**e**); and lidar data (**f**).

4. Results

With the multiplicative power model and our feature selection procedure, the final plot-level AGB prediction model is a simple power model based on the 90th percentile height:

$$\hat{B}_{plot} = 13.94 * H_{90th}^{0.86} \tag{7}$$

The model R^2 is 0.47 and the RMSE is 60.4 Mg/ha (see Figure 4a). Note that the mean AGB of all plots is 105.2 Mg/ha, which means that the coefficient of variation (CV) or relative prediction error is almost 60%. In particular, we found that plots 10 and 11 have relatively large residuals. Plot 11 is in an American oil palm plantation. The large residual of plot 11 is mainly caused by the unique DBH-H relationship of American oil palm trees, which implies that the AGB of American oil palm needs to be modeled separately from other vegetation types (see Section 5.1). However, we had only one plot of American oil palm, which eliminates the possibility of independently assessing the model prediction errors. Thus, plot 11 was excluded in our analysis, and the corresponding AGB model (see Figure 4b) developed after feature selection was:

$$\hat{B}_{plot} = 8.28 * H_{80th}^{1.06} \tag{8}$$

where H_{80th} is the 80th percentile height of lidar points.

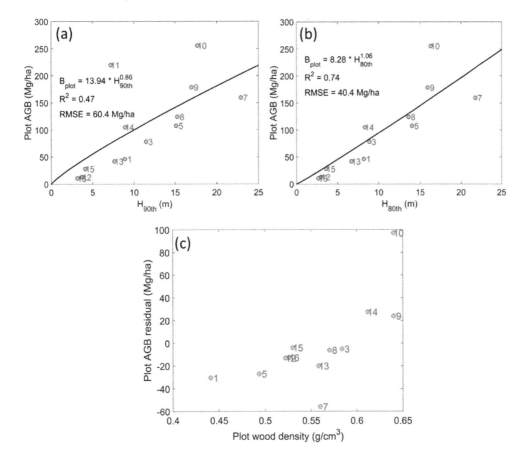

Figure 4. Models for predicting plot-level AGB with the American oil palm plot included (**a**) and excluded (**b**); and the relationship between residual and wood density at the plot level (**c**).

Note that when the allometric model for non-palm trees (Equation (2)) was used to estimate AGB, wood density is an important variable. As shown in Table 2, the teak plantation plots (plots 9 and 10) have the highest plot-level wood density. Plot 14 also has high average wood density because some of the species within it, such as ameixa and goiaba have high wood density (see Tables 1 and 2). The large plot-level wood density of these three plots can explain why they all have positive residuals in AGB prediction models (Figure 4b).

The patterns revealed in the residual errors of plot-level AGB model prediction (*i.e.*, plots with large wood density led to large positive residuals) (Figure 4c) suggest that the residuals are not statistically independent, *i.e.*, an obvious violation of the basic assumption of the ordinary least squares (OLS) regression models. A statistical approach that can naturally consider the correlation within individual observations is mixed-effects models. The idea of mixed-effects models is to stratify the observations (*i.e.*, plots in this study) into different groups and allow the model parameters vary as realizations of random variables across the groups. By doing so, the residuals of model prediction within individual groups will become more statistically independent. Mixed-effects models provide a trade-off between fitting all data points with one model and fitting models for each group independently. Hence it is well suited for handling data with few observations within groups [50].

The use of mixed-effects models requires the grouping of individual plots. The relatively large wood density of plots 9, 10, 14, and their positive residuals in the fixed-effects model imply that these three plots can be put into one group. The teak plantations (plots 9 and 10) not only have high wood density but also have the potential of being mapped over large area from remotely sensed data. Thus, we also tried to put only plots 9 and 10 into one group. As a result, we consider two different grouping scenarios as shown in Table 3. As shown in Equation (8), the developed fixed-effects model was a simple power model (in the form of $y = a*x^b$, where a and b are model parameters) with the 80th percentile height of lidar points as the predictor. In all of our mixed-effects modeling experiments, we found that only the parameter b had statistically significant random effects. Therefore, in the mixed-effects models, only the parameter b varies with vegetation groups. For grouping scheme A, the model parameters $a = 8.63$, $b_1 = 1.15$, and $b_2 = 0.98$, where b_1 and b_2 are the parameters for groups 1 and 2, respectively. For grouping scheme B, the model parameters $a = 7.23$, $b_1 = 1.22$, and $b_2 = 1.03$. Note that here the mixed-effects model for a given grouping scheme is equivalent to two simple power models for the two vegetation groups within the scheme (Figure 5).

Table 3. Different schemes of grouping plots for developing mixed-effects models.

Scheme	Group#1	Group#2
A	Teak plantation (plot 9, 10)	Non-teak (other plots)
B	High wood density plots (plot 9, 10, 14)	Other (other plots)

When we used the same plots for calibration and validation (*i.e.*, re-substitution), the R^2 was much higher and RMSE was much smaller when we used mixed-effects models in comparison to fixed-effects models (Table 4). The increase of goodness-of-fit in these statistics was expected because mixed-effects models are more complex (*i.e.*, more parameters) than fixed-effects models. AIC is a statistic that penalizes the increase of model fitness due to the use of more complex models [51]. When we fitted mixed-effects models based on two groups, the AIC values decreased slightly for scheme A and by about four for scheme B (note that lower AIC values are preferred for modeling). Figure 5 shows the lidar-based AGB models developed from the field plots as well as their AGB prediction in comparison to the field-based estimates.

We also used leave-one-out cross-validation for calculating R^2 and RMSE to further assess the models' prediction performance (Table 4). We found that the fixed-effects model had R^2 as low as 0.38, which was increased to 0.64 and 0.75 when mixed-effects models with scheme A and scheme B were used. This confirms the importance of stratification and mixed-effects modeling for predicting AGB.

Table 4. Comparison of fixed-effects and mixed-effects models for plot-level AGB estimation.

Model Type	Re-Substitution			Cross-Validation	
	R^2	RMSE (Mg/ha)	AIC	R^2	RMSE (Mg/ha)
Fixed-effects model	0.74	40.4	122.6	0.38	56.4
Mixed-effects model					
Scheme A	0.91	25.9	122.0	0.64	42.9
Scheme B	0.94	21.6	118.7	0.75	35.9

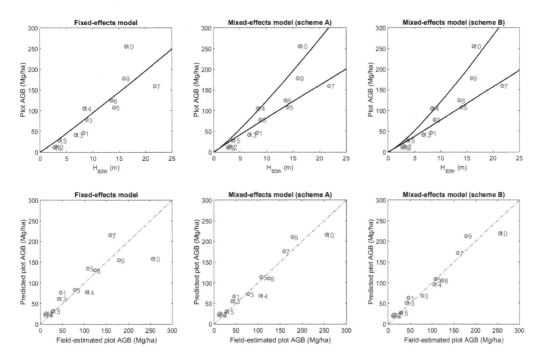

Figure 5. Comparison of fixed-effects and mixed-effects models for estimating plot AGB. The solid lines in the first-row figures are model curves. The dashed lines in the second-row figures are 1:1 line.

Figure 6 shows vegetation distribution 6a and the estimated AGB 6b and 6c over the whole agroforestry area using mixed- and fixed-effects models, respectively. The mean AGB densities estimated from fixed- and mixed-effects models were 57.0 Mg/ha and 51.7 Mg/ha, respectively. So, the average AGB density estimate from fixed-effects models was about 10% higher than the one from mixed-effects model. Moreover, by checking the spatial pattern of their difference at the pixel level (Figure 6d), we can see that the teak plantation AGB was underestimated by up to ~70 Mg/ha (or 33%) while the other agroforestry AGB were overestimated by up to ~60 Mg/ha (or 25%) using fixed-effects models in comparison to mixed-effects models.

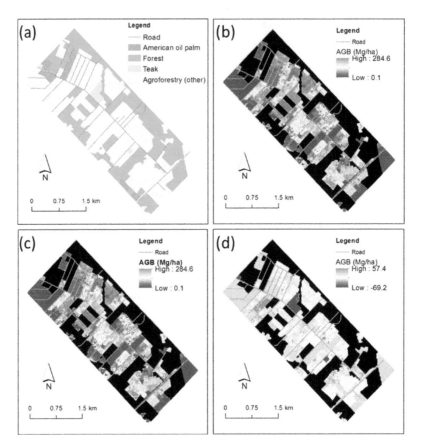

Figure 6. Maps of vegetation type (**a**); and AGB predicted with mixed-effects model (**b**); and fixed-effects model (**c**); and the difference between AGB predicted with fixed- and mixed-effects models (**d**). Black color indicates the area masked for analysis.

5. Discussion

5.1. Tree DBH-H Relationship

Note that, according to the allometric models used in this study (*i.e.*, Equations (1) and (2)), DBH is the only predictor for estimating palm trees' AGB and the major factor for predicting other trees' AGB. However, lidar is more suited for estimating vegetation height than for DBH because the latter usually has to be indirectly estimated from airborne lidar data via its statistical relationship with lidar height metrics. When such allometric models are used, the capability of predicting AGB from lidar is governed by the strength of tree-level DBH-H relationship.

Figure 7 shows the relationship between DBH and tree height for all species based on the field measurements. It is clear that the tree-level data points can be separated into two groups. The group at the right includes 13 American oil palm trees, all from plot 11. Although the other tree species in this study exhibits positive relationship between DBH and height, the American oil palms do not show such a pattern. This might be due to the fact that the plants grow by increasing its diameter rather than height after reaching a certain size. As a result, the American oil palms have unique DBH-H relationship compared to most other tree species in this study. Note that the plot-level AGB model (*i.e.*, Equation (7)) was calibrated to fit AGB from all plots while American oil palms exists in only plot 11. Hence, it is not surprising that the model (see Figure 4a) has large prediction error for the American oil palm plot. The unique DBH-H relationship of American oil palm suggested that it should be treated separately in our modeling. However, we were concerned that the model was not reliable with just one such plot. As a result, we decided to exclude the American oil palm plot in our further analysis.

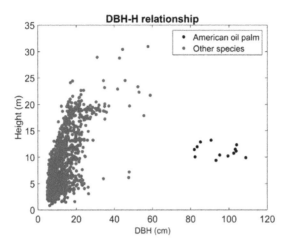

Figure 7. DBH and tree height (H) relationships for American oil palm and other tree species.

5.2. Allometry

One of the biggest uncertainties in our analysis is the tree-level AGB allometric models, especially for palms. For example, Goodman *et al.* [52] developed an allometric model for *Euterpe precatoria*, which is the allometric model that we could find and is the closest to our acai palm species (*Euterpe oleracea*). However, when the model ($\hat{B}_{tree}$ $(kg) = 13.589 * H_s - 108.81$) was applied to our acai palms, 235 out of 287 trees had negative AGB estimates. That allometric model was developed with 8 trees with a limited DBH range of 12–19 cm. Nevertheless, 242 Acai trees in this study have DBH < 12 cm. The small number and the limited range of the size for trees that were used to calibrate the model [52] can explain the failure of applying such a model into our study area. Unfortunately, no other allometric models specifically for acai palms were found in the literature. We do not have quantitative information about the errors of applying general allometric models (*i.e.*, Equation (1)) to predict the AGB of the palm species in our study. More than 2000 species of palm trees occur in the world. Palm is the most abundant arborescent plant family in western Amazonian forest [52]. Thus, improving our knowledge about the AGB allometry for palms trees, in addition to dicotyledonous trees, is critical for accurate estimation of the AGB and carbon stock in Amazon.

Tree AGB allometric models are key to quantify the errors of remotely sensed AGB. It has been increasingly recognized that it is crucial to characterize the uncertainty of AGB estimates at the pixel and larger spatial scales [36,53–55]. The RMSE based on leave-one-out cross validation only characterized the errors related to the lidar-based AGB models, without considering errors related to tree AGB allometry, model parameters, field measurements, and lidar data. Chen *et al.* [36] did a comprehensive study of different error sources of AGB prediction when airborne lidar data and field data were combined to model and map AGB in an African tropical forest. It was found that errors related to tree AGB allometry and lidar-based regression models were the two major error sources in pixel-level AGB estimates. One of the cornerstones of that study is the availability of the destructively measured tree AGB for characterizing the error of tree AGB allometric models. Although we can apply that methodology to characterize AGB prediction errors for dicotyledonous trees in this study, palms need separate allometric models [52]. Unfortunately, the lack of access to the destructive AGB measurements for palm trees prevents us from fully assessing the errors of agroforestry AGB in this study.

5.3. Mixed-Effects Modeling

A limitation of this study is the small sample size ($n = 12$) for plot-level AGB modeling. We used mixed-effects models to address the dependence of lidar-AGB relationship on vegetation type and that left us with small numbers of plots per vegetation type. We should use the developed models for AGB prediction for other areas with caution, especially when there are out of the range of the input variables. In the future, more field plots ought to be collected, especially for American oil palm and teak plantations, to develop AGB models for larger areas. The one-year time difference between the times of lidar and field data acquisitions could also contribute to the errors in AGB prediction.

The use of the mixed-effects models for spatially-explicit AGB mapping requires the classification of the agroforestry fields over the whole study area based on their corresponding grouping schemes. Although we found that the grouping scheme B had the best prediction accuracy (Table 4), this model is more challenging to use in our study area because it requires the identification of high wood density areas. The calculation of the average wood density over an area depends on mapping of individual tree species, which is difficult for polyculture agroforestry fields that have multilayer canopy structure. Instead, we chose grouping scheme A for vegetation type classification because teak plantations can be more easily recognized from remotely sensed data owing to the relatively uniform and simple canopy structure (Figure 3a–c). Preliminary results from a supervised classification using 5-m RapidEye satellite imagery (unpublished data) indicated that teak plantations could be classified with a user accuracy and producer accuracy of 97.7% and 96.4%, respectively. Therefore, our methodology, when combined with digital classification of remotely sensed data, has potential to be applied and extended over large areas. One future research direction is to apply object-based image analysis (OBIA) approach and/or hyperspectral imagery to improve the classification of agroforestry fields [56–58].

We excluded forests to stay focused in the study of agroforestry. As shown earlier, it was difficult to separate agroforestry field and forests when an agroforestry field is dense and mixed with many tree species (see Figure 2c,d). A remaining question is whether it is necessary to separate agroforestry and forests in such cases. In other words, it is unclear whether AGB for forests and such agroforestry fields can be fit using just one model. One cold answer this question with comparable field and lidar data over forest and agroforestry plots.

6. Conclusions

Agroforestry has great potential for carbon sequestration while providing multiple benefits to individual farmers and communities. This study examined the potentials and challenges of using airborne lidar data for estimating AGB in an agroforestry system in the Brazilian Amazon. Several significant lessons were learned.

(1) We found strong evidence to support the stratification of vegetation types in agroforestry fields for AGB modeling and mapping. This is in contrast to the widespread use of statistical models with no awareness of different vegetation types in most studies of tropical forest AGB using airborne lidar. Different from forests where trees are selected or adapted via natural processes, the trees in agroforestry are selected and planted by people to maximize economic and other benefits on purpose. Thus, the trees in agroforestry systems usually show more spatially-regularized patterns with a few co-occurring species. The species and structural diversities within agroforestry fields are usually smaller than the ones within forests while the diversities across different agroforestry fields may be larger. Therefore, the need for stratifying vegetation types in agroforestry is stronger than in forests for AGB studies.

(2) This study analyzed the residual errors resulted from regular fixed-effects models and found that the errors have a pattern related to the variations in plot-level wood density. Based on this pattern, agroforestry fields were classified for lidar-based AGB modeling. This is an improvement over previous studies (e.g., [50]) that used existing vegetation classification schemes not specifically developed for lidar-based AGB modeling and mapping.

(3) This study reinforced the utility of mixed-effects models for biomass modeling and mapping. Mixed-effects models can naturally incorporate how different species or groups have different wood densities and thus distinct lidar height—tree AGB relationships. Mixed-effects models also can elegantly cope with the issue of small sample size via adjustment of model parameters as a combination of fixed and random effects. With the classification of agroforestry into teak plantations and other types, we found the mixed-effects models improved the R^2 of AGB prediction from 0.38 to 0.64 and reduced the RMSE from 56.4 Mg/ha to 42.9 Mg/ha in comparison to fixed-effects models. We expect this study will encourage the further use of this under-investigated tool in the community of remote sensing of biomass and carbon.

This study highlighted the lack of reliable AGB allometry models, especially for palms, to comprehensively quantify the errors of lidar-based AGB estimates. This study also suggested the need to collect more field data over larger areas for automated mapping of plantation types and improved AGB modeling and mapping.

Acknowledgments: This study was financially supported by the China's National Natural Science Foundation (No# 41571411) and the Zhejiang A&F University's Research and Development Fund for the talent startup project (No# 2013FR052). Keller, dos-Santos, and Bolfe acknowledge the support from CNPq LBA. Data were acquired by the Sustainable Landscapes Brazil project supported by the Brazilian Agricultural Research Corporation (EMBRAPA), the US Forest Service, and USAID, and the US Department of State.

Author Contributions: Qi Chen developed the analytical framework, did the analysis, and wrote the manuscript. Dengsheng Lu contributed to the analytical framework and data analysis. Michael Keller contributed to the analytical framework and together with Edson Bolfe and Maiza Nara dos-Santos designed the field data collection plan. Edson Bolfe and Maiza Nara dos-Santos organized the field data. Yunyun Feng and Changwei Wang mapped vegetation types. All authors contributed to the editing of the manuscript.

Conflicts of Interest: The authors declare no conflict of interest.

References

1. Nair, P.R. *An Introduction to Agroforestry*; Kluwer Academic Publishers: Dordrecht, The Netherlands, 1993.
2. Negash, M.; Kanninen, M. Modeling biomass and soil carbon sequestration of indigenous agroforestry systems using CO2FIX approach. *Agric. Ecosyst. Environ.* **2015**, *203*, 147–155. [CrossRef]
3. Zomer, R.J.; Trabucco, A.; Coe, R.; Place, F. *Trees on Farms: Analysis of the Global Extent and Geographical Patterns of Agroforestry*; ICRAF Working Paper No. 89; World Agroforestry Centre: Nairobi, Kenya, 2009.
4. Mbow, C.; van Noordwijk, M.; Luedeling, E.; Neufeldt, H.; Minang, P.A.; Kowero, G. Agroforestry solutions to address food security and climate change challenges in Africa. *Curr. Opin. Environ. Sustain.* **2014**, *6*, 61–67. [CrossRef]
5. Albrecht, A.; Kandji, S.T. Carbon sequestration in tropical agroforestry systems. *Agric. Ecosyst. Environ.* **2003**, *99*, 15–27. [CrossRef]
6. Thorlakson, T.; Neufeldt, H.; Dutilleul, F.C. Reducing subsistence farmers' vulnerability to climate change: Evaluating the potential contributions of agroforestry in western Kenya. *Agric. Food Secur.* **2012**, *1*, 1–13. [CrossRef]
7. Nguyen, Q.; Hoang, M.H.; Öborn, I.; van Noordwijk, M. Multipurpose agroforestry as a climate change resiliency option for farmers: An example of local adaptation in Vietnam. *Clim. Change* **2013**, *117*, 241–257. [CrossRef]
8. McNeely, J.A. Nature *vs.* nurture: Managing relationships between forests, agroforestry and wild biodiversity. *Agrofor. Syst.* **2004**, *61*, 155–165.
9. Jose, S. Agroforestry for conserving and enhancing biodiversity. *Agrofor. Syst.* **2012**, *85*, 1–8. [CrossRef]
10. Ramos, N.C.; Gastauer, M.; de Cordeiro, A.A.C.; Meira-Neto, J.A.A. Environmental filtering of agroforestry systems reduces the risk of biological invasion. *Agrofor. Syst.* **2015**, *89*, 279–289. [CrossRef]
11. Anderson, S.H.; Udawatta, R.P.; Seobi, T.; Garrett, H.E. Soil water content and infiltration in agroforestry buffer strips. *Agrofor. Syst.* **2009**, *75*, 5–16. [CrossRef]
12. Hernandez, G.; Trabue, S.; Sauer, T.; Pfeiffer, R.; Tyndall, J. Odor mitigation with tree buffers: Swine production case study. *Agric. Ecosyst. Environ.* **2012**, *149*, 154–163. [CrossRef]

13. Asbjornsen, H.; Hernandez-Santana, V.; Liebman, M.; Bayala, J.; Chen, J.; Helmers, M.; Ong, C.; Schulte, L. Targeting perennial vegetation in agricultural landscapes for enhancing ecosystem services. *Renew. Agric. Food Syst.* **2014**, *29*, 101–125. [CrossRef]

14. Pandey, D.N. Carbon sequestration in agroforestry systems. *Clim. Policy* **2002**, *2*, 367–377. [CrossRef]

15. Sharrow, S.H.; Ismail, S. Carbon and nitrogen storage in agroforests, tree plantations, and pastures in western Oregon, USA. *Agrofor. Syst.* **2004**, *60*, 123–130. [CrossRef]

16. Kirby, K.R.; Potvin, C. Variation in carbon storage among tree species: Implications for the management of a small-scale carbon sink project. *For. Ecol. Manag.* **2007**, *246*, 208–221. [CrossRef]

17. Nair, P.K.R. Carbon sequestration studies in agroforestry systems: A reality-check. *Agrofor. Syst.* **2012**, *86*, 243–253. [CrossRef]

18. Lorenz, K.; Lal, R. Soil organic carbon sequestration in agroforestry systems: A review. *Agron. Sustain. Dev.* **2014**, *34*, 443–454. [CrossRef]

19. Smith, P.; Bustamante, M.; Ahammad, H.; Clark, H.; Dong, H.; Elsiddig, E.A.; Haberl, H.; Harper, R.; House, J.; Jafari, M.; *et al.* Agriculture, Forestry and Other Land Use (AFOLU). In *Climate Change 2014: Mitigation of Climate Change. Contribution of Working Group III to the Fifth Assessment Report of the Intergovernmental Panel on Climate Change*; Edenhofer, O., Pichs-Mdruga, R., Sokana, Y., Minx, J.C., Farahani, E., Kadner, S., Seyboth, K., Adler, A., Baum, I., Brunner, S., *et al*, Eds.; Cambridge University Press: Cambridge, UK; New York, NY, USA, 2014.

20. Skole, D.L.; Samek, J.H.; Chomentowski, W.; Smalligan, M. Forest, carbon, and the global environment: New directions in research. In *Land Use and the Carbon Cycle: Advances in Integrated Science, Management, and Policy*; Brown, D.G., Robinson, D.T., French, N.H.F., Reed, B.C., Eds.; Cambridge University Press: New York, NY, USA, 2013.

21. Lu, D.; Chen, Q.; Wang, G.; Liu, L.; Li, G.; Moran, E. A survey of remote sensing-based aboveground biomass estimation methods in forest ecosystems. *Int. J. Digit. Earth* **2014**. [CrossRef]

22. Bolfe, É.L.; Batistella, M.; Ferreira, M.C. Correlation of spectral variables and aboveground carbon stock of agroforestry systems. *Pesqui. Agropecu. Bras.* **2012**, *47*, 1261–1269. [CrossRef]

23. Czerepowicz, L.; Case, B.S.; Doscher, C. Using satellite image data to estimate aboveground shelterbelt carbon stocks across an agricultural landscape. *Agric. Ecosyst. Environ.* **2012**, *156*, 142–150. [CrossRef]

24. Mitchard, E.T.A.; Meir, P.; Ryan, C.M.; Woollen, E.S.; Williams, M.; Goodman, L.E.; Mucavele, J.A.; Watts, P.; Woodhouse, I.H.; Saatchi, S.S. A novel application of satellite radar data: Measuring carbon sequestration and detecting degradation in a community forestry project in Mozambique. *Plant Ecol. Divers.* **2013**, *6*, 159–170. [CrossRef]

25. Dube, T.; Mutanga, O. Investigating the robustness of the new Landsat-8 Operational Land Imager derived texture metrics in estimating plantation forest aboveground biomass in resource constrained areas. *ISPRS J. Photogramm. Remote Sens.* **2015**, *108*, 12–32. [CrossRef]

26. Chen, Q. Lidar remote sensing of vegetation biomass. In *Remote Sensing of Natural Resources*; Wang, G., Weng, Q., Eds.; CRC Press/Taylor and Francis: Boca Raton, FL, USA, 2013; pp. 399–420.

27. Montesano, P.M.; Nelson, R.F.; Dubayah, R.O.; Sun, G.; Cook, B.D.; Ranson, K.J.R.; Næsset, E.; Kharuk, V. The uncertainty of biomass estimates from LiDAR and SAR across a boreal forest structure gradient. *Remote Sens. Environ.* **2014**, *154*, 398–407. [CrossRef]

28. Margolis, H.A.; Nelson, R.F.; Montesano, P.M.; Beaudoin, A.; Sun, G.; Andersen, H.E.; Wulder, M. Combining Satellite Lidar, Airborne Lidar and Ground Plots to Estimate the Amount and Distribution of Aboveground Biomass in the Boreal Forest of North America. *Can. J. For. Res.* **2015**, *45*, 838–855. [CrossRef]

29. McRoberts, R.E.; Næsset, E.; Gobakken, T.; Bollandsås, O.M. Indirect and direct estimation of forest biomass change using forest inventory and airborne laser scanning data. *Remote Sens. Environ.* **2015**, *164*, 36–42. [CrossRef]

30. Latifi, H.; Fassnacht, F.; Koch, B. Forest structure modeling with combined airborne hyperspectral and LiDAR data. *Remote Sens. Environ.* **2012**, *121*, 10–25. [CrossRef]

31. Chen, Q. Modeling aboveground tree woody biomass using national-scale allometric methods and airborne lidar. *ISPRS J. Photogramm. Remote Sens.* **2015**, *106*, 95–106. [CrossRef]

32. Asner, G.P.; Clark, J.K.; Mascaro, J.; Galindo García, G.A.; Chadwick, K.D.; Navarrete Encinales, D.A.; Paez-Acosta, G.; Montenegro, E.C.; Kennedy-Bowdoin, T.; Duque, Á.; et al. High-resolution mapping of forest carbon stocks in the Colombian Amazon. *Biogeosciences* **2012**, *9*, 2683–2696. [CrossRef]

33. D'Oliveira, M.V.; Reutebuch, S.E.; McGaughey, R.J.; Andersen, H.E. Estimating forest biomass and identifying low-intensity logging areas using airborne scanning lidar in Antimary State Forest, Acre State, Western Brazilian Amazon. *Remote Sens. Environ.* **2012**, *124*, 479–491. [CrossRef]

34. Andersen, H.E.; Reutebuch, S.E.; McGaughey, R.J.; d'Oliveira, M.V.; Keller, M. Monitoring selective logging in western Amazonia with repeat lidar flights. *Remote Sens. Environ.* **2014**, *151*, 157–165. [CrossRef]

35. Vaglio Laurin, G.; Chen, Q.; Lindsell, J.A.; Coomes, D.A.; del Frate, F.; Guerriero, L.; Pirotti, F.; Valentini, R. Above ground biomass estimation in an African tropical forest with lidar and hyperspectral data. *ISPRS J. Photogramm. Remote Sens.* **2014**, *89*, 49–58. [CrossRef]

36. Chen, Q.; Laurin, G.V.; Valentini, R. Uncertainty of remotely sensed aboveground biomass over an African tropical forest: Propagating errors from trees to plots to pixels. *Remote Sens. Environ.* **2015**, *160*, 134–143. [CrossRef]

37. Yamada, M.; Gholz, H.L. An evaluation of agroforestry systems as a rural development option for the Brazilian Amazon. *Agrofor. Syst.* **2002**, *55*, 81–87. [CrossRef]

38. Rodrigues, T.E.; dos Santos, P.L.; Rolim, P.A.M.; Santos, E.; Rego, R.S.; da Silva, J.M.L.; Valente, M.A.; Gama, J.R.N.F. *Caracterização e Classificação dos Solos do Município de Tomé-Açu, Pará*; Embrapa Amazônia Oriental: Belém, Pará, Brazil, 2001; p. 49.

39. Homma, A.K.O. *História da Agricultura na Amazônia: Da era Pré-Colombiana ao Terceiro Milênio*; Embrapa Informação Tecnológica: Brasília, Distrito Federal, Brazil, 2003; p. 274.

40. Homma, A.K.O. Dinâmica dos sistemas agroflorestais: O caso da Colônia Agrícola de Tomé-Açu, Pará. *Rev. Inst. Estud. Super. Amazôn.* **2004**, *2*, 57–65.

41. Walker, W.; Baccini, A.; Nepstad, D.; Horning, N.; Knight, D.; Braun, E.; Bausch, A. *Field Guide for Forest Biomass and Carbon Estimation, Version 1.0*; Woods Hole Research Center: Falmouth, MA, USA, 2011.

42. Keller, M.; Palace, M.; Hurtt, G. Biomass estimation in the Tapajos National Forest, Brazil: Examination of sampling and allometric uncertainties. *For. Ecol. Manag.* **2001**, *154*, 371–382. [CrossRef]

43. Hunter, M.O.; Keller, M.; Victoria, D.; Morton, D.C. Tree height and tropical forest biomass estimation. *Biogeosciences* **2013**, *10*, 8385–8399. [CrossRef]

44. Nascimento, H.E.; Laurance, W.F. Total aboveground biomass in central Amazonian rainforests: A landscape-scale study. *For. Ecol. Manag.* **2002**, *168*, 311–321. [CrossRef]

45. Chave, J.; Réjou-Méchain, M.; Búrquez, A.; Chidumayo, E.; Colgan, M.S.; Delitti, W.B.; Duque, A.; Eid, T.; Fearnside, P.M.; Goodman, R.C.; et al. Improved allometric models to estimate the aboveground biomass of tropical trees. *Glob. Change Biol.* **2014**. [CrossRef] [PubMed]

46. Chave, J.; Coomes, D.; Jansen, S.; Lewis, S.L.; Swenson, N.G.; Zanne, A.E. Towards a worldwide wood economics spectrum. *Ecol. Lett.* **2009**, *12*, 351–366. [CrossRef] [PubMed]

47. Zanne, A.E.; Lopez-Gonzalez, G.; Coomes, D.A.; Ilic, J.; Jansen, S.; Lewis, S.L.; Miller, R.B.; Swenson, N.G.; Wiemann, M.C.; Chave, J. Global Wood Density Database. Dryad. Available online: http://hdl.handle.net/ 10255/dryad, 235 (accessed on 21 September 2015).

48. Fathi, L. Structural and Mechanical Properties of the Wood from Coconut Palms, Oil Palms and Date Palms. Ph.D. Thesis, University of Hamburg, Hamburg, Germany, 2014.

49. Chen, Q. Airborne lidar data processing and information extraction. *Photogramm. Eng. Remote Sens.* **2007**, *73*, 109–112.

50. Chen, Q.; Laurin, G.V.; Battles, J.J.; Saah, D. Integration of airborne lidar and vegetation types derived from aerial photography for mapping aboveground live biomass. *Remote Sens. Environ.* **2012**, *121*, 108–117. [CrossRef]

51. Chen, Q.; Gong, P.; Baldocchi, D.; Tian, Y.Q. Estimating basal area and stem volume for individual trees from lidar data. *Photogramm. Eng. Remote Sens.* **2007**, *73*, 1355–1365. [CrossRef]

52. Goodman, R.C.; Phillips, O.L.; del Castillo Torres, D.; Freitas, L.; Cortese, S.T.; Monteagudo, A.; Baker, T.R. Amazon palm biomass and allometry. *For. Ecol. Manag.* **2013**, *310*, 994–1004. [CrossRef]

53. Wang, G.; Oyana, T.; Zhang, M.; Adu-Prah, S.; Zeng, S.; Lin, H.; Se, J. Mapping and spatial uncertainty analysis of forest vegetation carbon by combining national forest inventory data and satellite images. *For. Ecol. Manag.* **2009**, *258*, 1275–1283. [CrossRef]

54. Lu, D.; Chen, Q.; Wang, G.; Moran, E.; Batistella, M.; Zhang, M.; Laurin, G.V.; Saah, D. Aboveground forest biomass estimation with Landsat and LiDAR data and uncertainty analysis of the estimates. *Int. J. For. Res.* **2012**. [CrossRef]

55. McRoberts, R.E.; Westfall, J.A. Effects of uncertainty in model predictions of individual tree volume on large area volume estimates. *For. Sci.* **2014**, *60*, 34–42. [CrossRef]

56. Lu, D.; Hetrick, S.; Moran, E. Land cover classification in a complex urban-rural Landscape with QuickBird imagery. *Photogramm. Eng. Remote Sens.* **2010**, *76*, 1159–1168. [CrossRef]

57. Bigdeli, B.; Samadzadegan, F.; Reinartz, P. A decision fusion method based on multiple support vector machine system for fusion of hyperspectral and LIDAR data. *Int. J. Image Data Fusion* **2014**, *5*, 196–209. [CrossRef]

58. Ghamisi, P.; Benediktsson, J.A.; Phinn, S. Land-cover classification using both hyperspectral and LiDAR data. *Int. J. Image Data Fusion* **2015**, *6*, 189–215. [CrossRef]

User Validation of VIIRS Satellite Imagery

Don Hillger [1,*], Tom Kopp [2], Curtis Seaman [3], Steven Miller [3], Dan Lindsey [1], Eric Stevens [4], Jeremy Solbrig [3], William Straka III [5], Melissa Kreller [6], Arunas Kuciauskas [7] and Amanda Terborg [8]

Academic Editors: Changyong Cao, Xiaofeng Li and Prasad S. Thenkabail

[1]   NOAA/NESDIS Center for Satellite Applications and Research (StAR), Fort Collins, CO 80523, USA; dan.lindsey@noaa.gov
[2]   The Aerospace Corporation, El Segundo, CA 90245, USA; Thomas.J.Kopp@aero.org
[3]   CIRA, Colorado State University, Fort Collins, CO 80523, USA; Curtis.Seaman@colostate.edu (C.S.); Steven.Miller@colostate.edu (S.M.); Jeremy.Solbrig@colostate.edu (J.S.)
[4]   Geographic Information Network of Alaska (GINA), Fairbanks, AK 99775, USA; eric@gina.alaska.edu
[5]   CIMSS, University of Wisconsin, Madison, WI 53706, USA; wstraka@ssec.wisc.edu
[6]   NWS, Fairbanks, AK 99775, USA; melissa.kreller@noaa.gov
[7]   NRL, Marine Meteorology Division, Monterey, CA 93943, USA; Arunas.Kuciauskas@nrlmry.navy.mil
[8]   Aviation Weather Center, NWS, Kansas, MO 64153, USA; amanda.terborg@noaa.gov
[*]   Correspondence: don.hillger@noaa.gov

Abstract: Visible/Infrared Imaging Radiometer Suite (VIIRS) Imagery from the Suomi National Polar-orbiting Partnership (S-NPP) satellite is the finest spatial resolution (375 m) multi-spectral imagery of any operational meteorological satellite to date. The Imagery environmental data record (EDR) has been designated as a Key Performance Parameter (KPP) for VIIRS, meaning that its performance is vital to the success of a series of Joint Polar Satellite System (JPSS) satellites that will carry this instrument. Because VIIRS covers the high-latitude and Polar Regions especially well via overlapping swaths from adjacent orbits, the Alaska theatre in particular benefits from VIIRS more than lower-latitude regions. While there are no requirements that specifically address the quality of the EDR Imagery aside from the VIIRS SDR performance requirements, the value of VIIRS Imagery to operational users is an important consideration in the Cal/Val process. As such, engaging a wide diversity of users constitutes a vital part of the Imagery validation strategy. The best possible image quality is of utmost importance. This paper summarizes the Imagery Cal/Val Team's quality assessment in this context. Since users are a vital component to the validation of VIIRS Imagery, specific examples of VIIRS imagery applied to operational needs are presented as an integral part of the post-checkout Imagery validation.

Keywords: VIIRS; DNB; NCC; imagery; validation; Alaska; KPP

1. Introduction

A major component in the overall strategy for the Imagery calibration and validation (Cal/Val) effort for the Visible Infrared Imaging Radiometer Suite (VIIRS) is to ensure that the Imagery is of suitable quality for effective operational use. Imagery of sufficient quality is often determined by the ability of human users to easily locate and discriminate atmospheric and ground features of interest. Such features include clouds and their type, especially convection and low clouds/fog, sea/lake ice edges, snow cover, volcanic eruptions, tropical cyclone structure, and dust storms [1–3]. In many of these cases, multi-spectral algorithms presented as false-color imagery are required to best identify land and atmospheric features over a given location. Pursuant to the Cal/Val strategy, VIIRS Imagery are analyzed to determine if it can be used to interpret the features noted above, considering both single and multi-spectral applications as appropriate.

The primary measurement obtained by VIIRS is digital counts (also known as digital numbers, or DN), designed to respond proportionally to the photons received by the detectors in the 22 bands that comprise the instrument. These DN values are converted into calibrated radiance, reflectance, and/or brightness temperatures based on the methodology of Cao et al. [4]. These calibrated data are distributed to the scientific community as Sensor Data Records (SDRs). A subset of the 22 bands is remapped to a satellite-relative Mercator projection and is distributed to the user community as Environmental Data Records (EDRs). Imagery produced from these data (both SDR and EDR) have been examined by the VIIRS EDR Imagery Team and collaboratively with the general user community to gauge performance and identify artifacts that characterize the overall quality of the imagery.

During the post-launch checkout, both single and multi-spectral images are analyzed to look for artifacts such as striping, banding, noise, geolocation errors, and collocation differences between bands that may plague multi-spectral imagery. In many cases, the root cause for such issues lies with the sensor and, as such, repairing or mitigating these artifacts are primarily the responsibility of the VIIRS SDR Cal/Val Team. The benefits of these ad-hoc software corrections are then inherited by the imagery EDRs produced from the SDRs. Much of this activity occurs before Imagery is declared to be an officially "validated" EDR by the Imagery Team. After this commissioning, VIIRS Imagery is made available to a wide audience of users who continue the Imagery validation process via assessments of operational utility, some examples of which are detailed herein.

Although S-NPP VIIRS visible and infrared capabilities are found to be clearly superior to heritage operational instrumentation (the Advanced Very High Resolution Radiometer, AVHRR), user validation is also tied to the unique features of VIIRS. These include the Day/Night Band (DNB) [5] that is not available from any current or near-future geostationary platform. The DNB, and the Near Constant Contrast (NCC) [6], a more user-friendly product derived from DNB, have found widespread use across NOAA, National Weather Service (NWS) and the U.S. Navy. NCC is capable of providing visual images at night, even under no moon conditions. While the quality varies with the amount of moonlight, day-night images have proven useful to numerous users [5,7,8].

2. VIIRS EDR Imagery

The Suomi National Polar-orbiting Partnership (S-NPP) satellite, which includes VIIRS as its principal imaging radiometer, was launched into a sun-synchronous, 1330 local time ascending node polar orbit on 28 October 2011. The VIIRS Imagery EDR is in fact comprised of three different sets of Imagery products. These are five Imaging-resolution or I-band products, six Moderate-resolution or M-band products, and Near Constant Contrast (NCC) Imagery derived from the DNB sensor. Each of these is remapped to a Ground Track Mercator (GTM) projection, the I-band with 400 m resolution and the others at 800 m resolution [9]. Details on the GTM mapping may be found in the VIIRS Imagery Products Algorithm Theoretical Basis Document (ATBD) [10] and are not discussed here.

The VIIRS Imagery EDR benefits from the many programs which preceded it, and many of the bands chosen to be created as Imagery products are based on heritage from such sensors as the Moderate Resolution Imaging Spectroradiometer (MODIS), AVHRR, and the Operational Line Scanner (OLS). Details on the VIIRS bands transformed into Imagery products by the JPSS ground system are provided in Tables 1 and 2. The Imagery EDR products in Table 1 include both the VIIRS bands explicitly spelled out in the Level 1 Requirements Documents (L1RD) [11] as well as VIIRS I3 band (1.61 μm, excellent in identifying snow/ice locations and ice-topped clouds) and the DNB, both of which became new Key Performance Parameter (KPP) bands in mid-2015. The products in Table 2 are the remaining EDR Imagery products created at present but not classified as KPPs at this time.

Table 1. Required Imagery Environmental Data Records (EDRs).

Imagery EDR Product	VIIRS Band	Wavelength (μm)	SDR Spatial Resolution Nadir/Edge-of-Scan (km)
Daytime Visible	I1	0.60–0.68	0.4/0.8
Short Wave IR (SWIR)	I3	1.58–1.64	0.4/0.8
Mid-Wave IR (MWIR)	I4	3.55–3.93	0.4/0.8
Long-Wave IR (LWIR)	I5	10.5–12.4	0.4/0.8
LWIR	M14	8.4–8.7	0.8/1.6
LWIR	M15	10.263–11.263	0.8/1.6
LWIR	M16	11.538–12.488	0.8/1.6
NCC	DNB	0.5–0.9	0.8/1.6

Table 2. Other Imagery EDRs.

Imagery EDR Product	VIIRS Band	Wavelength (μm)	Spatial Resolution Nadir/Edge-of-Scan (km)
Near Infrared (NIR)	I2	0.846–0.885	0.4/0.8
Visual	M1	0.402–0.422	0.8/1.6
Visual	M4	0.545–0.565	0.8/1.6
SWIR	M9	1.371–1.386	0.8/1.6

The required Imagery products shown in Table 1 primarily reflect user needs in the Alaskan weather theatre. Since no geostationary satellite properly covers the entire Alaskan region, user dependency on polar-orbiting products is much greater than any other US forecasting region. Of the explicit EDRs, bands I1 (0.64 μm), I4 (3.74 μm), and I5 (11.45 μm) comprise the basic building blocks for standard Imagery applications in the visible, MWIR, and LWIR portions of the radiative spectrum. The other three Moderate-resolution band images are LWIR bands that assist in locating clouds and in determining their composition (water or ice).

Use of the remaining bands in Table 2 varies according to the specific atmospheric and surface features of interest at a particular location. The I2 (0.86 μm) band complements the I1 (0.64 μm) band in the visible spectrum. Bands M1 (0.41 μm) and M4 (0.55 μm) assist in the creation of "true-color" and "natural-color" Imagery, while band M9 (1.378 μm) is superior to any other band at identifying thin cirrus during the day. These Imagery products are of practical use to various subsets of the user community.

The final EDR, the NCC as derived from the DNB, has found widespread use across NOAA, the NWS, and the U.S. Navy. NCC is capable of providing visible-wavelength images at night, even under no moon conditions [6]. While the quality varies with the amount of moonlight, NCC has proven useful at night in locating clouds, ice edge, snow cover, tropical cyclone centers (eyes), fires and gas flares, lightning, dust storms, and volcanic eruptions [5,12,13].

While Tables 1 and 2 encompass all of the EDR Imagery products created by the Ground System, they include only 6 of the 16 available M bands, implying that some of the VIIRS SDRs are not currently made available as Imagery EDRs. Some users work with the VIIRS SDRs from the 10 non-EDR rendered bands for their particular imagery needs. The Imagery Cal/Val Team is aware of these needs, and works with users to ensure that all spectral bands derived from VIIRS are of operational quality. DNB radiances, in particular, are used heavily in quantitative applications at the National Environmental Information Center (NEIC), Boulder CO [14] and at the National/Navy Ice Center (NIC).

3. VIIRS Imagery as a Key Performance Parameter (KPP)

Officially, only a subset of the global VIIRS Imagery products is designated as KPP. The specific verbiage of the VIIRS KPP reads as follows (updated to include the two new KPP bands): "VIIRS

Imagery EDR at 0.64 μm (I1), 1.61 μm (I3), 3.74 μm (I4), 11.45 μm (I5), 8.55 μm (M14), 10.763 μm (M15), 12.03 μm (M16), and Near Constant Contrast EDR for latitudes greater than 60°N in the Alaskan region."

The Cal/Val efforts place emphasis on high latitudes, as articulated by this verbiage. The overarching position taken by the Imagery Cal/Val Team is that the imagery requirements must be met first and foremost by the bands in Table 1 in the Alaskan region. Furthermore, the application of VIIRS Imagery must meet the user's expectations. This assessment relies heavily on user engagement and identification of representative use-cases.

Because Imagery is a KPP for VIIRS, it also is required to meet "Minimum Mission Success" in the Post Launch Test (PLT) time frame. How this will be accomplished for JPSS-1 is now spelled out in the newly-revised JPSS Cal/Val Plan [15] for the VIIRS Imagery Product. For JPSS-1 the PLT time frame ends at launch +85 days (L + 85). The objective of the Imagery Cal/Val Team is to show Imagery indeed meets the KPP criteria at L + 85, with the caveat understood that the time frame and season may limit the completeness of the Alaskan data set used for this assessment. For example, if the L + 85 period for JPSS-1 occurs primarily during the Northern Hemisphere winter, little visual data (Imagery) would be available to analyze north of 60° latitude. In contrast, if the period occurs primarily during the Northern Hemisphere summer season, when areas north of 60° latitude are bathed in sunlight, there will be few opportunities to assess the nocturnal component of the DNB's NCC product. Furthermore, certain key atmospheric events that drove the KPP for Alaska, such as volcanic ash, may not occur in the PLT time frame over Alaska. In these cases, the Imagery Cal/Val Team will use appropriate Imagery from other locations to show VIIRS Imagery products is sufficient or better than heritage imagery. Such use of alternative non-high-latitude locations complements the assessment over Alaska, and in most cases is adequate to show by proxy that VIIRS Imagery will achieve Minimum Mission Success.

The requirements for Imagery, as stated in the L1RD-Supplement, are simply spatial resolution requirements, as opposed to requirements that quantitatively address the quality of the EDR Imagery radiances and reflectances. Because of the lack of such specifications, Imagery validation rides heavily upon user feedback, such that the Cal/Val Team's work is complemented by users who are well versed in the use of Imagery applications.

4. Imagery Validation

During the early orbit instrument checkout phase, Imagery products were created as soon as the SDRs became available. In these early stages, VIIRS data are retrieved and evaluated from several sources: the Government Resource for Algorithm Verification Independent Testing and Evaluation (GRAVITE), the Comprehensive Large Array-data Stewardship System (CLASS), the Product and Evaluation and Test Element (PEATE), or from direct-broadcast line-of-sight reception sites. Post-checkout operational users have access to VIIRS data and derived products via NOAA's NPP Data Exploitation (NDE), as well as the NWS Advanced Weather Interactive Processing System (AWIPS) satellite data distribution system.

4.1. Validation Tools

There are many tools available designed to exploit the VIIRS Imagery. These tools, employed by the Cal/Val Team, are the Man-computer Interactive Data Access System (McIDAS-V in particular) [16], TeraScan processing software from SeaSpace Corporation [17], and various data processing and display tools using the Interactive Data Language (IDL) [18]. These tools are made available to various users during the Cal/Val process for Imagery. The primary operational user, the NWS, has its own display tools that are integrated into AWIPS [19]. Direct broadcast reception sites may display Imagery using the Community Satellite Processing Package (CSPP) [20] or the International Polar Orbiting Processing Package (IPOPP) [21]. Feedback from users of these systems is considered part of the extended validation process.

4.2. Intensive Calibration/Validation Phase

Among the basic qualities of Imagery are its spectral, spatial, temporal, and radiometric resolution. Spectral resolution is determined by the bands and bandwidths of VIIRS, as well as optical effects of the reflecting mirrors that direct light into the focal plane array. The spatial and radiometric resolutions of VIIRS are determined at the SDR level. The SDR and EDR radiances are unchanged, except for the special processing that goes into the DNB/NCC pair [6]. Table 3 compares the VIIRS SDRs and Imagery EDRs.

Table 3. Similarities and Differences between Visible/Infrared Imaging Radiometer Suite (VIIRS) Sensor Data Records (SDRs) *vs.* Imagery EDRs.

Characteristic	SDR	EDR
Solar reflective (visible) bands	Radiances and reflectances	Radiances and reflectances (same as SDR)
Infrared (thermal) bands	Radiances and brightness temperatures	Radiances and brightness temperatures (same as SDR)
Geo-spatial mapping	Satellite projection (with bowtie deletions and overlapping pixels)	Ground Track Mercator (GTM) projection (rectangular grid, no pixel deletions or pixel overlap)
Day/Night Band (DNB) imagery	DNB radiances (may vary by up to 7 orders of magnitude, depending on lunar and/or solar illumination)	NCC pseudo-albedos [6] (may vary by up to 3 orders of magnitude, to display features under conditions ranging from no moon to full solar illumination, as well as artificial lights)

The primary objective of the intensive Cal/Val phase is establishing in official sequence beta, provisional, and validated status for the Imagery products. The Imagery Cal/Val Team works with the VIIRS SDR Cal/Val Team to verify those requirements for Imagery that are tied to the SDR quality (e.g., radiance calibration accuracy, detector noise and striping, geolocation accuracy, *etc.*). The spatial resolution requirements are tied to the GTM projection, and are straightforward to verify. However, Imagery only passes through these validation stages as it is determined to be of quality for operational users.

The second component of Cal/Val user-oriented validation is operational user analysis of the Imagery. Band combinations targeting the characterization of clouds and cloud types, clouds phase, snow/ice on the surface, and dust storms are emphasized. Given the KPP emphasis on the Alaska region, this user base has a direct say in determining when validation has been reached. This determination is made via evaluation of VIIRS Imagery applications in the day-to-day operational environment.

The evaluation of NCC Imagery is a special case, due to its unique nature and its growing use for multiple purposes. DNB/NCC is used for such features ranging from ice edge at night to tropical cyclone center fixing. Some of the capabilities demonstrated at lower latitudes have analogous applications at the high latitudes. For example, the ability to peer through thin cirrus and reveal low-level circulation in tropical cyclones via moonlight also holds utility for peering through frontal cirrus and detecting the distribution of clouds or sea ice below (e.g., [5]). The Imagery Cal/Val Team helps users evaluate DNB/NCC for specialized applications such as gas flares, fishing boats, and auroras, while the operational applications of this nighttime imagery are the ultimate validation of its usefulness.

Mitigation of Non-Linearity for DNB on JPSS-1

The only significant difference with the input SDRs between S-NPP and JPSS-1 is with the DNB. It was discovered during routine laboratory testing that the DNB for JPSS-1 contains an anomalous non-linear response at high scan angles. The anomalies necessitated the design of DNB SDR post-processing software to mitigate the associated imagery artifacts. The method chosen impacts the spacing between DNB pixels towards the edge of the scan, as well as the location of nadir within the DNB SDR itself. The mitigation procedure is asymmetric, such that the DNB will actually extend in one direction (referred to as an "extended scene"). Furthermore, instead of preserving constant pixel size, the resolution of the DNB on JPSS-1 will degrade to approximately 1.2 km at the edge of the scan, and the spacing between pixels will not be as constant across the scan line as it is with S-NPP. The degradation in spatial resolution with the JPSS-1 DNB does not occur until a 49° viewing angle is reached.

Figure 1 shows the impact of these mitigation steps. These figures were produced using S-NPP DNB SDRs and executing the mitigation intended for JPSS-1. Figure 1A shows how the DNB SDR would look for JPSS-1. The blank space (blue) on the right side is the extended scene component. Nadir actually lies on the left side of the SDR image, as noted by the dashed line. In Figure 1B, which is the resulting NCC image, the NCC process shifts nadir (dashed line) to the center of the image, and truncates the extended scene portion of the DNB. This is intended, so NCC imagery from JPSS-1 will be similar to that from S-NPP, as well as match the GTM mapping for the VIIRS M-bands.

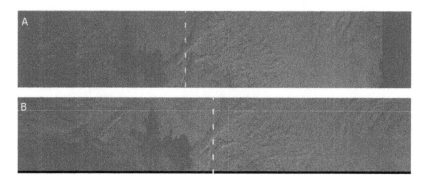

Figure 1. (A) DNB from Suomi National Polar-orbiting Partnership (S-NPP) used to display how DNB will look from JPSS-1, with the blue area on the right filled with extended scene imagery (currently missing in this simulation); **(B)** The DNB remapped into the GTM mapping used for Near Constant Contrast (NCC), showing that the NCC shifts the DNB imagery to the right, placing nadir at the center and ignoring the extended scene data on the right. In each image, the dashed line shows the approximate location of nadir.

4.3. Long-Term Monitoring Phase

Long-term monitoring of VIIRS Imagery is the responsibility of the VIIRS SDR and Imagery EDR Teams collectively. The task extends to the Imagery user base as well, who at any time may help the Teams to identify an imagery anomaly that may be either transient or recurring. Hence, long-term monitoring focuses on the ongoing value of the Imagery product quality as it applies to users.

With those Cal/Val basics as background, the rest of this article presents key examples of the uses of VIIRS Imagery that show its quality as being highly useful (even exceeding expectations) for many analysis and forecasting applications. In the cases presented, the Imagery or image products proved beneficial to the users.

5. Applications of VIIRS Imagery in Meteorological Operations

5.1. Alaska Examples

As noted in previous sections, Alaska users of VIIRS are specifically spelled out as primary users of VIIRS imagery since they are located on the northern edge of most geostationary satellite views and polar-orbiting data is best utilized in Polar Regions where there is a high frequency of satellite overpasses. The KPP explicitly spells out the Alaskan region as the most critical area for Imagery coverage and quality.

5.1.1. Visible and Longwave Imagery

VIIRS Imagery bands are displayable on NWS Alaska Region AWIPS as single-band products. Figure 2A shows an AWIPS screen capture of I1 (0.64 µm) visible imagery from 29 July 2015 at 0043 UTC (4:43 pm Alaska Daylight Time), 28 July 2015. Figure 2B shows the I5 (11.45 µm) longwave IR imagery. Each of these single-band products has its strengths and weaknesses. For example, consider the deck of low stratus over the ocean near the Bering Strait. This stratus is obvious in the VIIRS I1 visible imagery. But clouds like these typically develop in the lower troposphere and have temperatures similar to nearby sea surface temperatures, with the result that the distinction between the stratus deck and clear skies over the ocean (in the yellow circle) cannot be made using the VIIRS I5 (11.45 µm) longwave IR imagery alone. The I5 imagery helps forecasters identify colder convective clouds more easily than can be done in the I1 visible imagery. The colder clouds inland on the right side of these figures (noted by yellow arrows) represent typical summertime afternoon convection over the rough terrain of the Nulato Hills. These convective clouds are easy to identify in the I5 imagery because of their colder temperatures, but these clouds might not be immediately identified as convective in the I1 visible imagery alone, since they have the same white color as the low stratus over the nearby marine areas. Figure 2C is the result of combining the different advantages of the VIIRS I1 visible and I5 longwave IR imagery into a single product, thereby allowing the forecaster to gain the meteorological insights contained in two different bands by looking at just one product. In this case, the two Imagery bands are simply overlaid and blended together. This approach is simplistic, but effective. A more sophisticated and much more helpful approach is not to overlay different products but to combine two or more single bands into a multi-spectral product as described in the following section.

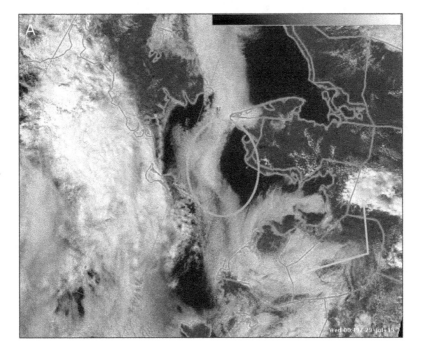

Figure 2. *Cont.*

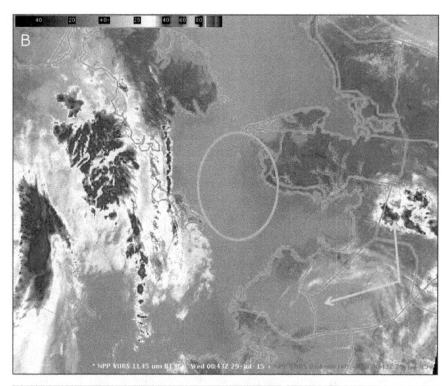

Figure 2. Annotated Advanced Weather Interactive Processing System (AWIPS) screen capture on 29 July 2015 at 0043 UTC centered over the Bering Strait of (**A**) VIIRS I1 (0.64 μm) band visible imagery; (**B**) VIIRS I5 (11.45 μm) band longwave infrared imagery; and (**C**) the two bands overlaid in AWIPS into a VIIRS Imagery product.

5.1.2. Fog/Low Cloud

Beginning with the GOES-8 imager, when the 3.9 μm band first became available, the 11 μm–3.9 μm brightness temperature difference (BTD) has traditionally been used for fog/low cloud detection [22]. At night, the 3.9 μm and I5 (11.45 μm) bands detect differences in radiometric temperature (due to spectral emissivity differences between the two bands) rather than thermodynamic temperature. These emissivity differences are related to the size of the particles, meaning that small droplets (such as fog) can be distinguished from larger droplets as well as cloud free surfaces. While these emissivity effects are present during the day, they are overwhelmed by the solar signal in the 3.9 μm band. Thus, this product is more useful during the night time. Fog is of critical importance in the Alaska region due to the large amount of private and commercial aviation and maritime traffic. During the northern hemisphere winter months, much of the Alaska region is in "night" or near-terminator illumination conditions, allowing for the application of the 11 μm–3.9 μm BTD for fog detection. Owing to the availability of timely direct broadcast S-NPP data, provided by the Geographic Information Network of Alaska (GINA, located at the University of Alaska, Fairbanks), there have been several instances where the NWS Weather Forecasting Office (WFO) in Fairbanks, Alaska, has been able to gain information on fog/stratus along the North Slope of Alaska in near real-time.

Figure 3 shows a color-enhanced 11 μm–3.9 μm BTD for the northern Alaska forecast area including the North Slope. The orange colors indicate areas of low cloud/fog, the light grays indicate higher clouds, and the black areas indicate thin cirrus. To accompany this figure, portions of the NWS Northern Alaska Forecast Discussion (AFD) for 11 March 2013 issued at 1258 pm (which follows) mention the use of VIIRS imagery (with yellow highlighting emphasizing the S-NPP VIIRS Imagery fog product). The AFD points out that the fog was evident in the VIIRS Imagery, surface observations, and MODIS products; but the higher spatial resolution of the VIIRS Imagery captured the fog in more detail than MODIS, and certainly more detail than surface observation; as well as VIIRS providing additional imagery at times other than MODIS overpasses.

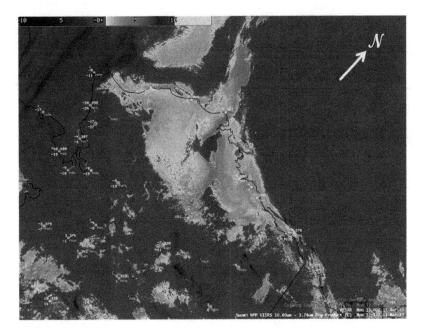

Figure 3. Color-enhanced 11 μm–3.9 μm BTD for 11 March 2013 at 1252 UTC, along with METAR observations at 1300 UTC. The oranges indicate areas of low cloud/fog, where the light grays indicate higher clouds and the black indicates thin cirrus.

"The Suomi NPP VIIRS satellite fog product was indicating a decent layer of stratus along the North Slope. Observations across the area generally indicated 1 to 2 miles (1.5 to 3 km) in visibility with flurries and fog. The IFR conditions align very well with the higher probabilities of MODIS IFR product. There are some very isolated pockets of higher probabilities of the MODIS LIFR conditions. These conditions should remain through Tuesday evening or Wednesday morning as the surface high pressure remains within the area. By Wednesday morning the surface pressure gradient begins to tighten providing an increase in winds and perhaps a break in some of the fog."

Fog and stratus have large impacts year round along the North Slope, particularly to the aviation community. The fog/stratus example in Figure 4 highlights the critical utilization the VIIRS imagery of the I3 (1.61 μm) band at the NWS Fairbanks WFO in the preparation of the Terminal Aerodrome Forecast (TAFs) issuance. The timely high-resolution imagery at 0004 UTC on 26 April 2015 allowed forecasters to identify and forecast the mesoscale circulation feature moving to the northwest of the Nuiqsut AK and showed improving visibility conditions for air flights. Meanwhile it was apparent that the visibility at the Prudhoe Bay airport was still ¼ mile (400 m) in freezing fog and would remain IFR conditions as well as along the Arctic Coast.

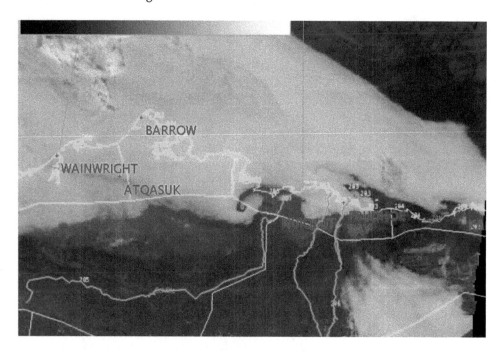

Figure 4. The Suomi National Polar-orbiting Partnership (S-NPP) VIIRS I3 (1.61 μm) band along the North Slope on 26 April 2015 at 0004 UTC. The white colors indicate areas of low clouds and fog, where the darker gray indicates the clear conditions. The light blue lines are the rivers with the green lings indicating the boundaries for the zone forecast areas.

5.1.3. Multi-Spectral Imagery

The large number of VIIRS I-bands and M-bands (5 and 16 bands, respectively) offers a wide variety of possible combinations into multi-spectral, or Red, Green, Blue (RGB) composite, image products. Figure 5A shows a true-color RGB from 9 July 2015 at 2302 UTC (3:02 pm Alaska Daylight Time), that combines the M5 (0.67 μm), M4 (0.55 μm), and M3 (0.49 μm) bands into a true-color image that represents what the human eye would see if we could ride along on the S-NPP satellite. A true-color image has great advantages for meteorological surveillance, in part because it is comparatively intuitive to interpret. Every multi-spectral Imagery product has its strengths and weaknesses, and while a strength of the true-color RGB is its ease of interpretation, a particular weakness (from the Alaskan perspective) is its inability to offer a distinction between clouds and sea ice over the Arctic Ocean because both clouds and ice appear white in true-color imagery. Figure 5B is a natural-color RGB composite image that assigns the VIIRS I3 (1.61 μm) band to the red component, the I2 (0.86 μm) band to the green component, and the I1 (0.64 μm) band to the blue component. This product takes advantage of the fact that incoming sunlight is absorbed or reflected to various degrees by different surfaces at these three wavelengths. Most helpfully, the resulting multi-spectral product depicts sea ice and ice clouds as cyan and liquid-based clouds as pink, making the separation of low clouds from ice comparatively straightforward. (Sea ice and ice clouds can normally be easily discriminated by texture and context.) The number of multi-spectral products available to NWS meteorologists in Alaska on their AWIPS workstations has increased substantially over recent years, with RGB composite images now being built routinely from VIIRS and MODIS data received locally in Alaska via direct broadcast antennas.

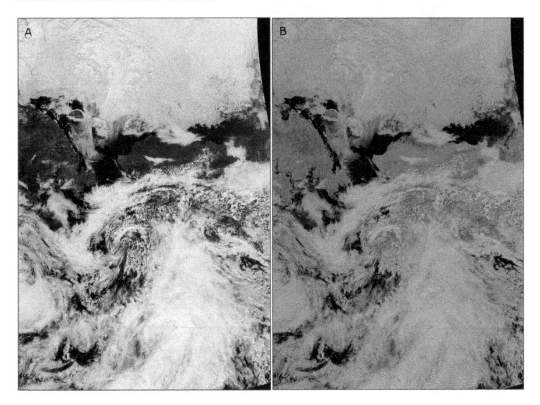

Figure 5. (**A**) VIIRS True-color Red, Green, Blue (RGB) combining the I1 (0.64 μm) band with the M3 (0.49 μm) and M4 (0.55 μm) bands; and (**B**) VIIRS natural-color RGB composite image combining the VIIRS I3 (1.61 μm), I2 (0.86 μm), and I1 bands, both from 9 July 2015 at 2302 UTC.

5.1.4. The Day/Night Band at Night

The VIIRS DNB has proven particularly useful to NWS forecasters during the extended periods of darkness during the Alaskan winter, because the DNB offers forecasters the ability to analyze meteorological and terrain features in the visible portion of the spectrum when ambient light levels are too low for conventional visible satellite imagery to be helpful. Figure 6 is an AWIPS screen capture of DNB imagery from 24 January 2013 at 1354 UTC (4:54 am Alaska Standard Time), a period of total darkness over all of Alaska. The scaling of DNB imagery needs to be appropriate to the lunar illumination at the time of the image, as the DNB radiances can vary by 7 orders of magnitude, from full daylight to nearly total darkness under new-moon conditions. Terrain and cloud features are very evident, as are the lights from the cities of Anchorage and Fairbanks. The raw S-NPP data used to produce this image were received via a direct broadcast antenna at GINA on the campus of the University of Alaska Fairbanks, then processed at GINA with CSPP software, and finally delivered to the NWS in Alaska for display in AWIPS via the Local Data Manager (LDM).

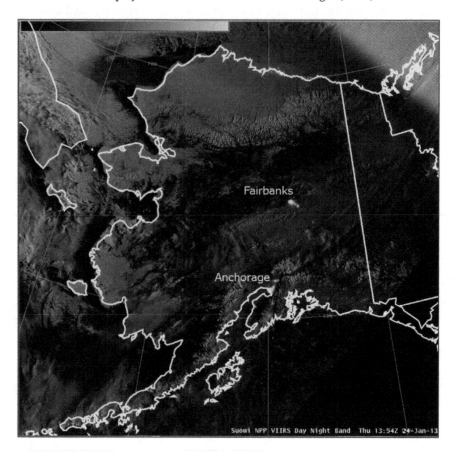

Figure 6. Example of VIIRS DNB imagery for 24 January 2013 as displayed on a National Weather Service (NWS) AWIPS workstation. An adjustable gray-scale appropriate to the lunar (or solar) illumination at the time is used to enhance the DNB radiances to best reveal cloud and surface features.

5.1.5. Volcanic Ash

Volcanic ash is also a common hazard in Alaska, not only from current eruptions, but even for ash from long-dormant eruptions as in the following example.

The 6–8 June 1912 eruption of Novarupta, one of the largest volcanic eruptions in recorded history, expelled ash with an estimated column height exceeding 23 km and deposited a layer of ash exceeding 100 m thick in the nearby Valley of Ten Thousand Smokes on the Alaska Peninsula [23]. During periods of high winds, this volcanic ash may become re-suspended in the atmosphere, posing a hazard to aviation [24] and human health [25]. VIIRS imagery was able to capture one such incident that occurred on 30–31 October 2012. According to local media reports [26], ash was lofted up to 1.2 km in altitude and resulted in the diversion and cancellation of flights in the vicinity of Kodiak Island during the high wind event. The NWS Alaska Aviation Weather Unit (AAWU) issued a Significant Meteorological event warning (SIGMET) for the re-suspended volcanic ash. Portions of the initial SIGMET [27] issued at 2216 UTC on 30 October 2012 is included below.

"Volcano: Novarupta 1102-18

Eruption details: NO Eruption. Resuspended ASH.

RMK: Resuspended ash due to high winds in area. Not from eruption."

Figure 7 shows the VIIRS DNB image (collected at night) along with the true-color RGB (Red/Green/Blue) composite image (collected during the following afternoon) from 30 October 2012. The plume of volcanic ash is visible in both the DNB and true-color images. In each image, the location of Novarupta is indicated by the yellow arrows. The ash plume extends from Novarupta to the southeast, across the Shelikof Strait and Kodiak Island. While the initial SIGMET was based on analysis of MODIS imagery, this case highlights the importance of satellite imagery for volcanic ash detection and demonstrates the utility of VIIRS imagery for this purpose.

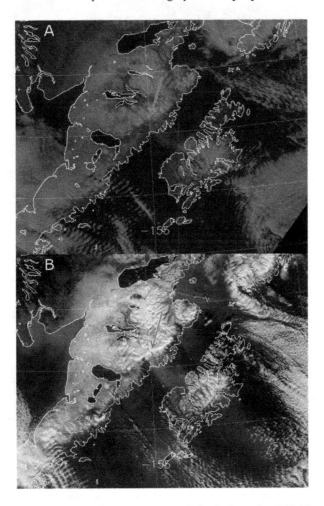

Figure 7. (**A**) VIIRS DNB image showing re-suspended ash from the 1912 Novarupta eruption (1411 UTC 30 October 2012) (**B**) VIIRS true-color image of the ash plume (2223 UTC 30 October 2012). In each image, the location of Novarupta is indicated by a red arrow. The ash plume extends from the volcano to the southeast across the Shelikof Strait and Kodiak Island.

5.1.6. Fire Weather

The VIIRS imagery with the I4 (3.74 μm) and I1 (0.64 μm) band is extremely beneficial to NWS forecasters for identifying hotspots and areas of smoke. The I4 brightness temperatures allow for the identification of the fire "hotspots", with the I1 visible band overlaid identifying the smoke plumes. The combination of these two products as in Figure 8 is heavily utilized during the Alaska fire weather season from late June through August. The 2015 fire season was extreme, and a record number of acres were burned. The area burned, 5.148 million acres (20.8 km^2), puts the season in second place (out of 66 years) behind the extreme 2004 season. During the active fire season three Incident meteorologists were dispatched to the wildfire complexes in the Fairbanks warning area [28] and one IMET was dispatched to the office to help with the number of wildfire spot forecasts (909 Fire Weather Spot forecasts were issued for the 2015 Fire weather season).

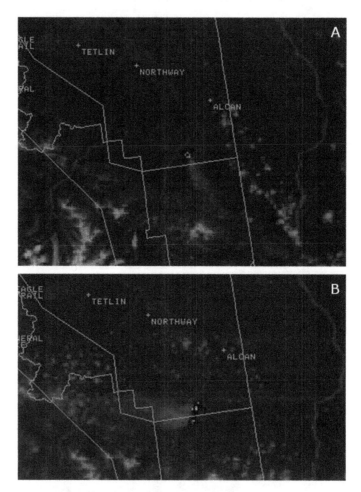

Figure 8. (**A**) The VIIRS I4 (3.74 μm) band brightness temperature and I1 (0.64 μm) band visible satellite imagery overlaid at 2211 UTC on 17 June 2013 in which you can see the smoke plume moving towards the southeast and a fire "hotspot" from the Chisana River, Eagle Creek, and Bruin Creek wildfires; (**B**) Image from 18 June at 2013 UTC after the winds shifted from the northwest to the southeast and moved the smoke plume.

With the active fire weather period, it was also important not only to identify these hot spots but also to inform the public on smoke hazards. The NWS Fairbanks WFO typically informs the public via social media, as in Figure 9 for 6 July 2015 regarding the Aggie Creek wildfire located northwest of Fairbanks. The reports show (top) the pyrocumulus images from the Aggie Creek fire, and (bottom) a side-by-side comparison of the VIIRS false-color imagery at 2312 UTC with the radar reflectivity image from the Pedro Dome (PAPD) radar on 6 July 2015. In this case the VIIRS

Imagery supports the radar, webcam, and other fire-related information, adding to the validity of the information presented to the public, which benefits from analyses such as this one compiled from multiple sources by NWS forecasters.

(a)

(b)

Figure 9. Two social media reports from 6 July 2015 about the Aggie Creek fire located northwest of Fairbanks: (a) as seen from the NWS Fairbanks office on the University of Fairbanks campus; (b) the VIIRS False-Color satellite imagery at 2312 UTC and the radar reflectivity image from the Pedro Dome radar are shown side by side.

5.1.7. Sea Ice

The determination of the sea ice state is important in Alaska in all seasons. VIIRS Imagery, when cloud-free, can provide a good view of sea ice, which is valuable input for a sea ice analysis. These ice analyses are particularly valuable when lives and properly are at stake, such as in the following example when VIIRS imagery allowed the NWS to help the US Coast Guard (USCG) rescue a mariner in need.

On 10 July 2014 the USCG was called upon to rescue a solo mariner in a small boat attempting to sail the Northwest Passage. The mariner had become stuck in the pack ice north of Barrow, Alaska, and the USCG coordinated with the NWS Ice Program based in Anchorage, Alaska. The NWS used multi-spectral VIIRS imagery, to analyze the sea ice in the area of concern and provide the USGC guidance regarding the best path to approach the mariner. The annotated VIIRS-based sea ice analysis for this case is shown in Figure 10. As a result, the Coast Guard successfully rescued the stranded sailor [29].

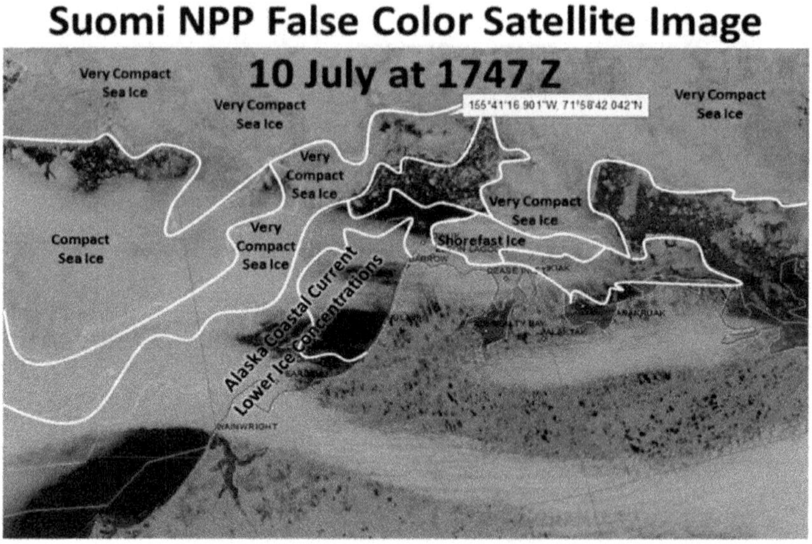

Figure 10. Annotated VIIRS RGB composite image from the NWS Ice Program in Anchorage AK, including the position of the boat stranded in the ice. (Image courtesy of Mary-Beth Schreck.)

5.2. Examples Outside of Alaska

Because S-NPP is a polar-orbiting satellite with worldwide coverage every 12 h, there are many opportunities for the use of VIIRS Imagery around the world, not just in the polar regions that specifically define VIIRS as a KPP.

5.2.1. Tropical Storm Centering at Night

It has become common practice at the Joint Typhoon Warning Center to consult an RGB composite image product that combines the VIIRS DNB with the I5 (11.45 μm) band during analysis of tropical cyclones. This imagery, made available by Naval Research Laboratory in Monterey CA (NRL-MRY) via the Automated Tropical Cyclone Forecasting System (ATCF) and the NRL Tropical Cyclone Webpage [30], has proven useful for distinguishing between high and low clouds in subjective analysis. Additionally, this band combination can allow analysts to discern situations where low cloud can be seen through optically-thin high cloud. RGB composite images have aided analysts in improving their estimates of storm center on multiple occasions.

To create the RGB composite image, the DNB lunar reflectance (following the model of Miller and Turner [31], or via NCC pseudo-albedo following Liang *et al.* [6]) is assigned to the red and green components, and inverted I5 band brightness temperatures are assigned to the blue component. This results in a false-color image product where: open ocean and optically-thin low cloud appear black; optically-thick low cloud appears yellow; optically-thin high cloud appears blue; optically-thick high cloud appears white; and optically-thin high cloud overlaying optically-thick low cloud also appears white, but can often be discerned from optically-thick high cloud by context.

In cases of extremely sheared storms, the RGB composite image can enable an analyst to observe the exposed low-altitude circulation center; however, even in less sheared cases, the imagery can provide useful information about the low-level circulation. An example of this is shown in Figure 11 where the left-hand panel shows I5 band brightness temperatures and the right-hand panel shows an RGB composed of the DNB and I5 band brightness temperatures. Both images depict Typhoon Linfa on 3 June 2015 at 1752 UTC. The I5 band image provides information about the storm's convective tops, but provides little information about the low-altitude circulations. Based solely on this I5 band image, analysts at the Joint Typhoon Warning Center (JTWC) would likely have determined that the storm center was at the center of the convective region (marked with a §).

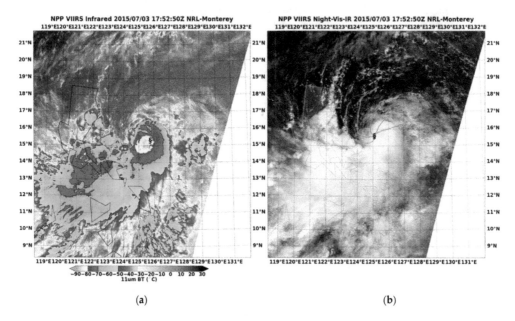

(a)   (b)

Figure 11. VIIRS (a) I5 (11.45 μm) band brightness temperature and (b) RGB composite imagery of Typhoon Linfa from 3 July 2015 at 1752 UTC. The RGB composite is composed of the DNB in the red and green bands and inverted I5 brightness temperatures in the blue. The RGB composite was used by Joint Typhoon Warning Center (JTWC) analysts to correctly determine the center of the typhoon by tracing the low-altitude cloud lines (orange) to the center of circulation.

The RGB composite image product provides a significant amount of additional information. The imagery allows analysts to readily discriminate low altitude patterns from high altitude patterns. JTWC stated that "without the VIIRS DNB image, the TC position would have been derived from IR only, which would have placed the center further southeast about 20 miles (30 km) with the assistance of the VIIRS DNB image, the forecaster correctly placed the best track position, which in turn improved model initialization and subsequent forecast accuracy."

In another example, the VIIRS DNB proved useful for identifying low cloud features in the presence of higher overriding cirrus clouds. The event occurred east of Hawaii on 29 July 2013. Tropical Storm Flossie was east of the Big Island of Hawaii, moving generally to the west-northwest. The Central Pacific Hurricane Center (CPHC) faced the challenge of issuing guidance on possible landfall over the nighttime hours when infrared data gave misleading information on the circulation center. After appealing to DNB imagery, the CPHC issued the following statement (with yellow highlighting noting the use of VIIRS nighttime visible imagery): "The center of Flossie was hidden by high clouds most of the night before VIIRS nighttime visual satellite imagery revealed an exposed low level circulation center farther north than expected. We re-bested the 0600 UTC position based on the visible data" [32].

Figure 12A shows the VIIRS DNB image at 1103 UTC that was being referred to, and Figure 12B shows the corresponding I5 (11.45 μm) infrared image. High cirrus clouds with brightness temperatures around −30 °C can be seen in the infrared image to the northwest of the deepest convection, but in the DNB some low clouds can be seen underneath. CPHC inferred the center of circulation based on the spiral structure of these low level clouds, and their 1200 UTC analysis of the center location is denoted in the figure with a maroon dot. First-light daytime visible imagery from GOES-W confirmed the DNB moonlight-based guidance. The infrared imagery alone would not have been useful in locating the center.

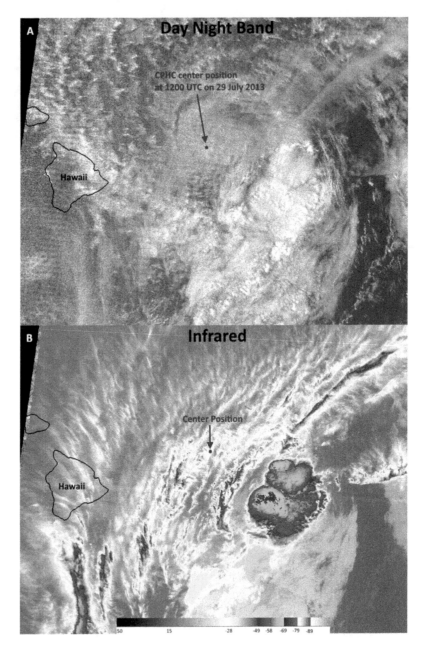

Figure 12. VIIRS (**A**) DNB and (**B**) I5 (11.45 μm) image showing Tropical Storm Flossie east of Hawaii on 29 July 2013 at 1103 UTC. The analyzed position by the Central Pacific Hurricane Center of the center of the storm at 1200 UTC is denoted by a maroon dot in both images. The units of brightness temperature in (b) are degrees C. Striping in the DNB becomes more apparent as the signal level decreases under lowlight illumination.

5.2.2. Puerto Rico Dust from Saharan Air Layer

During the spring through autumn months, the greater Caribbean region is susceptible to major dust outbreaks due to passages of dust-laden Saharan Air Layer (SAL) that can significantly degrade air quality and suppress local convection. SAL occurrence increases the potential for massive wildfire outbreaks throughout Puerto Rico and the surrounding West Indies. Additionally, the public in the greater Caribbean region suffers some of the world's highest rates of asthma, which has been postulated by health officials to be attributable to long-term exposure to high dust concentrations and the bio-chemical content of this dust [33–35]. The NWS in San Juan, Puerto Rico (NWS-PR) is responsible for issuing hazardous warning alerts to other agencies and the public. The resources available to the NWS-PR are limited with respect to making accurate and effective assessment and predictions of SAL events.

S-NPP VIIRS products are actively being used by the NWS-PR in monitoring the SAL via the NexSat website [36] from NRL-MRY. Through a wealth of remote sensing datasets and image products, in-situ observations, and an operational global dust model, NRL-MRY provides NWS-PR with enhanced capabilities to detect, assess, and predict SAL events as they propagate across the forecast area of responsibility (AOR). Included among the suite of products provided to NWS-PR through NRL-MRY's NexSat website are products derived from the S-NPP VIIRS sensor. These products include visible, infrared, true-color, dust, Aerosol Optical Depth (AOD), and DNB imagery.

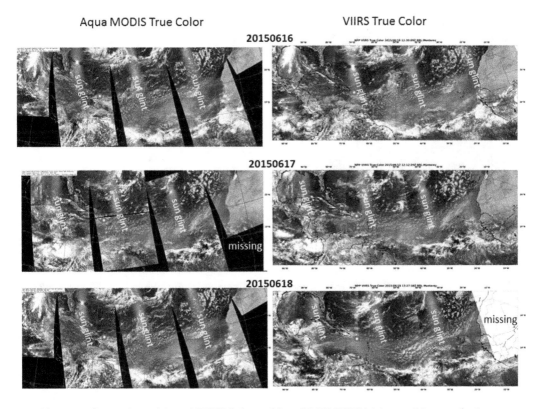

Figure 13. Comparison of Aqua MODIS (left panels) *vs.* S-NPP VIIRS (right panels) true-color image products while monitoring a SAL event across the north tropical Atlantic basin during 16–18 June 2015. The increased swath within VIIRS reduces the "guesswork" in tracking dust. Sun glint regions indicate enhanced levels of reflection from the ocean surface.

During the May–July 2015 timeframe, an almost continuous stream of strong African dust events propagated across the north tropical Atlantic basin, eventually impacting the greater Caribbean region. These events produced high counts of particulate matter ($PM_{2.5}$ and PM_{10}), high AOD values, and a rash of wildfires due to dry subsiding air, particularly within the forested regions of Puerto Rico. Figure 13 compares true-color imagery from Aqua MODIS and S-NPP VIIRS at the height of a SAL event that occurred during the period of 16–18 June 2015. The wider swath width and increased resolution at scan edge allows VIIRS (right-hand column) to provide a more continuous and more easily-analyzed product than MODIS (left-hand column), reducing the "guesswork" in tracking dust across the Atlantic basin. Figure 14 presents VIIRS-derived blue-light dust products [37] (left-hand column) during the approach of the same SAL-borne dust plume towards Barbados (marked with a red/yellow dot) for the period of 16–18 June. In this imagery, dust appears in shades of pink, while clouds are shown in cyan.

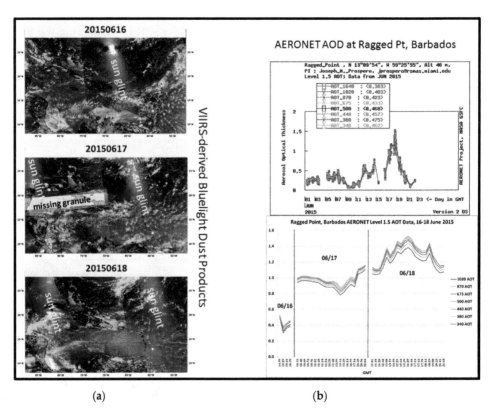

(a)   (b)

Figure 14. (a) A series of VIIRS-derived bluelight dust products during 16–18 June 2015 showing the approach and eventual impact of the SAL plume covering the greater Caribbean region. Dust appears in shades of pink, whereas clouds are in shades of cyan. The island of Barbados is annotated in red/yellow circles; (b) AERONET AOD measurements taken at Ragged Point, Barbados. The top-right panel indicates the monthly profile, with a spike during the SAL event (17–19 June). The bottom-right panel highlights the SAL event peaking during 14–17 UTC on 18 June.

Surface-based measurements of AOD indicate that the VIIRS imagery correctly identifies dust for subjective analysis. The plot in the top right of Figure 14 shows AOD measurements from the Aerosol Robotic Network (AERONET) site at Ragged Point, Barbados over the period from 1–23 June 2015. AOD values spike between the 17th and 19th, indicating the impact of the SAL-borne dust on the region. The figure in the lower right shows the same AOD data, but focusing on the period of 16–18 June, showing that the event peaks between 14 and 17 UTC on 18 June. These observations of elevated AOD values coincide with the approach of the dust plume as observed by VIIRS. In this case, VIIRS was able to provide forecasters with advanced warning of the SAL event and allowed forecasters to disseminate that information to the public.

5.2.3. Blowing Dust in the Great Plains

Another use of VIIRS multi-spectral Imagery is exemplified by the 29 April 2014 dust case over the Texas Panhandle. That day ended a three-day-long dust event associated with an intense upper-level system that lumbered across the western and central U.S. The placement and movement of the system kept a very strong jet over western Colorado and the Texas Panhandle, resulting in widespread severe winds gusting from 55 mph (90 km/h) to even greater than 65 mph (105 km/h) at times. Sustained drought over the region provided an ample supply of dust, which was immediately lofted by strong surface winds. Smaller regional airports in the Texas Panhandle reported visibilities at or near zero for the duration of this event. These conditions occurred on three consecutive afternoons from the 27 to the 29 April, and necessitated the issuance of multiple blowing dust SIGMET advisories.

On the afternoon of the 29 April, Aviation Weather Center (AWC) forecasters examined the VIIRS and MODIS dust enhancements as they issued several SIGMETs related to this event. Figure 15a is a VIIRS true-color image from 2000 UTC. There were no obvious signs of dust in the scene. However, the corresponding dust enhancement (Figure 15b), based on the multi-spectral algorithm [37], highlights in pink a large swath of blowing dust crossing into the Texas Panhandle from southwestern Colorado.

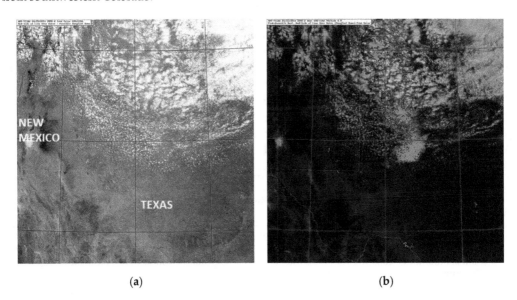

(a)   (b)

Figure 15. (a) VIIRS true-color imagery for the Texas Panhandle at 2000 UTC on 29 April 2014 with no obvious signs of dust in the scene; (b) VIIRS dust-enhancement imagery at the same time, highlighting the blowing dust in pink below the overlying clouds in cyan.

A blowing dust SIGMET was issued at 1700 UTC, valid until 2100 UTC. Because of the limited temporal resolution of polar imagery, this SIGMET was issued without the use of the dust enhancement. However, an extension of this SIGMET was issued just after 2100 UTC after forecasters examined the VIIRS dust enhancement imagery, and it was extended further south and east into central Texas. In this case the dust enhancement proved a very useful situational awareness tool for amending the SIGMET, which will only improve when geostationary assets begin carrying the bands required for implementing this algorithm.

6. Discussion

Although the initial post-launch Cal/Val checkout of VIIRS Imagery is shared between the EDR Imagery Team and the VIIRS SDR Team, many Imagery applications are validated during the intensive Cal/Val period ending with an 'official' validation status, signifying that Imagery is ready for use by a wide range of users worldwide. Each step involves users, but not to the extent that users become involved in the post-checkout timeframe. The post-launch checkout is followed by the long-term monitoring phase that continues for the life of VIIRS on each satellite. It is during this stage that users become heavily involved in VIIRS validation. Users are sought who benefit from the advantages that VIIRS has over heritage satellite instrumentation, particularly the DNB/NCC imagery not available from geostationary orbit.

The validation examples presented, most with direct and specific feedback from operational users, are the crux of the post-launch validation of VIIRS Imagery. Emphasis has been on the Alaska operations, since the Alaska region is specifically mentioned in the KPP statement for VIIRS. Alaska examples include identification of fog, convection, and the ability to distinguish between sea ice and clouds over the Arctic Ocean and Bering Sea. However, not all VIIRS validation necessarily takes place in the Polar Regions. Examples of tropical cyclone re-centering and Saharan dust in the Caribbean and dust on the Great Plains were presented as applications of VIIRS DNB imagery that has potential for worldwide application.

7. Conclusions

These cases present an overall picture of wide use and multiple-user benefits gained from VIIRS Imagery, both as individual bands and as band combinations and RGB composite image products suitable to specific atmospheric and land phenomena. The fact that numerous users are pleased with VIIRS is the "user" validation that goes beyond the "official" validation accomplished by the Imagery Team alone.

Future plans for VIIRS Imagery validation include continuing/routine checkout as operational ground systems (hardware and software) change, to make sure Imagery remains at consistent high quality. Then, the validation process begins over with follow-on satellites, JPSS-1 to be launched in 2016 or 2017, and similarly for JPSS-2 and beyond.

Acknowledgments: The authors would like to thank other on the VIIRS Imagery and Visualization Team [38] who are not listed as co-authors for their contributions: Stan Kidder, Debra Molenar, Steve Finley, Renate Brummer, Chris Elvidge, Kim Richardson, and Bill Thomas. The authors would also like to thank Capt. Brian Decicco and TSgt. Ricky Frye of the Joint Typhoon Warning Center for their contributions. Funding for this work was provided by the JPSS Program Office, NOAA/NESDIS/StAR, and the Naval Research Laboratory (Grant #N00173-14-G902), the Oceanographer of the Navy through office at the PEO C4I & Space/PMW-120 under program element PE-0603207N. The views, opinions, and findings contained in this article are those of the authors and should not be construed as an official National Oceanic and Atmospheric Administration (NOAA) or U.S. Government position, policy, or decision.

Author Contributions: Don Hillger and Tom Kopp, as VIIRS Imagery Team co-leads, wrote most of the text. The rest of the authors contributed the Imagery examples and the text to accompany those examples.

Conflicts of Interest: The authors declare no conflict of interest.

References

1.   Hillger, D.; Kopp, T.; Lee, T.; Lindsey, D.; Seaman, C.; Miller, S.; Solbrig, J.; Kidder, S.; Bachmeier, S.; Jasmin, T.; *et al.* First-light imagery from Suomi NPP VIIRS. *Bull. Am. Meteor. Soc.* **2013**. [CrossRef]

2.   Hillger, D.; Seaman, C.; Liang, C.; Miller, S.D.; Lindsey, D.; Kopp, T. Suomi NPP VIIRS imagery evaluation. *J. Geophys. Res.* **2014**, *119*, 6440–6455. [CrossRef]

3.   Kuciauskas, A.; Solbrig, J.; Lee, T.; Hawkins, J.; Miller, S.; Surratt, M.; Richardson, K.; Bankert, R.; Kent, J. Next-generation satellite meteorology technology unveiled. *Bull. Amer. Meteor. Soc.* **2013**, *94*, 1824–1825. [CrossRef]

4. Cao, C.; Shao, X.; Xiong, X.; Blonski, S.; Liu, Q.; Uprety, S.; Shao, X.; Bai, Y.; Weng, F. Suomi NPP VIIRS sensor data record verification, validation, and long-term performance monitoring. *J. Geophys. Res. Atmos.* **2013**, *118*, 11664–11678. [CrossRef]

5. Miller, S.; DStraka, W.; Mills, S.P.; Elvidge, C.D.; Lee, T.F.; Solbrig, J.; Walther, A.; Heidinger, A.K.; Weiss, S.C. Illuminating the capabilities of the Suomi NPP VIIRS Day/Night Band. *Remote Sens.* **2013**, *5*, 6717–6766. [CrossRef]

6. Liang, C.K.; Hauss, B.I.; Mills, S.; Miller, S.D. Improved VIIRS Day/Night band imagery with near constant contrast. *IEEE TGRS* **2014**, *52*, 6964–6971. [CrossRef]

7. Miller, S.D.; Mills, S.P.; Elvidge, C.D.; Lindsey, D.T.; Lee, T.F.; Hawkins, J.D. Suomi satellite brings to light a unique frontier of environmental imaging capabilities. *Proc. Nat. Acad. Sci. USA* **2012**, *109*, 15706–15711. [CrossRef] [PubMed]

8. Solbrig, J.E.; Lee, T.E.; Miller, S.D. Advances in remote sensing: Imaging the Earth by moonlight. *Eos* **2013**, *94*, 349–350. [CrossRef]

9. Seaman, C.; Hillger, D.; Kopp, T.; Williams, R.; Miller, S.; Lindsey, D. *Visible Infrared Imaging Radiometer Suite (VIIRS) Imagery Environmental Data Record (EDR) User's Guide*; NOAA Technical Report NESDIS: Washington, WA, USA, 2015.

10. Algorithm Theoretical Basis Documents (ATBD). Available online: http://www.star.nesdis.noaa.gov/jpss/Docs.php (accessed on 21 December 2015).

11. JPSS Level 1 Requirements Document Final. Available online: http://www.jpss.noaa.gov/pdf/L1RD_JPSS_REQ_1001_final_v1.8–1.pdf (accessed on 21 December 2015).

12. Miller, S. A satellite sensor that can see in the dark is revealing new information for meteorologists, firefighters, search teams and researchers worldwide. *Sci. Am.* **2015**, *312*, 78–81. [CrossRef] [PubMed]

13. Straka, W.J.; Seaman, C.J.; Baugh, K.; Cole, K.; Stevens, E.; Miller, S.D. Utilization of the Suomi National Polar-Orbiting Partnership (NPP) Visible Infrared Imaging Radiometer Suite (VIIRS) Day/Night band for arctic ship tracking and fisheries management. *Remote Sens.* **2015**, *7*, 971–989. [CrossRef]

14. Elvidge, C.D.; Zhizhin, M.; Hsu, F.-C.; Baugh, K.E. VIIRS nightfire: Satellite pyrometry at night. *Remote Sens.* **2013**, *5*, 4423–4449. [CrossRef]

15. JPSS Imagery CVP. *JPSS Cal/Val Plan for Imagery Product*; JPSS Program Office: Washington, WA, USA, 2015.

16. McIDAS-V. Available online: http://www.ssec.wisc.edu/mcidas/software/v/ (accessed on 21 December 2015).

17. SeaSpace Corporation. Available online: http://www.seaspace.com/software.php (accessed on 21 December 2015).

18. Interactive Data Language. Available online: http://exelisvis.com/ProductsServices/IDL.aspx (accessed on 21 December 2015).

19. Advanced Weather Interactive Processing System. Available online: http://www.nws.noaa.gov/ops2/ops24/awips.htm (accessed on 21 December 2015).

20. Community Satellite Processing Package. Available online: http://cimss.ssec.wisc.edu/cspp/ (accessed on 21 December 2015).

21. International Polar Orbiting Processing Package. Available online: https://directreadout.sci.gsfc.nasa.gov/?id=dspContent&cid=68 (accessed on 21 December 2015).

22. Ellrod, G.P. Advances in the detection and analysis of fog at night using GOES multispectral infrared imagery. *Wea. Forecast.* **1995**, *10*, 606–619. [CrossRef]

23. Fierstein, J.; Hildreth, W. The plinian eruptions of 1912 at Novarupta, Katmai National Park, Alaska. *Bull. Volcanol.* **1992**, *54*, 646–684. [CrossRef]

24. Casadevall, T. The 1989–1990 eruption of redoubt volcano, Alaska—Impacts on aircraft operations. *J. Volcanol. Geotherm. Res.* **1994**, *62*, 301–316. [CrossRef]

25. Horwell, C.J.; Baxter, P.J. The respiratory health hazards of volcanic ash: A review for volcanic risk mitigation. *Bull. Volcanol.* **2006**, *69*, 1–24. [CrossRef]

26. Ash from century-old Novarupta Volcanic Eruption Sweeps over Kodiak Island. Available online: http://www.adn.com/article/ash-century-old-novarupta-volcanic-eruption-sweeps-over-kodiak-island (accessed on 21 December 2015).

27. SIGMET. Available online: http://www.volcanodiscovery.com/archive/vaac/latest-reports-2012.html (accessed on 21 December 2015).

28. Alaska Wildfire Season Worst on Record So Far: NOAA Providing On-The-Scene Assistance. Available online: http://www.noaa.gov/features/03_protecting/070215-alaska-wildfire-season-worst-on-record-so-far.html (accessed on 21 December 2015).

29. USCG Rescues Man Trapped in Arctic Ice. Available online: http://navaltoday.com/2014/07/14/uscg-rescues-man-trapped-in-arctic-ice/ (accessed on 21 December 2015).

30. NRL Tropical Cyclone Webpage. Available online: http://www.nrlmry.navy.mil/TC.html (accessed on 21 December 2015).

31. Miller, S.D.; Turner, R.E. A dynamic lunar spectral irradiance dataset for NPOESS/VIIRS Day/Night band nighttime environmental applications. *IEEE Trans. Geosci. Remote Sens.* **2009**, *47*, 2316–2329. [CrossRef]

32. CPHC Statement. Available online: http://www.prh.noaa.gov/cphc/tcpages/archive/2013/TCDCP1.EP062013.019.1307291511 (accessed on 21 December 2015).

33. Akinbami, O.J.; Moorman, J.E.; Liu, X. *Asthma Prevalence, Health Care Use, and Mortality: United States, 2005–2009*; US Department of Health and Human Services, Centers for Disease Control and Prevention, National Center for Health Statistics: Centers for Disease Control and Premention: Washington, WA, USA, 2011.

34. Anstey, M.H. Climate change and health—What's the problem? *Global Health* **2013**, *9*. [CrossRef] [PubMed]

35. NexSat. Available online: http://www.nrlmry.navy.mil/NEXSAT.html (accessed on 21 December 2015).

36. Cadelis, G.; Molinie, J. Short-term effects of the particulate pollutants contained in Saharan Dust on the visits of children to the emergency department due to asthmatic conditions in Guadeloupe (French Archipelago of the Caribbean). *PloS ONE* **2014**, *6*. [CrossRef] [PubMed]

37. Miller, S.D. A consolidated technique for enhancing desert dust storms with MODIS. *Geophys. Res. Lett.* **2003**, *30*, 2071–2074. [CrossRef]

38. VIIRS Imagery and Visualization Team. Available online: http://rammb.cira.colostate.edu/projects/npp/ (accessed on 21 December 2015).

Use of SSU/MSU Satellite Observations to Validate Upper Atmospheric Temperature Trends in CMIP5 Simulations

Lilong Zhao [1,2], Jianjun Xu [2,*], Alfred M. Powell [3], Zhihong Jiang [1] and Donghai Wang [4]

Academic Editors: Xuepeng Zhao, Wenze Yang, Viju John, Hui Lu, Ken Knapp, Richard Gloaguen and Prasad S. Thenkabail

[1] Key Laboratory of Meteorological Disaster of Ministry of Education, Collaborative Innovation Center on Forecast and Evaluation of Meteorological Disasters, Nanjing University of Information Science and Technology, Najing 210044, China; zhaoyinuo8@gmail.com (L.Z.); zhjiang@nuist.edu.cn (Z.J.)

[2] Global Environment and Natural Resources Institute (GENRI), College of Science, George Mason University, Fairfax, WV 22030, USA

[3] NOAA/NESDIS/STAR, College Park, ML 20740, USA; al.powell@noaa.gov

[4] China State Key Laboratory of Severe Weather Chinese Academy of Meteorological Sciences, Beijing 100081, China; d.wang@hotmail.com

* Correspondence: jxu14@gmu.edu

Abstract: The tropospheric and stratospheric temperature trends and uncertainties in the fifth Coupled Model Intercomparison Project (CMIP5) model simulations in the period of 1979–2005 have been compared with satellite observations. The satellite data include those from the Stratospheric Sounding Units (SSU), *Microwave Sounding Units (MSU), and the Advanced Microwave Sounding Unit-A (AMSU)*. The results show that the CMIP5 model simulations reproduced the common stratospheric cooling (-0.46–-0.95 K/decade) and tropospheric warming (0.05–0.19 K/decade) features although a significant discrepancy was found among the individual models being selected. The changes of global mean temperature in CMIP5 simulations are highly consistent with the SSU measurements in the stratosphere, and the temporal correlation coefficients between observation and model simulations vary from 0.6–0.99 at the 99% confidence level. At the same time, the spread of temperature mean in CMIP5 simulations increased from stratosphere to troposphere. Multiple linear regression analysis indicates that the temperature variability in the stratosphere is dominated by radioactive gases, volcanic events and solar forcing. Generally, the high-top models show better agreement with observations than the low-top model, especially in the lower stratosphere. The CMIP5 simulations underestimated the stratospheric cooling in the tropics and overestimated the cooling over the Antarctic compared to the satellite observations. The largest spread of temperature trends in CMIP5 simulations is seen in both the Arctic and Antarctic areas, especially in the stratospheric Antarctic.

Keywords: climate change; SSU/MSU satellite observation; upper atmospheric temperature; CMIP5 simulation

1. Introduction

As an important aspect of climate change, the vertical structure of temperature trends from the troposphere to stratosphere has received a great deal of attention in the climate change research community [1–9]. The World Climate Research Program (WCRP) made an incredible effort to understand the variability in the stratosphere based on climate model simulations [10]. Compared to the third phase of the Coupled Model Intercomparison Project (CMIP3), many models of the fifth phase of the Coupled Model Intercomparison Project (CMIP5) increased the model top above

1 hPa to improve the representation of atmospheric change in the upper layers [10,11] in the coupled climate models. However, the reliability of these new simulations in the middle troposphere and stratosphere is still not completely clear [12]. Also, many of the climate models do not include all the physical and chemical processes necessary for simulating the stratospheric climate [11]. Santer and his co-authors compared CMIP5 model simulations with satellite observations to conduct attribution studies on atmospheric temperature trends [13,14], and they found that CMIP5 models underestimate the observed cooling of the lower stratosphere and overestimate the warming of the troposphere. As a consequence, evaluating the climate model results with integrated observational data sets is necessary to understand the model capabilities and limitations in representing long term climate change and short term variability.

The satellite observations from the Stratospheric Sounding Units (SSU), Microwave Sounding Units (MSU) and the Advanced Microwave Sounding Unit-A (AMSU-A) provide key assessments of climate change [15–19] and the performance of climate model simulations [7,14,20]. The National Oceanic and Atmospheric Administration (NOAA) Center for Satellite Applications and Research (STAR) developed both MSU/AMSU-A and SSU temperature time series [17–19,21] that can be used to validate climate model simulations. This is the only data available that can provide near-global temperature information over multidecadal periods from the middle troposphere up to the upper stratosphere (50 km).

In this study, an intercomparison of the temperature trends from the middle troposphere to the upper stratosphere between satellite observations and the CMIP5 simulations was conducted. The goal is to understand the uncertainties and deficiencies in estimating temperature trends in the CMIP5 simulations. Section 2 describes the data sets and methodologies. The temporal analysis of the global mean temperature and the spatial variation of the global temperature trend are presented in Sections 3 and 4 respectively. Lastly, in Sections 5 and 6 multiple linear regression analysis is performed to discuss the results, and conclusions are drawn.

2. Data and Methodology

In this study, the temperature trends from the middle troposphere to the upper stratosphere in the CMIP5 climate model simulations are assessed by the SSU and MSU temperature data records. All data sets spanned the period from 1979 through 2005.

2.1. SSU/MSU Data Sets

The NOAA/STAR SSU Version 2 dataset developed by Zou [16] is used in this study. The SSU is a three-channel infrared (IR) radiometer designed to measure temperatures in the middle to upper stratosphere (Figure 1) in which SSU1 (channel 1) peaks at 32 km, SSU2 (channel 2) peaks at 37 km, and SSU3 (channel 3) peaks at 45 km. The MSU/AMSU-A temperature dataset was created in STAR by Zou [15]. Three of the MSU/AMSU-A channels extend from the middle troposphere to the lower stratosphere (Figure 1): MSU2 (MSU channel 2 merged with AMSU-A channel 5) peaks at 6 km, MSU3 (MSU channel 3 merged with AMSU-A channel 7) peaks at 11 km, and MSU4 (MSU channel 4 merged with AMSU-A channel 9) peaks at 18 km. Similar to the previous study [3], the model temperatures on pressure levels are converted to the SSU/MSU layer-averaged brightness temperatures by applying weighting functions to the averaged vertical profile at each model grid point. The SSU/MSU weighting functions are normalized as shown in Figure 1. It is worth noting that a single weighting function over the globe shows some limitations because the weighting functions significantly depend on the latitudes, and the peak levels over the poles are different from the tropics. The best way is probably to use a fast radiative transfer model to transfer the pressure data to the SSU/MSU layers. However, based on previous studies [3,11,12], these limitations are not critical for the estimation of the temperature trends. In addition, one main problem with these satellite data is the discontinuities in the time series, due to that data from 13 different satellites have been used since 1979. Several correction have been made to compensate radiometric differences, tidal effects associated with orbit drifts [22,23], change of the vertical weighting functions due to atmospheric CO_2 changes [24], and long-term drift in the local time of measurements. So, errors associated with trend estimates are due to the uncertainties in the successive SSU adjustments and time continuity.

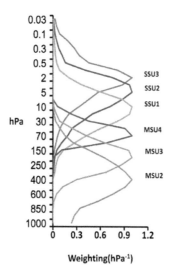

Figure 1. The weighting functions for the satellite Microwave Sounding Unit (MSU) and the Stratospheric Sounding Unit (SSU).

2.2. CMIP5 Simulations

We use the IPCC models at the Program for Climate Model Diagnosis and Intercomparison (PCMDI) [8]. The datasets used for this study are the historical run of 35 available models with 11 high-top models (model tops above 1 hPa), enabling the comparisons with the highest altitude SSU data for the period of 1979–2005. Table 1 lists information for all 35 models used in this study. Further details, together with access information, can be obtained at the following website [25]. The "historical" (HIS) run (1850–2005) is forced by past atmospheric composition changes (reflecting both anthropogenic and natural sources) including time evolving land cover. In addition, three types of experiments including pre-industrial control runs (PI), Greenhouse gases (GHG) only runs and natural forcing only (NAT) runs have been analyzed.

Table 1. The CMIP5/IPCC Data Sets and Selected Information.

IPCC I.D.	Pressure Level (hPa)	Center and Location	Forcing
CanESM2 CanCM4	1–1000	Canadian Centre for Climate Modelling and Analysis	GHG, SA, Oz, BC, OC, LU, Sl, Vl
HadGEM2-CC HadGEM2-AO HadGEM2-ES HadCM3	0.4–1000 10–1000	Met Office Hadley Centre	GHG, SA, Oz, BC, OC, LU, Sl, Vl
CESM1-WACCM CESM1-BGC CESM1-CAM5 CESM1-FASTCHEM CCSM4	0.4–1000 10–1000	NSF/DOE NCAR (National Center for Atmospheric Research) Boulder, CO, USA	Sl, GHG, Vl, SS, Ds, SD, BC, MD, OC, Oz, AA ,LU
MIROC-ESM-CHEM	0.03–1000	Japan Agency for Marine-Earth Science and Technology	GHG, SA, Oz, BC, OC, LU, Sl, Vl, MD
MIROC5	10–1000	Atmosphere and Ocean Research Institute, The University of Tokyo, Chiba, Japan National Institute for Environmental Studies, Ibaraki, Japan Japan Agency for Marine-Earth Science and Technology, Kanagawa, Japan	GHG, SA, Oz, LU, Sl, Vl, SS, Ds, BC, MD, OC

Table 1. *Cont.*

IPCC I.D.	Pressure Level (hPa)	Center and Location	Forcing
CMCC-CESM	0.01–1000	CMCC-Centro Euro-Mediterraneo peri Cambiamenti Climatici, Bologna, Italy	Nat, Ant, GHG, SA, Oz, Sl
MPI-ESM-LR MPI ESM-MR MPI-ESM-P	10–1000	Max Planck Institute for Meteorology	GHG, SD, Oz, LU, Sl, Vl
MRI-CGCM3 MRI-ESM1	0.4–1000	MRI (Meteorological Research Institute, Tsukuba, Japan)	GHG, SA, Oz, BC, OC, LU, Sl, Vl,
ACCESS1-0 ACCESS1-3	10–1000	Commonwealth Scientific and Industrial Research Organisation, Bureau of Meteorology, Australia	GHG, Oz, SA, Sl, Vl, BC, OC
BNU-ESM	10–1000	GCESS, BNU, Beijing, China	Nat, Ant
GISS-E2-H-CC GISS-E2-HGISS-E2-R-CC GISS-E2-R	10–1000	NASA Goddard Institute for Space Studies	GHG, LU, Sl, Vl, BC, OC, SA, Oz
IPSL-CM5A-LR IPSL-CM5B-LR IPSL-CM5B-MR	10–1000	Institut Pierre Simon Laplace, Paris, France	Nat, Ant, GHG, SA, Oz, LU, SS, Ds, BC, MD, OC, AA
NorESM1-M NorESM1-ME	10–1000	Norwegian Climate Centre	GHG, SA, Oz, Sl, Vl, BC, OC
inmcm4	10–1000	Institute for Numerical Mathematics, Moscow, Russia	N/A
CSIRO-Mk3-6-0	5–1000	Australian Commonwealth Scientific and Industrial Research Organization, Marine and Atmospheric Research, Queensland Climate Change Centre of Excellence	Ant, Nat (all forcings)
CNRM-CM5	10–1000	Centre National de Recherches Meteorologiques Centre Europeen de Recherches et de Formation Avancee en Calcul Scientifique	GHG, SA, Sl, Vl, BC, OC
FGOALS-g2	10–1000	Institute of Atmospheric Physics, Chinese Academy of Sciences, Tsinghua University	GHG, Oz, SA, BC, Ds, OC, SS, Sl, Vl

Notes: BC (black carbon); Ds (Dust); GHG (well-mixed greenhouse gases); LU (land-use change); MD (mineral dust); OC (organic carbon); Oz (tropospheric and stratospheric ozone); SA (anthropogenic sulfate aerosol direct and indirect effects); SD (anthropogenic sulfate aerosol, accounting only for direct effects); SI (anthropogenic sulfate aerosol, accounting only for indirect effects); Sl (solar irradiance); SO (stratospheric ozone); SS (sea salt); TO (tropospheric ozone); Vl (volcanic aerosol); Ant (anthropogenic forcing).

2.3. Methodology

To facilitate the inter-comparison study, all data are first interpolated to the same horizontal resolution of 5-degrees in longitude and latitude, then the temperature of the pressure levels in CMIP5 are converted to the equivalent brightness temperatures of the six SSU/MSU layers (SSU3, SSU2, SSU1, MSU4, MSU3, MSU2) based on the vertical weighting function of the SSU/MSU measurements in Figure 1 (the weighting functions were from the NOAA/STAR website). It is clear that only the 11 high-top models can be compared to the highest layers (SSU3, SSU2) of satellite data. Similar to the processing in the previous study [4], the six SSU/MSU channels represent the temperature for broad layer from the middle tropospheric MSU2 (peak at 6 km) to the upper stratospheric SSU3 (peak at 45 km).

Taylor's [26] diagram and ratio of signal to noise are used to evaluate the performance of the models. The fitting of linear least squares is used to estimate the temperature trend. The model trend uncertainty is measured by the ensemble spread, which is defined by the standard deviation among CMIP5 climate model simulations. Multiple linear regression was performed for the analysis of model performance in stratosphere and troposphere.

3. Temporal Characteristics of Global Mean Temperature

3.1. Global Mean Temperature

Figure 2 shows the time series of global mean temperature anomalies of the satellite observations (thick blue line) and CMIP5 simulations at the SSU/MSU six layers. The SSU channel (Figure 2a–c) indicates cooling temperature trend at a rate of approximately −0.85 K/decade in the upper stratosphere (SSU3). MSU channel 4 (MSU4) shows −0.38 K/decade in the lower stratosphere (MSU4). It is clear that all three SSU and MSU4 channels demonstrate strong anomalies during the 1982–1983 and 1991–1992 periods, which is attributed to the volcanic eruptions of El Chichón (1982) and Mt. Pinatubo (1991).

In comparison, the troposphere (Figure 2e,f) shows a weak warming at a rate of +0.07 K/decade in MSU3, and +0.14 K/decade in the middle troposphere (MSU2).

For the CMIP5 simulations, all the models are able to capture the observed temperature variability in the upper and low stratosphere (Figure 2a–d) except some models demonstrate a larger, short-lived warming than observations due to the Mount Pinatubo volcano in 1991–1992. The overestimation of the temperature response is mainly due to the fact that the observed decreases in ozone concentrations following the major volcanic eruptions were not included in the forcing data set, which is why most models lead to overestimation of the stratospheric temperature response to volcanic eruptions, especially Pinatubo [11].

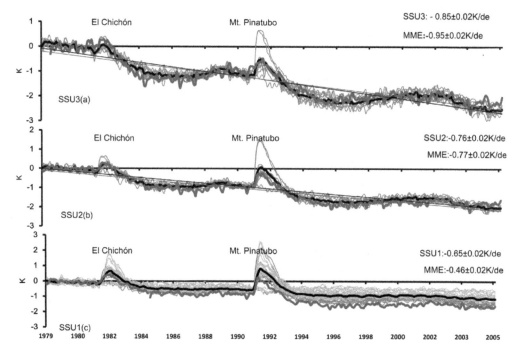

Figure 2. *Cont.*

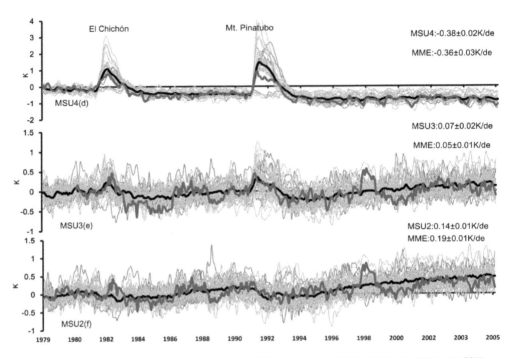

Figure 2. Global temperature anomalies time series (K) in the period of 1979–2005 at (**a**) SSU3, (**b**) SSU2; (**c**) SSU1. Note SSU1~SSU3 represent the SSU observational layers. Blue thick line represents the STAR observation. Gray and orange lines indicate the low-top and high-top CMIP5 model, respectively, Black thick line represents model ensemble mean. Figure 2b is the same as Figure 2a except for the MSU observation (**d**) MSU4; (**e**) MSU3; (**f**) MSU2. Note MSU2-MSU4 represents the MSU observational layers. MME is multiple model ensemble mean.

The model ensemble mean (MME: black thick line) is very similar to observations in the stratosphere. The major differences between the model ensemble mean and the observation is that the model ensemble mean overestimates the cooling in SSU3 channels and underestimates that in SSU1 where the difference reaches 0.19 K/decade. It should be noted that some models, IPSL-CM5A-LR, IPSL-CM5A-MR, IPSL-CM5B-LR, CMCC-CESM and INMCM4 even cannot reproduce strong volcanic eruptions anomalies in the stratosphere during the 1982–1983 and 1991–1992 periods, because they did not include volcanic aerosols.

In contrast, in the comparison of stratosphere, the CMIP5 models show an obvious discrepancy from the MSU observations (Figure 2e,f) where the multi-model ensemble mean cannot produce the temperature variability. Some models even show an opposite phase of variability to the MSU observations, but there is no large difference in the global mean temperature trend between models and observations.

3.2. Consistencies between Simulations and Observations

The evaluation of the global mean temperatures in CMIP5 simulations in comparison with the SSU/MSU observations is accomplished through the Taylor-diagram. The Taylor diagram is a convenient way of evaluating different model's performance with observation using three related parameters: correlation with observed data, centered root-mean-square (RMS), and standard deviation. Models with as much variance as the observations' largest correlation and with the least RMS error are considered best performers in the Taylor diagram. In the stratosphere (Figure 3), the correlation coefficient between SSU/MSU and CMIP5 climate models have a large value ranging from 0.60 in the lower stratosphere to 0.95 in the middle to upper stratosphere. This reflects the strong consistency in the global mean stratospheric temperature between the CMIP5 climate model simulations and observation. The results showed that the 11 high-top models (triangle symbol) have higher correlation coefficients and a more centralized distribution than the low-top models in the low stratosphere. It is worth noting that the high-top model CMCC-CESM does not include volcanic aerosol in the stratosphere during the 1982–1983 and 1991–1992 periods, but the high correlation coefficient in SSU3, SSU2, SSU1 and MSU4 indicated the better agreement with observation in the stratosphere, which is totally different from the poor simulations in the low top model discovered in many previous studies [27–29]. It is obvious that the model lid height plays a very crucial role in the simulation of stratospheric temperature variability.

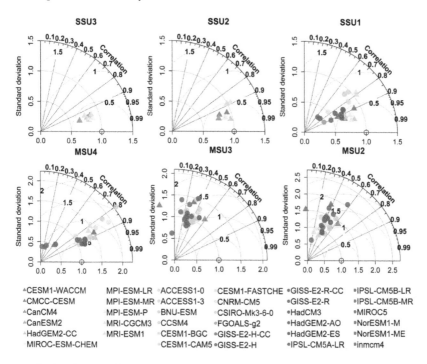

Figure 3. Taylor diagram for time series of observed and simulated global mean temperature. Normalized Standard deviation, Correlation coefficient, and The Root Mean Squared Deviation (RMSD) are presented in one diagram. An ideal model has a standard deviation ratio of 1.0 and a correlation coefficient of 1.0. Triangle symbols represent high-top models (correlations greater than 0.14 are statistically significant at the 99% confidence level).

In contrast, the correlations of the CMIP5 model simulations with the MSU observations in the troposphere are sharply reduced. This demonstrates a lower correlation with the MSU in the CMIP5 simulations, which indicates that the CMIP5 simulations in the troposphere are in worse agreement than those in the stratosphere. The model CMCC-CESM shows a negative correlation with the MSU observations in the two MSU channels, and CanCM4 model simulations in the troposphere show much better agreement in the troposphere than their counterpart climate models. Additionally, models have a more centralized distribution in the stratosphere than troposphere implying that most of the models have better agreement in the stratosphere.

Results from the Taylor diagram suggest that models show better agreement with observation and smaller intermodel discrepancy in the stratosphere than in the troposphere, especially for the high-top models. The lower correlation in the troposphere is generally because the temperature variations in the troposphere demonstrate mostly unforced internal variability, and free-running coupled-ocean-atmosphere models will never capture the timing of that variability. Further research is therefore needed to understand the forcing of internal variability and the principle of the coupled ocean–atmosphere system to improve the performance of the CMIP5 models in the troposphere.

3.3. Uncertainty Analysis in Model Simulations

To quantitatively reveal the spread and convergence of the climate models in reproducing the Stratospheric and Tropospheric Temperature, the signal-to-noise ratio is evaluated using the method of Zhou [30].

Ensemble mean is given by the ensemble average of all model simulations, that is

$$x_e(t) = \frac{1}{N} \sum_{n=1}^{N} x(n,t) \tag{1}$$

where N is the total number of models, x is the global mean time series, $x(n, t)$ represents the simulation of nth model at year t of a time length T. Climatological mean is defined by σ_e.

$$x_c = \frac{1}{NT} \sum_{n=1}^{N} \sum_{t=1}^{T} x(n,t) = \frac{1}{T} \sum_{t=1}^{T} x_e(t) \tag{2}$$

The standard deviation of the $x_e(t)$ is used to measure modelled temperature trends in response to external signal. The dispersion of the simulation (measured by σ_i the standard deviation of $x(n, t)$) indicates intermodel variability, which is noise for the climate reproduction. From the definition of the two averages, we have

$$\sigma_e^2 = \frac{1}{T} \sum_{t=1}^{T} [x_e(t) - x_c]^2 \tag{3}$$

$$\sigma_i^2 = \frac{1}{T} \sum_{t=1}^{T} \left\{ \frac{1}{N} \sum_{n=1}^{N} [x(n,t) - x_e(t)]^2 \right\} \tag{4}$$

The signal to noise ratio (S/N) is defined as S/N = $\lambda = \frac{\sigma_e^2}{\sigma_i^2}$.

Here we compute that S/N ratio for the period of 1979–2005 from stratosphere to troposphere. Analysis of S/N (Table 2) indicates that the stratosphere stands out as having much higher S/N (3.53–57.98) than the troposphere (0.48–1.36), where the forced signal is much larger than intermodal noise, especially for SSU3. The S/N ratio reduced sharply from SSU2 (28.31) to SSU1 (5.26), which is partially due to the inclusion of low-top models increasing intermodal noise.

Table 2. Signal-to-noise ratio (S/N) for stratospheric and tropospheric air temperature.

Layer	SSU3	SSU2	SSU1	MSU4	MSU3	MSU2
S/N	57.98	28.31	5.26	3.53	0.48	1.36

3.4. Trend Changes with Vertical Level

The vertical profile of the global mean temperature trends shows (Figure 4) that the CMIP5 model's (Figure 4a) temperature trend cooling rates in the upper stratosphere are less than the SSU observations, especially in the SSU1 channel. Also, the low-top models further underestimate the cooling trend than the high-top models in the stratosphere.

The crossover points identify a transition from tropospheric warming to stratospheric cooling. It is obvious that the most of the low-top models are higher than the corresponding crossover point from the MSU observations and high-top models. On the other hand, the ensemble spread among the CMIP5 model simulations (Figure 4b) is generally between 0.04 and 0.1 K/decade from the middle troposphere to the upper stratosphere, with the maximum spread appearing at the point of the SSU1 level, which is mainly due to the low-top models.

The above results clearly show that high-top models have better consistency with the SSU/MSU observations than low-top models in global mean temperature variability.

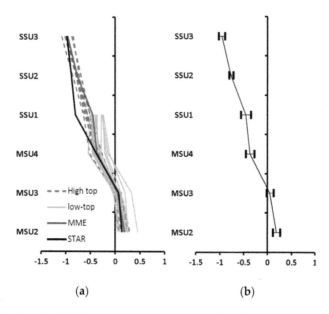

Figure 4. Vertical profile of global mean temperature trends and spread among the data sets for CMIP5 model simulations. Note SSU1-3 and MSU2-4 represent the SSU/MSU observational layers (unit K/decade). (a) Vertical profile temperature trends; (b) spread among models.

4. Spatial Pattern of Temperature Trends

4.1. Trend Changes with Latitude

To better understand the spatial pattern of the CMIP5 model simulations, the latitude profiles of the temperature trend have been plotted to facilitate comparing the differences between the datasets. The results indicate (Figure 5) that the linear trends are highly sensitive to the latitude of interest. All exhibit predominant cooling in the stratosphere and warming in the troposphere except for the southern high latitudes. Also, there is an extremely strong cooling trend in the three SSU channel observations over the tropics and arctic.

For the upper stratosphere (Figure 5a–c), a distinguishing difference in the CMIP5 simulations from SSU observations is found over the tropics, where the cooling rates get up to −1.2 K/decade in the SSU2 layer (Figure 5b). This is approximately −0.5 K/decade lower than the value in the CMIP5 models. In addition, the cooling trend shows a sharp gradient from high to low latitude in the SSU observations.

For the layer from the middle troposphere to the lower stratosphere (Figure 5d–f), the MSU data shows a consistent trend with the CMIP5 models, except some models show more cooling than that observed at MSU4 over Antarctica. An important fact worth noting is that cooling was found in the troposphere over Antarctica, with the maximum cooling trend being approximately −1.8 K/decade in the upper troposphere. At the same time, warming trends have been observed over the tropics and the whole Northern hemisphere. This layer also displayed a substantial temperature difference between the Antarctic and the rest of the areas.

It is obvious that the cooling trends of the stratospheric temperature change markedly with latitude and the largest trend is found in the tropical and arctic latitudes. In contrast, the warming trend increases with latitude from south to north in the troposphere, but the spread retains a small value except for both polar areas. Conversely, the south to north latitudinal cooling trend in the stratosphere decreased. To a first order (linear), the tropospheric warming by latitude is offset by stronger latitudinal cooling in the stratosphere indicating the atmosphere is adjusting to surface and tropospheric heating to maintain radiative balance.

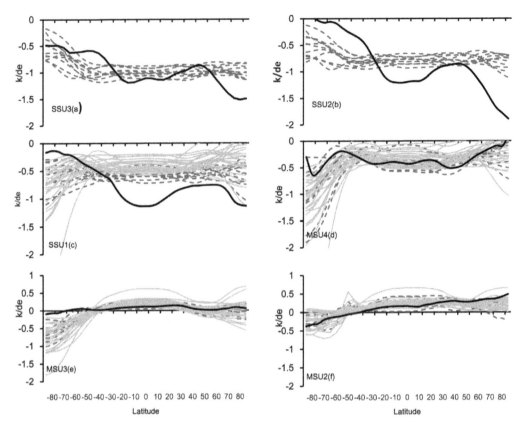

Figure 5. Temperature trend change with latitude for the layers (**a**) SSU3; (**b**) SSU2; (**c**) SSU1; (**d**) MSU4; (**e**) MSU3; (**f**) MSU2. Note SSU1~3 and MSU2~4 represent the SSU/MSU observational layers. Heavy black line indicates the SSU/MSU observation, gray and red thin lines represent the low-top and high top CMIP5 model, respectively.

4.2. Distribution of Longitude-Latitude Trend Spread

Figure 6 displays the latitudinal-longitudinal variation of global temperature trend spreads for 1979–2005 for the layers from the middle troposphere (6 km height) to the upper stratosphere (45 km height). The results indicate (Figure 6) that the spreads are highly sensitive to the latitude of interest, from 0.05 K/decade in tropical and subtropical areas to 0.5 K/decade in the southern polar region. The largest layer spread locations change with vertical level; the largest spread in the lower stratosphere (SSU1, MSU4) exceeds 0.5 K/decade over Antarctic areas, which is mainly due to some models significantly overestimating the cooling.

In contrast, there are some remarkable discrepancies in the tropical regions in the MSU4 layer, especially in the central Pacific region. It is worth noting that the spread of the CMIP5 simulations in the middle troposphere (MSU2) remains a relatively small value at all latitudes. The smaller spread in the MSU2 reflects the high consistency at all latitudes in CMIP5 simulations.

To summarize, the cooling trends of the stratospheric temperature markedly change with latitude, and the largest trend is found in the tropics-subtropics but the largest spread is found in the South Polar Region. In contrast, the warming trend increases with latitude from south to north in the troposphere, but the spread retains a small value except for both polar areas. Conversely, the cooling trend decreases with latitude from south to north in the stratosphere consistent with latitudinal radiative balance.

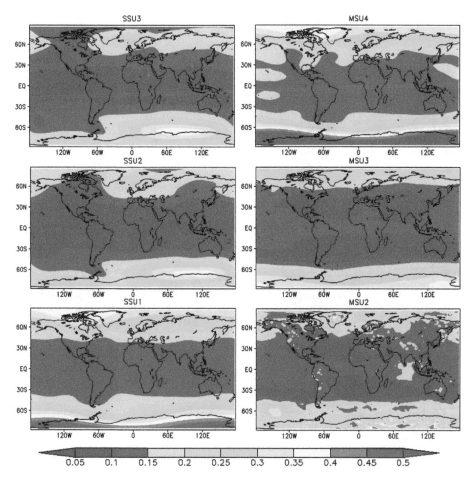

Figure 6. Spatial distribution of spread of temperature trends in CMIP5 model simulations. Note SSU1-3 and MSU2-4 represent the SSU/MSU observational layers (unit K/de).

5. Discussion

According to the above analysis, there is one point worth noting. All selected CMIP5 models showed a much higher correlation with SSU/MSU observations and higher intermodal consistency in the stratosphere compared to the global mean temperature in the troposphere.

In order to understand the possible reasons for the difference between the stratosphere and troposphere, the model's response to different forcings and their internal variability are investigated. In particular, three types of simulations are analyzed: pre-industrial control runs (piControl: PI), Greenhouse gases (GHG) only runs and natural forcing only (historicalNat: NAT) runs. Only two out of the 11 high-top models MIROC-ESM-CHEM and the MRI-CGCM3 included all three types of simulations and thus are analyzed here.

The time series of global mean temperature anomalies for SSU3 and MSU2 observational layers in GHG-only experiment, natural-only forcing run and preindustrial control run are shown in Figure 7. In SSU3, there is no significant trend in the natural-only forcing run and preindustrial control run, stratospheric cooling can only be reproduced in the model with anthropogenic forcing. Both models underestimated the cooling in the GHG-only experiment (0.91 ± 0.02 K/de) compared to all-forcing historical simulation (MRI-CGCM3 0.98 ± 0.02 K/de and MIROC-ESM-CHEM 0.96 ± 0.02 K/de) (ozone forcing is not included due to the dataset being unavailable). Solar variability and volcanic eruptions can be easily found in the natural-only forcing run. Similar results can be obtained in other stratospheric observation layers (Figure 8).

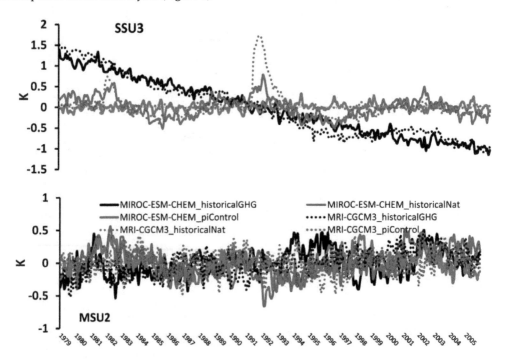

Figure 7. Global mean temperature anomalies for SSU3 and MSU2 observational layers in GHG-only historical experiment, Natural-only forcing run and preindustrial control run.

In comparison, the warming trend only can be detected from the GHG-only experiment, but the difference in the trend (MRI-CGCM3 0.15 ± 0.02 K/de and MIROC-ESM-CHEM 0.09 ± 0.03 K/de)is bigger compared to simulation results from SSU3, and also no significant trend in natural-only forcing run and preindustrial control run was observed.

To quantitatively link the different model performances in the stratosphere and troposphere with the internal variability of the climate model and the model response to a given external forcing, a multiple linear regression analyses [31] were performed for the all forcing historical runs using the GHG-only experiment, natural-only forcing run and preindustrial control run as regressors. Multiple linear regression attempts were made to model the relationship between two or more explanatory variables and a response variable by fitting a linear equation.

$$Y = b_0 + b_1 X_1 + b_2 X_2 + \cdots + b_p X_p \tag{5}$$

Y is the dependent variable, (x1, x2 ... xp) is a set of p explanatory variables. b_0 is the constant term and b_1 to b_p are the coefficients relating the p explanatory variables. A regression coefficient which is significantly greater than zero indicates a detectable response to the forcing concerned. Here, we assume that model output may be written as a linear sum of the simulated responses to individual forcing (GHG, NAT) and internal variability, each scaled by a regression coefficient plus residual variability.

The regression coefficients are shown in Figure 9. The multiple linear regression results with GHG, NAT and PI can accurately reproduce the amplitude and phase of historical runs in the stratosphere (Figure 10). In the troposphere, the multiple linear regression results have their own amplitude and phase variability which does not match historical runs in any detail.

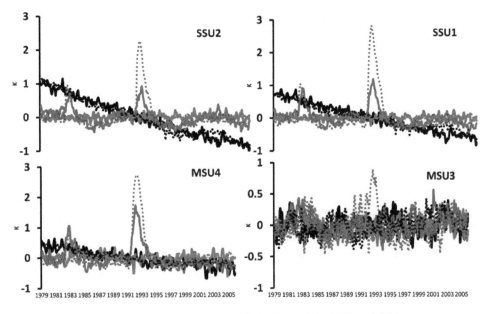

Figure 8. Same as Figure 7 except for MSU4, MSU3, SSU2, and SSU1.

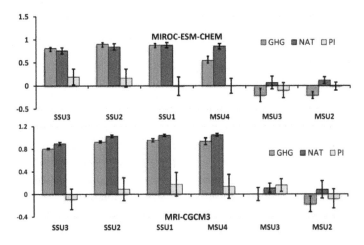

Figure 9. Regression coefficients derived from the regression of historical global mean temperatures' time series with greenhouse gases forced run (GHG) natural forced run (NAT) and pre-industrial control run (PI) (error bar: 95% confidence intervals for the coefficient estimates).

The results suggest that the stratospheric temperature is dominated by external forcing and response [32]. This is different to the troposphere where the temperature variability is driven by both internal variability and external forcing, and the model response to external forcing is nonlinear.

That is why all selected CMIP5 models showed a much higher correlation with SSU/MSU observations and higher intermodal consistency in the stratosphere compared to the global mean temperature in the troposphere.

According to the above analysis, a final important point to emphasize is that all selected CMIP5 model simulations showed high correlation with SSU/MSU observations in the stratosphere compared to the global mean temperature in the troposphere, but these models failed to reproduce the latitude-longitude pattern of the temperature trends. On the other hand, most of the CMIP5 models do not have some of the necessary physical processes. For example, most of the selected CMIP5 models do not include a chemistry model in the stratosphere; only the MIROC-ESM-CHEM and CESM1-WACCM includes chemistry and the chemistry model was recognized as a very important component to reproduce the true atmosphere [33,34].

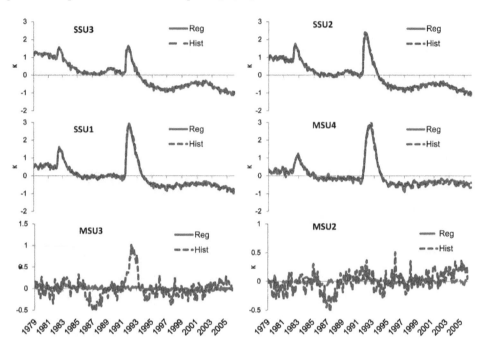

Figure 10. Global mean temperature anomalies simulated by MRI-CGCM3 at SSU1-SSU3 and MSU2-4. Red dash line represents the historical run (Hist). Blue line indicates the multiple linear regression results (Reg) with greenhouse gases forced run (GHG) natural forced run (NAT) and pre-industrial control run (PI).

6. Conclusions

Based on the satellite SSU and *MSU* temperature observations from 1979 through 2005, the trends and uncertainties in CMIP5 model simulations from the middle troposphere to the upper stratosphere (5–50 km) have been examined. The results are summarized as follows:

The CMIP5 model simulations reproduced a common feature with cooling in the stratosphere and warming in the troposphere, but the trend exhibits a remarkable discrepancy among the selected models. The cooling rate is higher than the SSU3 measurements at the upper stratosphere and less than SSU at the lower stratosphere.

Regarding the temporal variation of the global mean temperature, the CMIP5 model simulations significantly reproduced the volcanic signal and were highly correlated with the SSU measurements in the upper stratosphere during the study period. However, these models have lower temporal correlation with observations in the middle-upper troposphere.

Regarding the regional variation of the global temperature trends, the CMIP5 simulations displayed a different latitudinal pattern compared to the SSU/MSU measurements in all six layers from the middle troposphere to the upper stratosphere.

Generally, the high-top models show better agreement with observations than the low-top model, especially in the lower stratosphere. The temperature trends and spread show marked changes with latitude; the greatest cooling is found in the tropics in the upper stratosphere and the greatest warming appears in the Arctic in the middle troposphere. The CMIP5 simulations underestimated the stratospheric cooling in the tropics compared to the SSU observations and remarkably overestimated the cooling in the Antarctic from the upper troposphere to the lower stratosphere (MSU3-SSU3). The largest trend spread among the CMIP5 simulations is seen in both the Arctic and Antarctic in the stratosphere and troposphere, and the CMIP5 simulations retain similar spread values in the tropics in both the troposphere and stratosphere.

Acknowledgments: The SSU and MSU data sets were obtained from the Center for Satellite Applications and Research (STAR).The first author was partially supported by a key program of the National Natural Science Foundation of China (Grant 41230528, 41305039, 91537213, 41375047) and Postdoctoral Science Foundation of Jiangsu province (1402166C). The authors would like to thank these agencies for providing the data and funding support. Thanks to the four anonymous reviewers giving the good suggestions to improve the presentation of manuscripts. This work was supported by the National Oceanic and Atmospheric Administration (NOAA), Center for Satellite Applications and Research (STAR). The views, opinions, and findings contained in this publication are those of the authors and should not be considered an official NOAA or U.S. Government position, policy, or decision.

Author Contributions: The concept and design of framework for Use of SSU/MSU Satellite Observations to Validate Upper Atmospheric Temperature Trends in CMIP5 Simulations were proposed by Jianjun Xu. Lilong Zhao performed the experiments and analyzed the data; Lilong Zhao and Jianjun Xu wrote the paper, Alfred M. Powell, Zhihong Jiang and Donghai Wang reviewed and edited the manuscript. All authors are in agreement with the submitted and accepted versions of the publication.

Conflicts of Interest: The authors declare no conflict of interest.

References

1. Eichelberger, S.J.; Hartmann, D.L. Changes in the strength of the brewer–dobson circulation in a simple AGCM. *Geophys. Res. Lett.* **2005**, *32*, L15807. [CrossRef]

2. Cordero, E.C.; Forster, P.M. Stratospheric variability and trends in models used for the IPCC AR4. *Atmos. Chem. Phys.* **2006**, *6*, 5369–5380. [CrossRef]

3. Seidel, D.J.; Angell, J.K.; Christy, J.; Free, M.; Klein, S.A.; Lanzante, J.R.; Mears, C.; Parker, D.; Schabel, M.; Spencer, R.; *et al.* Uncertainty in signals of large-scale climate variations in radiosonde and satellite upper-air temperature datasets. *J. Clim.* **2004**, *17*, 2225–2240. [CrossRef]

4. Seidel, D.J.; Gillett, N.P.; Lanzante, J.R.; Shine, K.P.; Thorne, P.W. Stratospheric temperature trends: Our evolving understanding. *Wiley Interdiscip. Rev. Clim. Chang.* **2011**, *2*, 592–616. [CrossRef]

5. Young, P.J.; Thompson, D.W.J.; Rosenlof, K.H.; Solomon, S.; Lamarque, J. The seasonal cycle and interannual variability in stratospheric temperatures and links to the brewer–dobson circulation: An analysis of MSU and SSU Data. *J. Clim.* **2011**, *24*, 6243–6258. [CrossRef]

6. Xu, J.; Powell, A. Ensemble spread and its implication for the evaluation of temperature trends from multiple radiosondes and reanalyses products. *Geophys. Res. Lett.* **2010**, *37*, L17704. [CrossRef]

7. Xu, J.; Powell, A. Uncertainty estimation of the global temperature trends for multiple radiosondes, reanalyses, and CMIP3/IPCC climate model simulations. *Theor. Appl. Climatol.* **2012**, *108*, 505–518. [CrossRef]

8. Jiang, Z.; Li, W.; Xu, J.; Li, L. Extreme precipitation indices over China in CMIP5 models. Part I: Model evaluation. *J. Clim.* **2015**, *28*, 8603–8619. [CrossRef]

9. Cai, J.; Xu, J.; Powell, A.; Guan, Z.; Li, L. Intercomparison of the temperature contrast between the arctic and equator in the pre- and post periods of the 1976/1977 regime shift. *Theor. Appl. Climatol.* **2015**. [CrossRef]

10. Taylor, K.E.; Stouffer, R.J.; Meehl, G.A. An overview of CMIP5 and the experiment design. *Bull. Am. Meteor. Soc.* **2012**, *93*, 485–498. [CrossRef]

11. Charlton-Perez, A.J.; Baldwin, M.P.; Birner, T. On the lack of stratospheric dynamical variability in low-top versions of the CMIP5 models. *J. Geophys. Res. Atmos.* **2013**, *118*, 2494–2505. [CrossRef]

12. Thompson, D.W.J.; Seidel, D.J.; Randel, W.J.; Zou, C.Z.; Butler, A.H.; Mears, C.; Osso, A.; Long, C.; Lin, R. The mystery of recent stratospheric temperature trends. *Nature* **2012**, *491*, 692–697. [CrossRef] [PubMed]

13. Santer, B.D.; Painter, J.F.; Bonfils, C.; Mears, C.A. Identifying human influences on atmospheric temperature. *Proc. Natl. Acad. Sci. USA* **2012**, *110*, 26–33. [CrossRef] [PubMed]

14. Santer, B.D.; Painter, J.F.; Bonfils, C.; Mears, C.A. Human and natural influences on the changing thermal structure of the atmosphere. *Proc. Natl Acad. Sci. USA* **2013**, *110*, 17235–17240. [CrossRef] [PubMed]

15. Christy, J.R.; Spencer, R.W.; Braswell, W.D. MSU tropospheric temperatures: Dataset construction and radiosonde comparisons. *J. Atmos. Oceanic Technol.* **2000**, *17*, 1153–1170. [CrossRef]

16. Mears, C.A.; Schabel, M.C.; Wentz, F.J. A reanalysis of the MSU channel 2 tropospheric temperature record. *J. Clim.* **2003**, *16*, 3650–3664. [CrossRef]

17. Zou, C.; Goldberg, M.; Cheng, Z.; Grody, N.; Sullivan, J.; Cao, C.; Tarpley, D. Recalibration of microwave sounding unit for climate studies using simultaneous nadir overpasses. *J. Geophys. Res.* **2006**, *111*, D19114. [CrossRef]

18. Zou, C.Z.; Qian, H.F.; Wang, W.H. Recalibration and merging of SSU observations for stratospheric temperature trend studies. *J. Geophys. Res.: Atmos.* **2014**, *119*, JD021603. [CrossRef]

19. Wang, W.H.; Zou, C.Z. AMSU-A-Only atmospheric temperature data records from the lower troposphere to the top of the stratosphere. *J. Atmos. Oceanic Technol.* **2014**, *31*, 808–825. [CrossRef]

20. Santer, B.D.; Hnilo, J.J.; Wigley, T.M.L.; Boyle, J.S.; Doutriaux, C.; Fiorino, M.; Parker, D.E.; Taylor, K.E. Uncertainties in observationally based estimates of temperature change in the free atmosphere. *J. Geophys. Res.* **1999**, *104*, 6305–6333. [CrossRef]

21. Wang, L.; Zou, C.Z.; Qian, H. Construction of stratospheric temperature data records from stratospheric sounding units. *J. Clim.* **2012**, *25*, 2931–2946. [CrossRef]

22. Keckhut, P.; Randel, W.J.; Claud, C.; Leblanc, T.; Steinbrecht, W.; Funatsu, B.M. Tidal effects on stratospheric temperature series derived from successive advanced microwave sounding units. *Q. J. R. Meteorol. Soc.* **2014**, *141*, 477–483. [CrossRef] [PubMed]

23. Nash, J.; Forrester, G.F. Long-term monitoring of stratospheric temperature trends using radiance measurements obtained by the TIROS-N series of NOAA spacecraft. *Adv. Sp. Res.* **1986**, *6*, 37–44. [CrossRef]

24. Shine, K.P.; Barnett, J.J.; Randel, W.J. Temperature trends derived from stratospheric sounding unit radiances: The effect of increasing CO_2 on the weighting function. *Geophys. Res. Lett.* **2008**, *35*, L02710. [CrossRef]

25. Detail Information about CMIP5 Model. Available online: http: //cmip -pcmdi. llnl. gov/cmip5 (accessed on 21 December 2015).

26. Taylor, K.E. Summarizing multiple aspects of model performance in a single diagram. *J. Geophys. Res.* **2001**, *106*, 7183–7192. [CrossRef]

27. Cagnazzo, C.; Manzini, E. Impact of the Stratosphere on the winter tropospheric teleconnections between ENSO and the North-Atlantic European region. *J. Clim.* **2009**, *22*, 1223–1238. [CrossRef]

28. Hardiman, S.C.; Butchart, N.; Hinton, T.J.; Osprey, S.M.; Gray, L.J. The effect of a well-resolved stratosphere on surface climate: Differences between CMIP5 simulations with high and low top versions of the met office climate model. *J. Clim.* **2012**, *25*, 7083–7099. [CrossRef]

29. Shaw, T.A.; Perlwitz, J. The impact of stratospheric model configuration on planetary scale waves in northern hemisphere winter. *J. Clim.* **2010**, *22*, 3369–3389. [CrossRef]

30. Zhou, T.; Yu, R. Twentieth-century surface air temperature over China and the globe simulated by coupled climate models. *J. Clim.* **2006**, *19*, 5843–5858. [CrossRef]

31. Gillett, N.P.; Akiyoshi, H.; Bekki, S.; Braesicke, P.; Eyring, V.; Garcia, R.R.; Karpechko, A.Y. Attribution of observed changes in stratospheric ozone and temperature. *Atmos. Chem. Phys.* **2011**, *11*, 599–609. [CrossRef]

32. McLandress, C.; Jonsson, A.I.; Plummer, D.A.; Reader, M.C.; Scinocca, J.F.; Shepherd, T.G. Separating the effects of climate change and ozone depletion. Part 1: Southern Hemisphere stratosphere. *J. Clim.* **2010**, *23*, 5002–5020. [CrossRef]

33. Guo, D.; Su, Y.; Shi, C.; Xu, J.; Powell, A.M. Double core of ozone valley over the Tibetan plateau and its possible mechanisms. *J. Atmos. Sol. Terr. Phys.* **2015**, *130*, 127–131. [CrossRef]

34. Meehl, G.A.; Arblaster, J.M.; Matthes, K.; Sassi, F.; Loon, H. Amplifying the Pacific climate system response to a small 11-year solar cycle forcing. *Science* **2009**. [CrossRef] [PubMed]

Investigation on the Weighted RANSAC Approaches for Building Roof Plane Segmentation from LiDAR Point Clouds

Bo Xu [1], Wanshou Jiang [1,*], Jie Shan [2], Jing Zhang [1] and Lelin Li [3]

Academic Editors: Devrim Akca, Zhong Lu and Prasad Thenkabail

[1] State Key Laboratory of Information Engineering in Surveying, Mapping and Remote Sensing, Wuhan University, Wuhan 430072, China; lmars_xubo@whu.edu.cn (B.X.); jing.zhang@whu.edu.cn (J.Z.)
[2] Lyles School of Civil Engineering, Purdue University, West Lafayette, IN 47907, USA; jshan@purdue.edu
[3] National-Local Joint Engineering Laboratory of Geo-Spatial Information Technology, Hunan University of Science and Technology, Xiangtan 411201, China; lilelin@hnust.edu.cn
[*] Correspondence: jws@whu.edu.cn

Abstract: RANdom SAmple Consensus (RANSAC) is a widely adopted method for LiDAR point cloud segmentation because of its robustness to noise and outliers. However, RANSAC has a tendency to generate false segments consisting of points from several nearly coplanar surfaces. To address this problem, we formulate the weighted RANSAC approach for the purpose of point cloud segmentation. In our proposed solution, the hard threshold voting function which considers both the point-plane distance and the normal vector consistency is transformed into a soft threshold voting function based on two weight functions. To improve weighted RANSAC's ability to distinguish planes, we designed the weight functions according to the difference in the error distribution between the proper and improper plane hypotheses, based on which an outlier suppression ratio was also defined. Using the ratio, a thorough comparison was conducted between these different weight functions to determine the best performing function. The selected weight function was then compared to the existing weighted RANSAC methods, the original RANSAC, and a representative region growing (RG) method. Experiments with two airborne LiDAR datasets of varying densities show that the various weighted methods can improve the segmentation quality differently, but the dedicated designed weight functions can significantly improve the segmentation accuracy and the topology correctness. Moreover, its robustness is much better when compared to the RG method.

Keywords: 3D point clouds; building reconstruction; building roof segmentation; weighted RANSAC

1. Introduction

Numerous studies have been conducted in 3D building reconstruction in the past two decades [1–3]. According to [4,5], reconstruction methods can be divided into two general categories: data-driven and model-driven. For high density point cloud data or complex roof structures, the task often converges on a data-driven process based on segmentation [2]. According to [6,7], there are three data-driven segmentation techniques: edge-based or region growing (RG), feature clustering, and model fitting.

Segmentation methods based on edge or region information [8–12] are relatively simple and efficient but are error-prone in the presence of outliers and incomplete boundaries. When the transitions between two regions are smooth, finding a complete edge or determining a stop criterion for RG becomes difficult [6]. Techniques using feature clustering for segmentation [6,13–18] experience problems in deciding the number of segments; and poor segmentation (over-, under-, no segmentation, or artifacts) can occur when small roof sub-structures exist or tree points close to the building roofs are not completely filtered beforehand. Compared with the above techniques, [7] suggested that model fitting methods can be more efficient and robust in the presence of noise and outliers. RANdom SAmple Consensus (RANSAC) [19] and Hough transform are two well-known algorithms for model fitting. The concept and implementation of the RANSAC method are simple. It simply iterates two steps: generating a hypothesis by random samples and verifying the hypothesis with the remaining data. Given different hypothesis models, RANSAC can detect planes, spheres, cylinders, cones, and tori [20]. Numerous variants have been derived from RANSAC; and a comprehensive review is available in the work of [21]. Those variants (*i.e.*, [22–24]) provide the possibility of improving the methods in both robustness and efficiency. Information like point surface normal [4,7] and connectivity [25] also can be incorporated in RANSAC for better results. Moreover, although the RANSAC method is an iterative process, reference [5] suggests that it is faster than the Hough transform.

LiDAR techniques generate ever increasingly high resolution data. This provides the possibility to recognize subtle roof details and rather complex structures; but in the meantime, it brings challenges to current RANSAC-based segmentation methods. A widely concerning problem is the spurious planes that consist of points from different planes or roof surface [4,6,7,26,27]. A detected plane overlapping multiple reference planes or a plane snatching parts of the points from its neighbor planes are frequent occurrences. Their misidentification and incorrect reconstruction may have a crucial effect on the understanding of the building structure (*i.e.*, topology of the building) [28,29]. To address this issue, many additional processes were designed and used in past studies, such as normal vector consistency validation [4,7], connectivity [26], and standard deviation of the point-plane distances. Those processes need careful fine tuning of their parameters in order to achieve the best performance (e.g., reference [30] suggests that the threshold should be in agreement with the segment scale). This is a difficult task and highly relies on prior knowledge of the data and scene as well as the experience of the operators. Therefore, a more accurate fitting method is needed to suppress the spurious planes.

Although no applications were found in building roof segmentation, the M-estimate SAC (MSAC) and the Maximum Likelihood SAC (MLESAC) in [31] provided a potential solution to the spurious planes problem. In these two methods, the contribution of a point to the hypothesis plane is no longer a constant 0 or 1, but rather a loss function (inversed to weight) according to the point-plane distance. Basically, a large distance is assigned to a large loss, and false hypotheses are suppressed because of the larger total loss. However, we argue that their loss functions were not sufficient to distinguish the spurious plans from the correct hypothesis plane for complex roof segmentation problems. Inclusion of other additional factors into the loss function, such as surface normal, would make the methods more adaptive and robust.

This paper implements the idea of loss function into the popular RANSAC method and proposes a weighted RANSAC framework for roof plane segmentation. In the framework of our new method, the hard threshold voting function which considers both the point-plane distance and the normal vector consistence is transformed into a soft threshold voting function based on two weight functions. New weight functions are introduced based on the error distribution between the proper and improper hypothesis planes. Different forms of weights were tested and compared, yielding a recommended weight form.

The remainder of this paper is organized as follows. Section 2 discusses the related work and the modification of the existing weighted RANSAC into the normalized forms. In Section 3, the design of an ideal weight function is discussed, and several different weight functions are proposed and evaluated. Experimental results are presented and analyzed in Section 4, followed by discussion and concluding remarks in Section 5.

2. Background

2.1. RANSAC-based Segmentation

Although the RANSAC-based segmentation methods have several variations, they consist of three steps [4]: preprocessing, RANSAC, and post-processing. The preprocessing step yields the surface normal for each LiDAR point. The roof points can be separated to a planar set and a nonplanar set (if so, the nonplanar points are excluded from the second step and be retrieved in the final step). The second step is a standard implementation of the RANSAC method [14]. It iteratively and randomly samples points to estimate the hypothesis plane and then tests the plane against the remainder of the dataset. A point is taken as an inlier if the point-plane distance and the angle between the point's normal vector and plane's normal vector (in [6]) are smaller than the given thresholds. After a certain number of iterations, the shape that possessed the largest percentage of inliers relative to the entire data is extracted. The method detects only one plane at a time from the entire point set. Thus, the process has to be implemented iteratively in a subtractive manner, which means that once a plane is detected, the points belonging to the plane are removed and the algorithm continues on the remainder of the dataset until no satisfactory planes are found. To be fast, the constraints of normal vectors [7,32] and local sampling [4,32] are used to avoid the meaningless hypotheses. A fast and rough clustering (or classification) process can be used to decompose the dataset [7,33]. To be robust, validations on normal vector consistency [4,7], connectivity [26], and standard deviation of point-plane distances are also adopted. The main task of post-processing is to refine the segmentation results, retrieve roof points from unsegmented point sets, find missing planes, and remove false spurious planes [27,32].

For classical RANSAC methods, the plane with the maximum inliers is generated when determining the most probable hypothesis plane $\hat{M}$:

$$\hat{M} = \underset{M}{\mathrm{argmax}} \left\{ \sum_{P_i \in U} T(P_i, M) \right\} \tag{1}$$

where U is the set of remaining points, N_U is the number of points in U, and $T(P_i)$ is the inlier indicator:

$$T(P_i, M) = \begin{cases} 1 & d_i < d_t \ and \ \theta_i < \theta_t \\ 0 & otherwise \end{cases} \tag{2}$$

where d_i is the point-plane distance, θ_i is the angle between point P_i's normal and plane's normal [7], and d_t and θ_t are the corresponding thresholds.

2.2. Spurious Planes

The problem of spurious planes is a widely discussed common problem that has yet to be resolved in RANSAC-based segmentation. Generally, the planes detected by the RANSAC methods may belong to different planes or roof surfaces. As shown in Figure 1a, suppose that the threshold d_t can be well estimated beforehand according to the precision of the point clouds, then a proper segmentation is achieved if the hypothesis planes π_1 and π_2 receive the largest inlier ratio. However, poorly estimated planes may be detected, such as plane π_3 in Figure 1b, whose point count is much larger than that of π_1 or π_2, thus leading to false segmentation. The RANSAC method extracts planes one after the other from LiDAR points so these mistakes may occur at plane transitions. The situation in Figure 1b can further intensify such competitions as the inaccurate hypothesis tends to generate more supports from roof points.

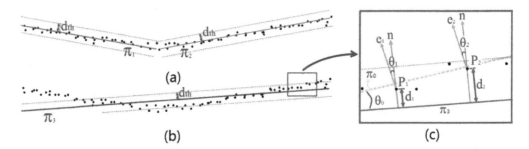

Figure 1. An example of spurious planes. (a) The well estimated hypothesis planes (π_1 and π_2); the two green parallel lines are the boundary of the point-to-plane distance threshold; (b) A spurious plane (π_3) is generated under the same thresholds; (c) A detail view of (b), where n is the normal vector of the plane π_3, and e_1 and e_2 are the point normal vectors. The d_1, d_2, θ_1, θ_2 are the corresponding observed values of point P_1 and P_2 in Equation (2). θ_0 is the angle between π_3 and the real roof surface (π_0).

2.3. Existing Weighted RANSAC Methods

Instead of using fixed thresholds in the determination of inliers, MSAC and MLESAC [31] use a loss function to count the contribution, which is actually a contribution loss, of the inliers based on the point-to-plane distance. The most probable hypothesis $\hat{M}$ is determined by minimizing the total loss of hypothesis M:

$$\hat{M} = \underset{M}{\operatorname{argmin}} \left\{ \sum_{P_i \in U} Loss\left(d(P_i, M)\right) \right\} \tag{3}$$

The MSAC adopts bounded loss as follows:

$$Loss\left(d\right) = \begin{cases} d^2 & |d| < d_t \\ d_t^2 & otherwise \end{cases} \tag{4}$$

MLESAC utilizes the probability distribution of error by inliers and outliers, models inlier errors as Gaussian distribution and outlier errors as uniform distribution:

$$Loss\left(d\right) = -\log\left(\gamma \frac{1}{\sqrt{2\pi}\sigma_d}\exp(-\frac{d^2}{2\sigma_d^2}) + (1-\gamma)\frac{1}{v}\right) \tag{5}$$

where γ is the prior probability of being an inlier, which is the inlier ratio in Equation (1), σ is the standard deviation of Gaussian noise ($\sigma_d = d_t/1.96$), and v is a constant which reflects the size of available error space.

For the so-called weighted RANSAC, the loss function is transformed as the normalized weight functions for testing points, having values from 0 to 1, so that different weight functions can be easily compared and further applied to more than one factor (*i.e.*, considering both distance and normal directions). As a result, the loss functions of MSAC and MLESAC are normalized as Equation (6).

$$weight\left(d\right) = \frac{loss\left(+\infty\right) - loss\left(d\right)}{loss\left(+\infty\right) - loss\left(0\right)} \tag{6}$$

3. Weighted RANSAC for Point Cloud Segmentation

For the weighted RANSAC methods, the weight value of an inlier reflects its consistency with the hypothesis plane. An ideal weight function is expected to suppress the spurious planes as far as possible without excessively penalizing the proper planes. In this work, the purpose is achieved by comparing the error distribution between the proper and improper hypotheses. In Section 3.1, we discuss the drawback of the existing weighted methods that form the design principle of the ideal weight function. Then, several new weight functions are defined based on the design principle in Section 3.2. In Section 3.3, adding the factor of normal vector errors into the weight functions is considered, and a joint weight function is designed via the multiplication of the two factors (distance and normal vector). Those new weight functions are compared and evaluated in Section 3.4, together with the existing weighted methods.

3.1. Improvements Consideration of the Weighted Function

Figure 2 (Bottom) provides examples of the point-to-plane distance distribution for both the proper hypothesis and the spurious plane. To clarify the discussion that follows, the distance range is divided into three regions, namely A, B, and C. Generally, the inliers of a proper-plane tend to focus on the region with a smaller distance, which follows the normal distribution in theory, while the distribution of the distances to a spurious plane tend to be more dispersed. For traditional RANSAC, the spurious planes are detected instead of the proper plane if there are too many points in region C. An intuitive solution to alleviate the problem is simply using a smaller distance threshold, *i.e.*, changing the threshold d_t to d'_t, which reduces the inliers count of the spurious plane (yellow region). However, a too small threshold will decrease the number of inliers (red region) and eventually result in over-segmentation.

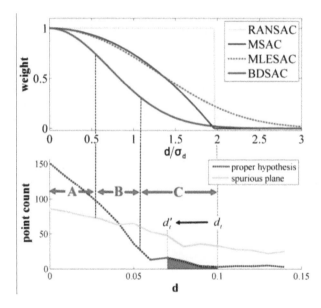

Figure 2. Comparison of point-to-plane distance distribution between proper and improper hypotheses. (**Top**) Plots of the weight functions ($d_t = 1.96\sigma_d$, MLESAC: $\gamma = 0.3$, $v = 3\sigma_d$); (**Bottom**) Examples of distance distribution for the proper hypothesis and the spurious plane. A, B, and C are a rough division of the distance range: A for regions where the proper planes are dominant in the point count, B for regions where the point counts are similar, and C for regions where spurious planes generate more inliers. The red region represents the lost roof points when using a stricter threshold and the yellow region indicates that more points are excluded from the spurious planes than the proper hypothesis. BDSAC is a newly designed weight curve.

Without changing the d_t threshold, MSAC and MLESAC suppress the spurious planes by assigning smaller weights to the inliers with larger distances so that the inliers in area C of Figure 2 will contribute less to the evaluation of the hypothesis plane. The inadequacy of MSAC and MLESAC are mainly caused by its slow decrease of the weight curves. Generally, the weighted methods are expected to suppress the spurious plane as far as possible without excessively penalizing the proper planes. Under such consideration, we expect the curve of the weight function to decrease rapidly in area B and gradually with small weight values in area C (*i.e.*, the curve of BDSAC in Figure 2). However, as shown in Figure 2, there are still a great deal of inliers that have large weight values and gradients in area C for MSAC and MLESAC. MSAC has the largest absolute gradient at the threshold boundary, and the MLESAC has a boundary weight value of over 0.2, which limit their suppressing to spurious planes. To overcome the drawbacks of these two methods, we attempted to modify the weight functions, and the improved versions of weight functions are shown in Section 3.2.

3.2. Modified Weight Functions and New Weight Functions

First, the weight functions of RANSAC, MSAC, and MLESAC were modified. Generally, after a hypothesis plane is accepted, it is expected that all the inliers should be excluded to avoid affecting the detection of other planes; while in the plane detection step, it is wished that as fewer outliers included as possible to decrease the possibility of false plane detection and the absence of minor inliers is acceptable. This reminds us to reduce the thresholds used in the weight functions and to keep the threshold unchanged for inlier exclusion. For such an objective, a reduction ratio μ was applied to the distance threshold d_t in the weight function. For example, the MSAC with a reduction ratio μ is expressed by (denoted by MSAC$_\mu$):

$$weight_\mu(d) = \begin{cases} 1 - \left(\dfrac{d}{\mu \cdot d_t}\right)^2 & |d| < (\mu \cdot d_t) \\ 0 & otherwise \end{cases} \tag{7}$$

Similarly, the reduction ratio μ also was adopted in classical RANSAC and MLESAC to generate the two modified versions, named RANSAC$_u$ and MLESAC$_u$. RANSAC$_u$ uses smaller threshold $\mu \cdot d_t$ for inlier determination; and the σ_d in Equation (5) of MLESAC$_u$ is reduced to $\mu \cdot d_t/1.96$.

As discussed in Section 3.1, two new weight functions stricter in theory can be designed, whose value is close to 1 in region A, close to 0 in region C, and rapidly decreasing in region B. One weight function is a piecewise-linear function, which linearly decreases in Region B (denote by LDSAC).

$$weight(d) = \begin{cases} 1 & |d| \leqslant d_1 \\ \dfrac{d_2 - |d|}{d_2 - d_1} & d_1 < |d| < d_2 \\ 0 & others \end{cases} \tag{8}$$

where d_1 and d_2 are the selected thresholds between 0 and d_t (*i.e.*, $0.2d_t$ and $0.7d_t$ in our test).

Another weight function is a smooth curve decreasing along the "bell" curve (denote by BDSAC):

$$weight(d) = \exp(-\dfrac{d^2}{\sigma_d{}^2}) \tag{9}$$

The curves of the weight functions and the absolute value of their gradients are illustrated in Figure 3. All the weight functions are inversely proportional to the point-to-plane distance d with a range of from 0 to 1, thus the most probable hypothesis plane $\hat{M}$ is decided similarly with classical RANSAC in Equation (1):

$$\hat{M} = \underset{M}{\operatorname{argmax}} \left\{ \sum_{P_i \in U} weight(d_i) \right\} \tag{10}$$

As the value of the weights is generated by simply mapping the value of d/σ_d into pre-calculated tables, the efficiency of all the methods are similar.

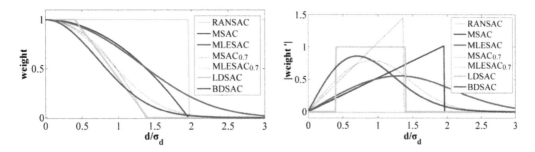

Figure 3. Weight functions of various RANSAC methods: **(Left)** Plots of the weight functions; **(Right)** Plots of the absolute value of gradient.

3.3. Joint Weight Function Regarding Angular Difference

The normal vectors of the inliers are generated by neighborhood analysis [6,10] and often have fine consistency for 2.5D roof surfaces. For poor hypotheses, a systematic deviation of the normal vectors (*i.e.*, θ_0 in Figure 1c) can exist between the hypotheses plane and the real roof surface. As the normal of the points turns out to be in accord with the real roof surface, this deviation will reflect in most roof points. As a result, the angular difference between the points and the hypothesis plane (θ in Equation (2)) has long been used to evaluate the quality of inliers, either as constant thresholds in [7] or as a normal vector consistency validation in [4]. It is very natural for us to consider adding the angular difference into the weight definition.

Suppose the distribution of angular difference θ is independent with the distance and also obeys the normal distribution with a standard deviation of σ_θ. Then, the weight of the angular difference can be defined by using the same form as the distance (simply replace d by θ). For instance, the weight functions of BDSAC for angular difference θ can be defined as:

$$w\left(\theta\right) = \exp(-\frac{\theta^2}{\sigma_\theta^2}) \tag{11}$$

Then, the final weight $weight\left(d_i, \theta_i\right)$, considering both the point-to-plane distances and the angular differences, can be defined as the product of the two weights:

$$weight\left(d_i, \theta_i\right) = w(d_i)w(\theta_i) \tag{12}$$

Similar to Equation (9), the most probable hypothesis plane $\hat{M}$ is determined by maximizing the total weight of all the hypothesis of M

$$\hat{M} = \underset{M}{\operatorname{argmax}} \left\{ \sum_{P_i \in U} weight\left(d_i, \theta_i\right) \right\} \tag{13}$$

To distinguish from methods considering point-to-plane distance only, a subscript of "$_{nv}$" is added to the methods that take the angular difference into account (e.g., $BDSAC_{nv}$ for the improved method of BDSAC).

3.4. Weight Function Evaluation

A proper weight function is expected to suppress the improper hypotheses as much as possible, without excessively penalizing the proper ones. Since the decreasing rates of the total weights for the hypothesis plane under different weight functions are different, an outlier suppression ratio is defined as the evaluation metric here:

$$ratio_{os} = \frac{W_{test}}{W_{ref}} \tag{14}$$

where W_{ref} stands for the total weight of the reference plane (the plane fitted by all the inliers), and W_{test} is the total weight of the test hypothesis plane.

The test hypothesis planes are randomly generated and manually marked as positive or negative, based on whether a correct segmentation can be generated. For a positive hypothesis, we expect that the ratio of a good weight function is stable, which should be close to 1. For negatives hypotheses, we need the ratio to be as small as possible, and a ratio over 1 indicates that a false hypothesis gains larger weights than the proper ones, leading to false segmentation.

To evaluate the weight functions defined in Sections 3.2 and 3.3 10 hypotheses planes are generated from the point cloud of the building in Figure 4, among which three hypotheses are positive and seven hypotheses are negative.

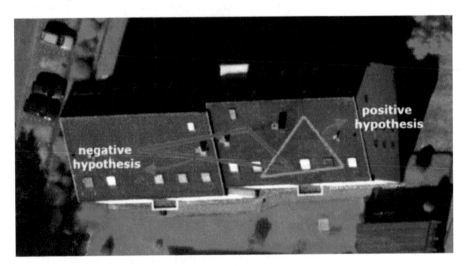

Figure 4. Buildings with both positive and negative hypotheses. The deep blue triangle is a negative hypothesis as it is athwart the two roof planes, and the cyan triangle is a positive hypothesis which can produce a correct segmentation.

The outlier suppression ratios of the 10 hypotheses are shown in Figure 5. As shown in Figure 5a, the ratios of eight methods considering only distances error are compared, and the mean ratio of the 10 hypotheses under different thresholds are illustrated in Figure 5c. As shown in Figure 5b,d, we compare the improvements of the methods after considering both the distance and angular difference in the weight function, corresponding to Figure 5a,c. Several conclusions can be made at this point:

(1)   For all the weighted methods, the evaluation of the positive hypotheses (planes 1, 2, and 3) are stable as the ratios in Figure 5a are close to 1.0 and the ratio reductions in Figure 5b are close to 0. Meanwhile, all the weighted methods can significantly decrease the ratios of the negative hypotheses when compared to RANSAC, but their suppressing ability are different.

(2)   By comparing the results between the modified weight functions and the original functions (*i.e.*, $MSAC_{0.7}$ and MSAC), it can be concluded that reduction of the inlier threshold can suppress the outliers effectively. The newly designed LDSAC and BDSAC functions have the best performances, which verifies our considerations in Section 3.1.

(3)   From Figure 5c, it can be seen that all the methods can be affected by the threshold in some degree, but the newly designed weighted methods are least influenced.

(4)   Figure 5b,d illustrate the improvements after taking the angular differences into the weight functions. All the weighted methods gain positive effects and the effects are not sensitive to the thresholds.

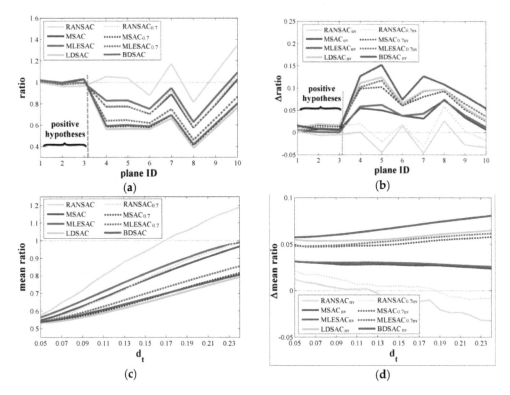

Figure 5. Suppressing ability and threshold sensitivity test. (a) Suppressing ratios for planes under different weight forms; (b) Ratio reductions after considering the angular difference (*i.e.*, the ratio reduction of BDSACnv is the ratio of BDSAC minus the ratio of BDSACnv); (c) Mean ratio of the ten planes under different dt thresholds; (d) Mean ratios reduction after considering the angular difference (the reduction approach is similar to (b)).

As the performances of the segmentation methods are greatly influenced by the complexity of the input data and the threshold parameters, we simulate the data in Figure 6 to test the robustness of the algorithm on a variety of conditions. The data consists of two adjacent horizontal planes, both 10 m × 5 m with an average point distance is 0.5 m. The height difference between the planes (Δd) and the added Gaussian noise (with a standard deviation of σ) are both changeable. The thresholds dt for the methods are tested from 0.02 m to 0.2 m, every 0.01 m a trail.

The difficulty of segmentation will obviously increase when Δd decreased or σ increased, which will influence the selection of the d_t thresholds. For data with a larger σ, the d_t needs to be larger in order to include all the plane inliers; otherwise, over-segmentation may occur. As a result, nearly all the methods fail when d_t is smaller than 2σ in Figure 6c (the value need to be even larger in real applications). The value of Δd reflects the separability of the two planes and stricter thresholds are needed for a successful separation. The setting of thresholds needs to consider both factors and find a proper value between the two limitations, finally forming the acceptable areas for different weighted methods in Figure 6. For classical RANSAC, the results are rather disappointing and a proper threshold is difficult to generate. However, for the weighted methods, as the spurious planes are suppressed, much looser thresholds are allowed which result in larger areas in Figure 6. It also can be seen that both adding new weight forms and considering the angular difference in the weights produce positive effects on the acceptable areas. This decreases the difficulty of threshold selection and allows the possibility of processing more complex data. For instance, when Δd equals 0.15 m and 0.2 m in Figure 6b or when σ equals 0.03 m and 0.04 m in Figure 6c, the classical RANSAC methods will always fail while our new weighted methods can produce a correct segmentation. Intuitively, a spurious plane that passes through the middle of the two planes will include all the points if d_t is larger than $\Delta d/2$ for classical RANSAC and cannot distinguish the two planes well when d_t is larger than $\Delta d/3$ in our experiments. In comparison, proper results are produced by $BDSAC_{nv}$ even when d_t is larger than $2\Delta d/3$.

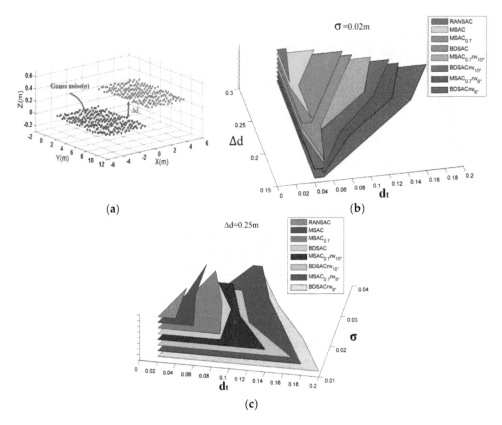

Figure 6. Data sensitivity test. (**a**) Simulated data, with changeable Δd and σ; (**b**) Segmentation results under different Δd; (**c**) Segmentation results under different σ. The colored regions depict the range of d_t that can produce a correct segmentation.

4. Experiments and Evaluation

After comparing the effects of different weight functions in suppressing spurious planes and their sensitivity to thresholds and input data, this section presents the stability and robustness segmentation results and the optimal weight recommendations. The various assessment metrics are introduced, and the experiments on various datasets to test the overall performance of the methods are presented.

4.1. Datasets and Fundamental Algorithm

The experiments utilized two datasets. The first dataset was collected in the city of Vaihingen, ISPRS dataset [34] and the other set, which has a higher point density, was collected on the Wuhan University campus, China. In the quantitative tests, the reference data were created manually based on the initial segmentation results and aerial images. Since our segmentation algorithms initiates from the classified building roof points, the points on the ground, walls, and vegetation were filtered beforehand and excluded from the quantitative tests. The results of a RG-based method [11] are also used for comparison purposes.

To evaluate the effects of the weighted methods, a fundamental RANSAC-based segmentation algorithm is needed. Since this paper focuses on the effects of the different weight functions, only a brief introduction to the algorithm implementation is provided here. The main framework of the algorithm follows the work of [4], but we also refer to the work of [7,25] (described in Section 2.1). In the pre-processing stage, the points normal are estimated through the tensor voting algorithm [10], which also divides the points into planar and nonplanar sets. In the second step (standard RANSAC stage), the density-based connectivity clustering is implemented [22] to ensure the spatial connectivity of the detected planes. Some speed-up techniques also are also utilized: a fast and rough connectivity clustering to decompose the integral data [35] (connectivity of the octree cells) and the ND-RANSAC [32] and Local RANSAC [4] to avoid meaningless hypotheses. The post-processing mainly included the following aspects: (1) completion of the roof plane by searching the points from the unsegmented points; (2) clustering of the remaining points set and an extra searching process to detect the lost segments; and (3) to avoid over-segmentation or detecting a plane twice, a region merging process [32] was adopted among the neighbor planes, which required the total weights of the merged plane to be larger than that of either single plane. Some basic specifications for the datasets are provided in Table 1, and the main parameters are shown in Table 2.

Table 1. Properties of the two datasets.

Site	Vaihingen	Wuhan
Acquisition Date	22 August 2008	22 July 2014
Acquisition System	Leica ALS 50	Trimble Harrier 68i
Fly Height	500 m	1000 m (cross flight)
Point Density	~4/m^2	>15/m^2

Table 2. Parameters used in the experiments.

	MinPt	MinLen	Angle	d_t	θ_t	Ncc	Dcc	P_0	NbPt1	NbPt2
Vaihingen	5	1 m	15°	0.15 m	10°	5	1.5 m	0.99	5	20
Wuhan	20	1 m	15°	0.20 m	10°	5	0.75 m	0.99	10	30

MinPt: the minimum number of points for a plane; MinLen: the minimum length of the detected edge. Angle: the angle threshold between the three sample points' normal and the plane's normal used in hypothesis generation (ND-RANSAC, see [31]). Ncc (least number of points) and Dcc (searching distance) are the parameters used in the density-based connectivity clustering [24]. P_0 is the confidence probability to select the positive hypotheses at least once. NbPt1 and NbPt1 are the two parameters (nearest n points) used in the tensor voting-based method [10] (two rounds of voting).

4.2. Evaluation Metrics

The evaluation metrics consisted of two parts: the object-level evaluation metrics provided in [36] and the quality of the roof ridges detected after segmentation. *Completeness* (*Comp*), *Correctness* (*Corr*) and *Quality* in [36] are used to assess the segmentation results:

$$comp = \frac{||TP||}{||TP|| + ||FN||}$$
$$corr = \frac{||TP||}{||TP|| + ||FP||} \qquad (15)$$
$$Quality = \frac{||TP||}{||TP|| + ||FN|| + ||FP||}$$

where *TP* (True Positive) is the number of objects found both in the reference and segmentation, *FN* (False Negative) is the number of reference objects not found in segmentation, and *FP* (False Positive) is the number of detected objects not found in the reference. Different from the metrics defined in [37], which are widely adopted for the ISPRS benchmark dataset, the metrics in [36] found the correspondences between the reference and the segmented data by using the "maximum overlap" instead of the "overall coverage". As they only establish one-to-one correspondences, the TP values in the reference and segmented data are always the same. This can be more convenient for distinguishing the segmentation errors when the relationships of one-to-many, many-to-one, or many-to-many occurred. For example, if one segmented plane corresponds to two reference planes (one-to-many), the two reference planes will all be taken as TPs for the metrics in [37] (fail to detect under-segmentation), while the smaller reference plane will be detected as FN in [36] instead.

Even a small number of incorrectly segmented points sometimes can have a very large influence on the identification of building structures (*i.e.*, false division of roof boundary points can affect the roof topology). Such errors may not be easily detected by the segmentation-based metrics as they only offer a quick assessment at plane level, (*i.e.*, a minimum overlap of 50% with the reference is required to be a TP). Consequently, a result-driven metrics is designed based on whether the segmentation results influences the extraction of roof ridges. The intersection line is calculated using the method and parameters provided in [28]. Considering that the intersections of roofs ridges in corners (*i.e.*, using the close-circle analysis in [38]) may cover up some mistakes in segmentation, only the original ridges are compared in the experiments.

Two ridge-based metrics are utilized in our experiments. One metric is based on the roof topology graph (RTG) which mainly considers the existing ridges between planes. In the above metrics, the one-to-one correspondences among the reference planes and roof planes have been established. A detected ridge is related to two extracted planes and is taken as a TP only when two correspondences between planes can be found and a reference ridge exists between the two planes. The second metric is much stricter and accepts a TP only when the corresponding ridges are similar enough. To achieve such a goal, the similarity between the reference ridges and the test ridges are defined (Figure 7), which consists of three aspects: distance consistence (*dc*), orientation consistence (*oc*), and projection consistence (*pc*):

$$dc = \exp\left\{ -\left(\frac{|CC_1| + |DD_1|}{2 \cdot dis_0} \right)^2 \right\}$$

$$oc = \exp\left\{ -\left(\frac{\alpha}{\alpha_0} \right)^2 \right\} \tag{16}$$

$$pc = \frac{|AB| \cap |C_1 D_1|}{|AB| \cup |C_1 D_1|} = \frac{|C_1 B|}{|AD_1|}$$

where α_0 and dis_0 are two previously established values (*i.e.*, 5° and 0.2 m). As the *oc*, *pc* and *dc* are values between 0 and 1, the large the better, the integral consistence is set as the product of the three values:

$$ic = oc \cdot dc \cdot pc \tag{17}$$

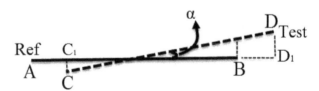

Figure 7. Definition of ridge similarity. Line **AB**: the reference ridge (Ref); line **CD**: the detected ridge (Test), where C1 and D1 are the corresponding projection points of C and D; and α is the intersect angle.

4.3. Experiments

In this section, the improvements from using the weighted approach experimentally are verified. First, the methods for typical scenes that are error-prone for classical RANSAC are presented. Then, the results for the Vaihingen and Wuhan University datasets are evaluated by the metrics given in Section 4.2.

4.3.1. Local Data

For the RANSAC-based methods, spurious planes that consist of points from several roof surfaces are easily generated when adjacent planes have very similar heights or normal orientations. As shown in Figure 8, we select eight typical buildings (a)–(h) to examine the new weighted methods, with the error-prone regions numbered from 1 to 12. In regions 1–2, 5, and 11, a detected plane is possibly be overlapping multiple reference planes; for regions 3–4 and 9–10, poor segmentation may occur when inaccurate hypothesis planes snatch points from neighbor planes; for regions 6–8, two planes are shown as merged into one; and the roof in region 12 is not complanate, thus the segmented results are likely fragmentized. Due to the limited space, we present only the segmentation results for RANSAC, BDSAC, and RG. The MSAC and MLESAC results are very similar to the conventional RANSAC results and can hardlyable to distinguish the poorly estimated planes. Further discussion will be provided in Figure 9.

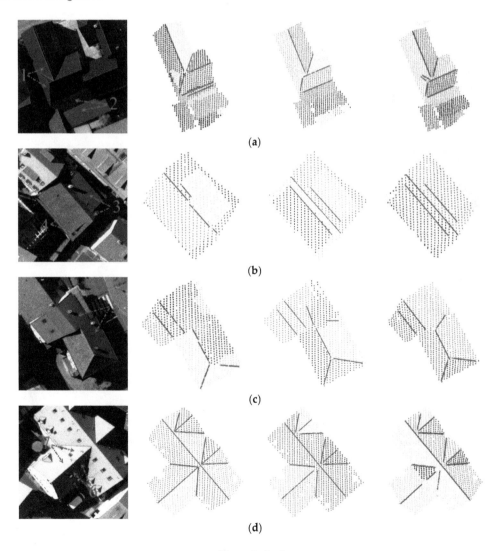

(a)

(b)

(c)

(d)

Figure 8. *Cont.*

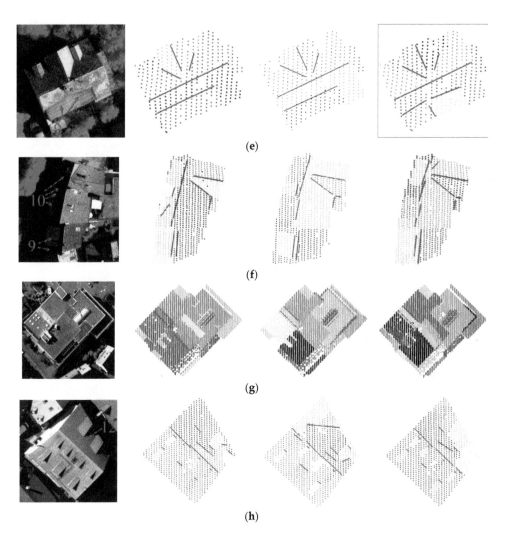

Figure 8. Results of segmentation and ridge detection for error-prone buildings. (**a–h**) are eight selected buildings containing error-prone regions. From left to right: reference images, results by classical RANSAC, results by RG, and results by BDSAC$_{nv}$.

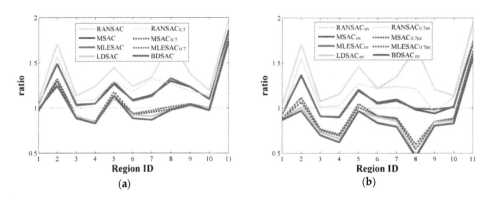

Figure 9. Suppressing ratios comparison. The spurious planes detected by RANSAC in Figure 8 (regions 1–11). (**a**) Suppressing ratio for methods that only consider point-plane in weight functions; (**b**) Ratios for methods considering both distance and angular difference.

As shown in Figure 8, our new weighted method significantly improved the segmentation and ridge detection results. In regions 1–9, most of the segmentation errors for the RANSAC method, which are also common for the RG method (regions 5–9), are properly solved. In regions 10–12, all the methods fail to create ideal results. The errors in region 10 are mainly caused by sparse data; and in region 11, the normal difference between planes A and B is about 3°and the height difference between B and C is only about 0.15 m, which are too small to distinguish under the current thresholds. The RG method successfully distinguished roofs A and B while it fail to separate B and C. For region 12, all the methods fail because the origin data is not complanate. As a result, our method, compared to the RG method, is slightly better in region 10 but worse in region 11. The quantitative results in Table 3 support our conclusion. Comparing the results of our method to classical RANSAC, the overall segmentation quality increases from 61.3% to 77.2%, and the two ridge-based metrics also increase from 51.8% to 81.7% and 41.6% to 69.3%. Meanwhile, our results are also better than the RG method by the metrics. It can be seen that for regions like 1–4 and 9–0, the incorrectly classified points may not be significant in point count but had strong influences on the identification of roof topology. Such errors are not distinguished by the segmentation-based metrics (*i.e.*, the three planes in region 9 are considered as TPs). Our ridge based metrics show more reasonable evaluation under such situations as the errors will damage the distinguishing of roof ridges. The metrics based on ridge similarity are stricter than those based simply on RTG and exclude some ambiguous or incomplete ridges, such as the ridges in building (b) in Table 3, and thus are more reasonable in some situations.

Table 3. Quality of segmentation results for data in Figure 8.

ID	nPls	nRidges	Method	Segmentation			Ridges (RTG)			Ridges (ic > 0.3)		
				%Cm	%Cr	%Qua	%Cm	%Cr	%Qua	%Cm	%Cr	%Qua
a	10	7	RANSAC	80	100	80	71.4	55.5	45.5	57.1	44.4	33.3
			RG	100	100	100	71.4	100	71.4	71.4	100	71.4
			BDSAC$_{nv}$	100	100	100	100	100	100	71.4	100	71.4
b	5	3	RANSAC	80	57.1	50	66.7	50.0	40.0	0	0	0
			RG	100	100	100	100	100	100	100	100	100
			BDSAC$_{nv}$	100	100	100	100	100	100	100	100	100
c	7	5	RANSAC	85.7	60	51.5	60.0	37.5	30.0	40.0	25.0	18.2
			RG	85.7	75	66.7	100	71.4	71.4	100	71.4	71.4
			BDSAC$_{nv}$	100	100	100	100	83.3	83.3	100	83.3	83.3
d	10	11	RANSAC	80.0	100	80.0	90.9	100	90.9	90.9	100	90.9
			RG	80.0	100	80.0	90.9	100	90.9	90.9	100	90.9
			BDSAC$_{nv}$	100	100	100	100	100	100	100	100	100
e	9	7	RANSAC	88.9	100	88.9	71.4	100	71.4	57.1	80	50
			RG	60	100	60	71.4	100	71.4	57.1	80	50
			BDSAC$_{nv}$	100	100	100	100	100	100	100	100	100
f	12	12	RANSAC	66.7	66.7	50	33.3	50.0	25.0	33.3	50	25.0
			RG	83.3	90.9	76.9	41.7	55.5	31.2	41.7	55.5	31.2
			BDSAC$_{nv}$	91.7	91.7	84.6	91.7	91.7	84.6	66.7	66.7	50.0
g	23	5	RANSAC	69.6	64.0	50.0	100	62.5	62.5	100	62.5	62.5
			RG	78.3	72.0	60.0	100	71.4	71.4	100	71.4	71.4
			BDSAC$_{nv}$	69.6	66.7	51.6	100	50.0	50.0	100	50.0	50.0
h	11	10	RANSAC	90.9	71.4	66.7	90.0	64.3	60.0	80.0	57.1	50.0
			RG	81.8	69.2	60.0	70.0	46.7	38.9	60.0	40.0	31.6
			BDSAC$_{nv}$	90.9	66.7	62.5	90.0	69.2	64.3	80.0	61.5	53.3
sum	87	60	RANSAC	78.2	73.9	61.3	71.7	65.2	51.8	61.7	56.1	41.6
			RG	82.8	83.7	71.3	75.0	73.8	59.2	71.7	70.5	55.1
			BDSAC$_{nv}$	89.7	84.8	77.2	96.7	84.1	81.7	86.7	77.6	69.3

Figure 9 depicts the performance results of the different weighted methods based on the data in Figure 8, via the outlier suppression ratio (Equation (14)). In each error-prone region, we utilize the largest spurious plane by RANSAC as the test hypothesis plane, whose total weight is W_{test}, and the total weight of the largest reference plane in the corresponding region is W_{ref}. Regions 1–11 in Figure 8 are evaluated. Region 12 is omitted because the roof surface is nonplanar.

Although the ratios are smaller for MSAC and MLESAC than for classical RANSAC, all of them are over 1; thus, the two methods will still accept all the spurious planes that result in false segmentation. As a result, simply using MSAC and MLESAC cannot improve the segmentation results. For our new method, both the new weight forms and the weights regarding angular difference have distinct positive effects on the final results. Considering only one factor may fail in some situations, such as regions 4 and 7 for BDSAC and $MSAC_{nv}$. Meanwhile, the extent of the improvements by angular difference in the weights may be different for the planes. For planes with distinct biases in both the distance and normal vectors, such as regions 2 and 8, the suppressing of the total weights can be larger than in other regions. Again, $BDSAC_{nv}$ provides the best results. In addition, although our methods fail in region 11, the ratio of the $BDSAC_{nv}$ is still smaller than the other weighted methods.

4.3.2. Vaihingen (Germany)

Figure 10 illustrates the segmentation results for the Vaihingen data; specifically, Figure 10a–c are the benchmark data of the "ISPRS Test Project on Urban Classification and 3D Building Reconstruction", in which A and B have been tested in Figure 8e,f, respectively. Other error-prone regions for the classical methods also are indicated in the Figures. For region E, the situation is similar to Figure 8b,f, where spurious planes overlapping multiple roof planes can be produced. In region G, several planes intersect at the same roof corner, which requires more accurate segmentation methods and neighbor competition in the post-processing to better divide the roof boundary points. Our weighted methods show advantages in those regions as well, as the planes with smaller distance errors are more likely to be accepted. Such differences can be detected by the ridge-based metrics. When the transitions between the neighbor regions are smooth, as shown in E, the RG-based methods may fail. Some of the errors are caused by the processing before roof segmentation; for example, the points in region C are classified as vegetation and parts of the points in region D are lost in the original data. All the methods fail in those regions and the related roof ridges are also lost. For E, F, and G, the segmentation results of the different methods are illustrated in Appendix I for comparison.

The quantitative results of the Vaihingen data are shown in Figure 11. It can be seen that our $BDSAC_{nv}$ method generates significant improvements compared to the traditional methods RANSAC and MSAC. Higher scores are achieved by our methods when using either the segmentation-based metrics or the two ridge-based metrics. The improvements of MSAC and MLESAC to RANSAC are not evident in the test data, and many spurious planes are still detected. It should be noted that some error-prone regions are also difficult for the RG method because regions with small angular or height differences often have very smooth transitions (e.g., B and E). Besides, the RG methods seem to be unstable in a few regions, such as the over-segmentation that unexpectedly occurs in F (see Appendix I).

 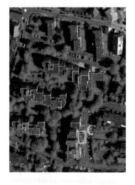 

Figure 10. *Cont.*

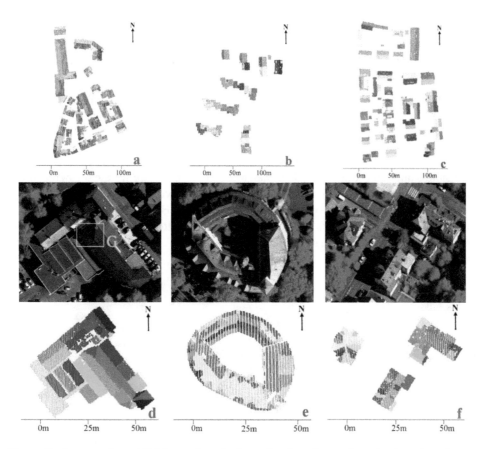

Figure 10. Segmentation of Vaihingen data. (**a–f**) are six selected areas from the data. (top: image, bottom: results of BDSAC$_{nv}$).

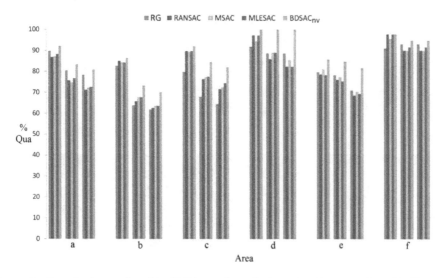

Figure 11. Quantitative results of the Vaihingen data. (**a–f**) are the six areas selected in Figure 10. Three metrics are used, from left to right: quality of segmentation and quality of two ridge based metrics.

4.3.3. Wuhan University (China)

The segmentation results of the Wuhan University data are illustrated in Figure 12. Some error-prone regions are designated. For L and M, spurious planes that overlap multiple roof planes can be produced. For J and O, the small roof planes or short roof ridges may be lost because of small point counts and the competition of roof points from large neighbor planes. In areas (g)–(l), there are many roof details (e.g., Figure 10e), including small windows, eaves, and even guard bars made of glazed tiles, which greatly increase the segmentation difficulties and ultimately results in small plane pieces and short false ridges. In K, a horizontal plane is produced passing through the four planes because the normal errors are not considered in the weight functions. The weighted methods demonstrate great robustness under those situations and therefore significantly improve the segmentation results. Our methods also encounter problems which are unable to resolve. For example, since our weighted methods are not yet adaptable to a curved surface, they divide H and N into several broken pieces. In addition, the RG-based methods fail in I because the points from the upper structure divide the bottom plane into many pieces. The segmentation results for I, K, L and M using different methods are also shown in Appendix I.

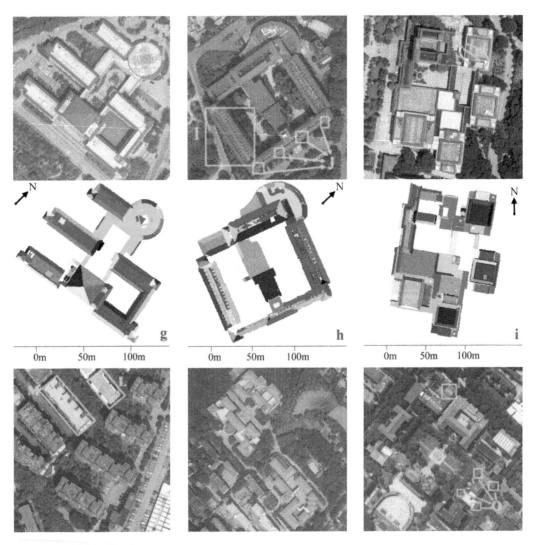

Figure 12. *Cont.*

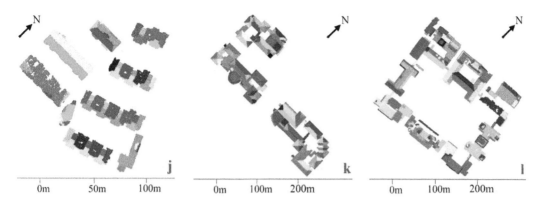

Figure 12. Segmentation results of Wuhan University data. (g–l) are six selected areas from the data. (top: image, bottom: results of $BDSAC_{nv}$).

The quantitative results of the Wuhan University data are illustrated in Figure 13. Similar to Figure 11, our method makes significant improvements compared to the other weighted methods. For areas (g) and (h), the RG method's performance is unsatisfactory because many broken fragments exist. For I in area (h) especially, the bottom planes become numerous broken fragments. The RANSAC-based method can be more robust in those situations. Since the RG method considers the roof slope, it can distinguish very small angular differences, which makes it better than the original RANSAC method results in L and M; and over-segmentation also occurred using RG in L and M. A detailed comparison is available in Appendix I.

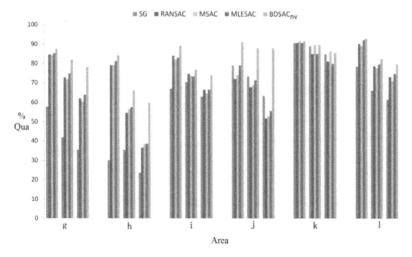

Figure 13. Quantitative results of the Wuhan University data. (g–l) are the six areas selected in Figure 12. Three metrics are used, from left to right: quality of segmentation and quality of two ridge based metrics.

The overall quantitative results are shown in Figure 14, which includes the results of Figures 8, 10 and 12. It can be seen that, while the improvements were not very obvious, the results of MSAC and MLESAC are slightly better than that of RANSAC. Our $BDSAC_{nv}$ generate significant improvements compared to both classical RANSAC and the existing weighted methods. Compared to RANSAC, $BDSAC_{nv}$ improves the overall segmentation quality from 85.7% to 90.1%, as well as the two ridge-based metrics from 75.9% to 83.6% and 68.9% to 80.2%. The quality of the RG method is lower than the RANSAC-based methods, mainly due to their instability in areas (c), (g) and (h).

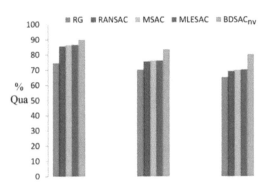

Figure 14. Integral quantitative results. Three metrics are used, from left to right: quality of segmentation and quality of two ridge based metrics.

5. Conclusions

A new weighted RANSAC algorithm for roof point cloud segmentation is introduced in this paper, in which the hard threshold voting function considering both the point-plane distance and the normal vector consistence is transformed into a soft threshold voting function based on two weight functions. Our method utilizes a new strategy to design the ideal weight functions based on the error distribution between the proper and improper hypotheses. Several different weight functions are defined using this strategy, and an outlier suppression ratio is put forward to compare the performance of different weight functions. Preliminary experiments comparing the suppression ratios of different weight functions demonstrated that the $BDSAC_{nv}$ method is able to effectively suppress the outliers from spurious planes. As a result, we chose $BDSAC_{nv}$ for the further experiments and compare its performance with other existing segmentation methods, including original RANSAC, MSAC, MLESAC, and a representative RG method. A set of local data with error-prone regions and two large area datasets of varying densities are used to evaluate the performance of the different methods. The quantitative results of both the segmentation-based metrics and the ridge-based metrics indicated that the different weighted methods improve the segmentation quality differently, but $BDSAC_{nv}$ significantly improve the segmentation accuracy and topology correctness. When compared with RANSAC, $BDSAC_{nv}$ improved the overall segmentation quality from 85.7% to 90.1%; and the two ridge-based metrics also improved from 75.9% to 83.6% and 68.9% to 80.2%. Moreover, the robustness of $BDSAC_{nv}$ is better compared to the RG method. As a result, we believe there is potential for the wide adoption of $BDSAC_{nv}$ as an upgrade to or replacement of classical RANSAC in roof plane segmentation.

However, our method has several limitations. First, although the weighted RANSAC approach is robust to parameters, a small amount of post-processing is still needed to avoid false segmentation or artifacts (see Section 4.1). Second, the weight definition of our method requires a robust estimate of point surface normal, which can be problematic for small buildings or when the point density is low with regard to the roof dimensions. Third, the issue of spurious planes is efficiently suppressed by our method but not completely solved; therefore, spurious planes still may occur in extreme conditions (*i.e.*, Figure 8g).

There are also some possible improvement directions for future work. The number of iterations for RANSAC increases rapidly when the inlier ratio decreases, thus a combination of cluster and fitting to decompose the input data step by step could greatly improve the algorithm's efficiency and robustness. Meanwhile, RANSAC is a one-at-a-time process so adopting the competition approach among neighbor planes could improve the accuracy of segmentation. Finally, only the segmentation of roof planes was considered in this paper, but applying the weighted methods to other roof shapes is possible as the methods mainly are concerned with the procedure of hypothesis verification and do not change the generation of the hypothesis.

Acknowledgments: This work was partially supported by the National Basic Research Program of China under Grant 2012CB719904 as well as the science and technology plan of the Sichuan Bureau of Surveying, Mapping and Geoinformation under Grant of J2014ZC02.   The Vaihingen dataset was provided by the German Society for Photogrammetry, Remote Sensing, and Geoinformation (DGPF) [Cramer, 2010]: http://www.ifp.uni-stuttgart.de/dgpf/DKEP-Allg.html. The segmentation results by the RG-based method were provided by Biao Xiong.

Author Contributions: Bo Xu, Wanshou Jiang and Jie Shan contributed to the study design and manuscript writing. Bo Xu and Wanshou Jiang conceived and designed the experiments; Bo Xu performed the experiments; Jing Zhang and Lelin Li contributed to the initial data, the analysis tools and partial codes of the algorithm.

Conflicts of Interest: The authors declare no conflict of interest.

Appendix I

Appendix I Results compare for some regions marked in Figures 10 and 12.

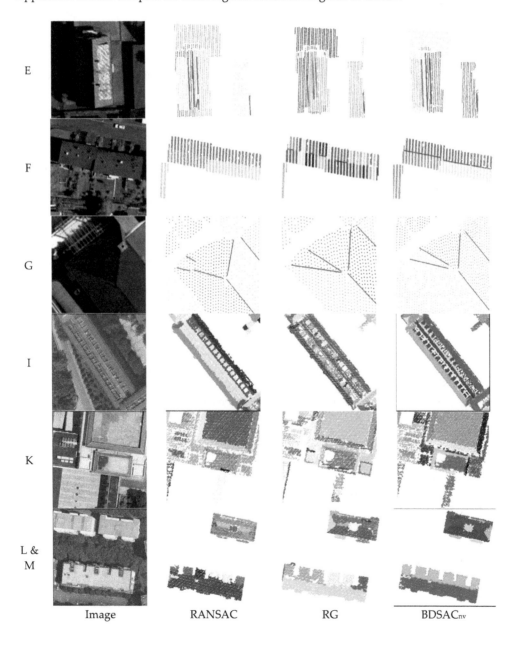

| | Image | RANSAC | RG | BDSACnv |

References

1. Baltsavias, E.P. Object extraction and revision by image analysis using existing geodata and knowledge: Current status and steps towards operational systems. *ISPRS J. Photogramm. Remote Sens.* **2004**, *58*, 129–151. [CrossRef]

2. Haala, N.; Kada, M. An update on automatic 3D building reconstruction. *ISPRS J. Photogramm. Remote Sens.* **2010**, *65*, 570–580. [CrossRef]

3. Rottensteiner, F.; Sohn, G.; Gerke, M.; Wegner, J.D.; Breitkopf, U.; Jung, J. Results of the ISPRS benchmark on urban object detection and 3D building reconstruction. *ISPRS J. Photogramm. Remote Sens.* **2014**, *93*, 256–271. [CrossRef]

4. Chen, D.; Zhang, L.Q.; Li, J.; Liu, R. Urban building roof segmentation from airborne Lidar point clouds. *Int. J. Remote Sens.* **2012**, *33*, 6497–6515. [CrossRef]

5. Tarsha-Kurdi, F.; Landes, T.; Grussenmeyer, P.; Koehl, M. Model-driven and data-driven approaches using Lidar data analysis and comparison. In Proceedings of the ISPRS, Workshop, Photogrammetric Image Analysis (PIA07), Munich, Germany, 19–21 September 2007.

6. Sampath, A.; Shan, J. Segmentation and reconstruction of polyhedral building roofs from aerial Lidar point clouds. *IEEE Trans. Geosci. Remote Sens.* **2010**, *48*, 1554–1567. [CrossRef]

7. Awwad, T.M.; Zhu, Q.; Du, Z.; Zhang, Y. An improved segmentation approach for planar surfaces from unconstructed 3D point clouds. *Photogramm. Rec.* **2010**, *25*, 5–23. [CrossRef]

8. Fan, T.J.; Medioni, G.; Nevatia, R. Segmented descriptions of 3-D surfaces. *IEEE Trans. Rob. Autom.* **1987**, *3*, 527–538.

9. Alharthy, A.; Bethel, J. Detailed building reconstruction from airborne laser data using a moving surface method. *Int. Arch. Photogramm. Remote Sens.* **2004**, *35*, 213–218.

10. You, R.J.; Lin, B.C. Building feature extraction from airborne Lidar data based on tensor voting algorithm. *Photogramm. Eng. Remote Sens.* **2011**, *77*, 1221–1231. [CrossRef]

11. Vosselman, G. Automated planimetric quality control in high accuracy airborne laser scanning surveys. *ISPRS J. Photogramm. Remote Sens.* **2012**, *74*, 90–100. [CrossRef]

12. Lagüela, S.; Díaz-Vilariño, L.; Armesto, J.; Arias, P. Non-destructive approach for the generation and thermal characterization of an as-built BIM. *Constr. Build. Mater.* **2014**, *51*, 55–61. [CrossRef]

13. Hoffman, R.; Jain, A.K. Segmentation and classification of range images. *IEEE Trans. Pattern Anal. Mach. Intell.* **1987**, *9*, 608–620. [CrossRef] [PubMed]

14. Filin, S. Surface clustering from airborne laser scanning data. *Int. Arch. Photogramm. Remote Sens.* **2002**, *34*, 119–124.

15. Filin, S.; Pfeifer, N. Segmentation of airborne laser scanning data using a slope adaptive neighborhood. *ISPRS J. Photogramm. Remote Sens.* **2006**, *60*, 71–80. [CrossRef]

16. Biosca, J.M.; Lerma, J.L. Unsupervised robust planar segmentation of terrestrial laser scanner point clouds based on fuzzy clustering methods. *ISPRS J. Photogramm. Remote Sens.* **2008**, *63*, 84–98. [CrossRef]

17. Dorninger, P.; Pfeifer, N. A comprehensive automated 3D approach for building extraction, reconstruction, and regularization from airborne laser scanning point clouds. *Sensors* **2008**, *8*, 7323–7343. [CrossRef]

18. Awrangjeb, M.; Fraser, C.S. Automatic segmentation of raw Lidar data for extraction of building roofs. *Remote Sens.* **2014**, *6*, 3716–3751. [CrossRef]

19. Fischler, M.A.; Bolles, R.C. Random sample consensus—A paradigm for model-fitting with applications to image-analysis and automated cartography. *Commun. ACM* **1981**, *24*, 381–395. [CrossRef]

20. Schnabel, R.; Wahl, R.; Klein, R. Efficient RANSAC for point-cloud shape detection. *Comput. Graph. Forum* **2007**, *26*, 214–226. [CrossRef]

21. Choi, S.; Kim, T.; Yu, W. Performance evaluation of RANSAC family. In Proceedings of the British Machine Vision Conference, London, UK, 7–10 September 2009.

22. Frahm, J.-M.; Pollefeys, M. RANSAC for (Quasi-) degenerate data (QDEGSAC). In Proceedings of the IEEE Computer Society Conference on Computer Vision and Pattern Recognition, New York, NY, USA, 17–22 June 2006.

23. Chum, O.R.; Matas, J.R.I. Matching with PROSAC-progressive sampling consensus. In Proceedings of the IEEE Computer Society Conference on Computer Vision and Pattern Recognition, Miami, FL, USA, 20–25 June 2005.

24. Chum, O.; Matas, J. Optimal randomized RANSAC. *IEEE Trans. Pattern Anal. Mach. Intell.* **2008**, *30*, 1472–1482. [CrossRef] [PubMed]

25. Berkhin, P. *A Survey of Clustering Data Mining Techniques*; Springer Berlin Heidelberg: New York, NY, USA, 2006.

26. Gallo, O.; Manduchi, R.; Rafii, A. CC-RANSAC: Fitting planes in the presence of multiple surfaces in range data. *Pattern Recognit. Lett.* **2011**, *32*, 403–410. [CrossRef]

27. Yan, J.X.; Shan, J.; Jiang, W.S. A global optimization approach to roof segmentation from airborne Lidar point clouds. *ISPRS J. Photogramm. Remote Sens.* **2014**, *94*, 183–193. [CrossRef]

28. Xiong, B.; Elberink, S.O.; Vosselman, G. A graph edit dictionary for correcting errors in roof topology graphs reconstructed from point clouds. *ISPRS J. Photogramm. Remote Sens.* **2014**, *93*, 227–242. [CrossRef]

29. Elberink, S.O.; Vosselman, G. Building reconstruction by target based graph matching on incomplete laser data: Analysis and limitations. *Sensors* **2009**, *9*, 6101–6118. [CrossRef] [PubMed]

30. Hesami, R.; BabHadiashar, A.; HosseinNezhad, R. Range segmentation of large building exteriors: A hierarchical robust approach. *Comput. Vis. Image Underst.* **2010**, *114*, 475–490. [CrossRef]

31. Torr, P.H.S.; Zisserman, A. Mlesac: A new robust estimator with application to estimating image geometry. *Comput. Vis. Image Underst.* **2000**, *78*, 138–156. [CrossRef]

32. Bretar, F.; Roux, M. Hybrid image segmentation using Lidar 3D planar primitives. In Proceedings of the ISPRS Workshop Laser Scanning, Enschede, The Netherlands, 12–14 September 2005.

33. López-Fernández, L.; Lagüela, S.; Picón, I.; González-Aguilera, D. Large-scale automatic analysis and classification of roof surfaces for the installation of solar panels using a multi-sensor aerial platform. *Remote Sens.* **2015**, *7*, 11226–11248. [CrossRef]

34. Wang, C.; Sha, Y. A designed beta-hairpin forming peptide undergoes a consecutive stepwise process for self-assembly into nanofibrils. *Protein Pept. Lett.* **2010**, *17*, 410–415. [CrossRef] [PubMed]

35. Girardeau-Montaut, D. Detection de Changement sur des Données Géométriques 3D. Ph.D. Thesis, Télécom ParisTech, Paris, France, 2006.

36. Awrangjeb, M.; Fraser, C.S. An automatic and threshold-free performance evaluation system for building extraction techniques from airborne Lidar data. *IEEE J. Sel. Top. Appl. Earth Observ. Remote Sens.* **2014**, *7*, 4184–4198. [CrossRef]

37. Rutzinger, M.; Rottensteiner, F.; Pfeifer, N. A comparison of evaluation techniques for building extraction from airborne laser scanning. *IEEE J. Sel. Top. Appl. Earth Observ. Remote Sens.* **2009**, *2*, 11–20. [CrossRef]

38. Perera, G.S.N.; Maas, H.G. Cycle graph analysis for 3D roof structure modeling: Concepts and performance. *ISPRS J. Photogramm. Remote Sens.* **2014**, *93*, 213–226. [CrossRef]

Cloud and Snow Discrimination for CCD Images of HJ-1A/B Constellation Based on Spectral Signature and Spatio-Temporal Context

Jinhu Bian [1,2,†], Ainong Li [1,*,†], Qiannan Liu [1,2,†] and Chengquan Huang [3]

Academic Editors: Richard Müller and Prasad S. Thenkabail

1   Institute of Mountain Hazards and Environment, Chinese Academy of Sciences, Chengdu 610041, China; bianjinhu@imde.ac.cn (J.B.); qnliu@imde.ac.cn (Q.L.)
2   University of Chinese Academy of Sciences, Beijing 100049, China
3   Department of Geography, University of Maryland, College Park, MD 20742, USA; cqhuang@umd.edu
*   Correspondence: ainongli@imde.ac.cn
†   These authors contributed equally to this work.

Abstract: It is highly desirable to accurately detect the clouds in satellite images before any kind of applications. However, clouds and snow discrimination in remote sensing images is a challenging task because of their similar spectral signature. The shortwave infrared (SWIR, e.g., Landsat TM 1.55–1.75 μm band) band is widely used for the separation of cloud and snow. However, for some sensors such as the CBERS-2 (China-Brazil Earth Resources Satellite), CBERS-4 and HJ-1A/B (HuanJing (HJ), which means environment in Chinese) that are designed without SWIR band, such methods are no longer practical. In this paper, a new practical method was proposed to discriminate clouds from snow through combining the spectral reflectance with the spatio-temporal contextual information. Taking the Mt. Gongga region, where there is frequent clouds and snow cover, in China as a case area, the detailed methodology was introduced on how to use the 181 scenes of HJ-1A/B CCD images in the year 2011 to discriminate clouds and snow in these images. Visual inspection revealed that clouds and snow pixels can be accurately separated by the proposed method. The pixel-level quantitative accuracy validation was conducted by comparing the detection results with the reference cloud masks generated by a random-tile validation scheme. The pixel-level validation results showed that the coefficient of determination (R^2) between the reference cloud masks and the detection results was 0.95, and the average overall accuracy, precision and recall for clouds were 91.32%, 85.33% and 81.82%, respectively. The experimental results confirmed that the proposed method was effective at providing reasonable cloud mask for the SWIR-lacking HJ-1A/B CCD images. Since HJ-1A/B have been in orbit for over seven years and these satellites still run well, the proposed method is helpful for the cloud mask generation of the historical archive HJ-1A/B images and even similar sensors.

Keywords: cloud; snow; HJ-1A/B; regional covariance matrix; spectral; spatio-temporal; texture; context

1. Introduction

Accurate detection of clouds in satellite images is critically important for a wide range of applications [1]. It is the fundamental pre-processing step for the land cover classification [2], change detection [3], image compositing [4,5], or biophysical variables inversion [6,7]. Generally, undetected clouds in satellite images can introduce serious positive bias in aerosol concentration, increase land surface albedo and result in identification of land cover change where none occurred [8,9].

Precise detection of clouds in the images is a quite challenging work not only because of the complexity of clouds themselves but also the difficulty in clouds and snow discrimination [1]. Regarding the complexity of clouds, many types of clouds exist and different types of clouds have various spectral signature. The differences are mainly depended on cloud properties such as optical thickness, height, particle effective radius and thermodynamic phase [10,11]. In particular, thin cirrus clouds are notoriously difficult to be detected due to their mixture spectral signature with land surface [12,13]. Regarding the clouds and snow discrimination, the spectral signature of snow and clouds are both visually bright in the visible wavelengths. This spectral similarity makes them very difficult to be distinguished in visible bands. Moreover, the spectral reflectance of snow can also vary greatly with its properties like grain size, amount of impurities, and thickness of snowpack. Sometimes snow can have very similar spectral signature to certain clouds [9].

Over the years, a number of automated cloud detection methods have been developed based on the spectral signature (hereafter called spectral-based methods) [5,9,10,12,14–17]. Among different spectral bands, the shortwave infrared (SWIR, e.g., Landsat TM 1.55–1.75 μm band) band is most widely used for the clouds and snow discrimination. The primary reason is that the reflectance of snow is usually lower than clouds and therefore snow is generally darker than clouds in the SWIR wavelengths [18]. The Normalized Difference Snow Index (NDSI) [19,20], a combination of visible and SWIR, is very effective for snow and clouds separation [1,10,16]. For instance, it was used for the generation of the internal cloud and snow masking algorithm in the Landsat Ecosystem Disturbance Adaptive Processing System (LEDAPS) [21,22]. However, for some newly launched sensors without SWIR band, such as CBERS (China-Brazil Earth Resources Satellite), HJ-1A/B (HuanJing (HJ)), which means environment in Chinese), the aforementioned spectral-based methods are no longer practical and the difficulty in discriminating cloud and snow for these sensors is even more serious.

Combining spectral signature with spatio-temporal context provides more complementary information for the clouds or snow detection process [1,8,23,24]. For temporal context, it has been used in the new algorithms development for a number of satellite sensors, including MODIS [8], Landsat [9], SPOT [25], and Sentinel-2 [23]. A direct and simple strategy using temporal context is to compare the spectral difference of cloudy images with a referenced clear image [23]. This kind of strategy based on the basic hypothesis that the presence of clouds will introduce high-frequency random changes, and by comparing with a clear-sky reference image, the cloud and snow can be easily separated (hereafter called temporal-based methods). However, there are also some limitations in the temporal-based methods. The main limitation comes from the revisit cycles of satellites, especially for the fine spatial resolution but long revisit cycle sensors like Landsat. Due to the long revisiting cycles for these satellites, the phenological variations of some land cover types in the revisit cycle, such as the deciduous forest, grassland, farmland, and the glacier may vary so greatly that the basic assumption that the change of reflectance was induced by cloud presentence might be invalid. Another limitation comes from the geometric location relationships between clouds and snow. If the clouds cover the same region of snow, the spectral differencing will weaken clouds spectral signature and consequently may cause omission errors. Moreover, SWIR band is usually still needed in most temporal-based algorithms [8,9,23], which limits their application for the sensors without SWIR bands.

For spatial contextual information, it can be defined as how the probability of presence of one object (or objects) is affected by its (or their) neighbors [26]. It is the relationship between the target object and synthesis information generated from the spatial environment [27,28]. Texture is the widely used spatial contextual information, and it refers to repeated local patterns and their regular arrangement of kinds of objects in the images [29]. The texture of clouds and snow are one of their important properties. Even though the texture element of clouds and snow is variable and unpredictable, they are still obviously different from the ground object texture features. On the basis of texture feature analysis, many scientists have done considerable works to improve the accuracy of cloud or snow detection [30–33]. However, most of the present texture-based algorithms assume that there is no snow in the cloudy images or no clouds in the snow covered images [33]. Besides, there are many types of texture features, and using more features does not naturally result in higher accuracy because of the Hughes effect [29]. Moreover, using more feature also result in a large computation cost [32].

The HJ-1A/B is a kind of new generation polar orbit constellation launched in September 2008 by China. It provides fine-resolution (30-m) images like Landsat, and dense observations (every two days revisiting), which is appropriate for capturing anthropogenic impacts and retrieving biophysical parameters over heterogeneous land surface [34]. However, for the lack of SWIR bands, the cloud detection method for HJ-1A/B CCD images is still under the exploration stage. It is noted that taking fully advantage of the spectral signature and spatio-temporal context information of HJ-1A/B might be a practical way to provide an accurate cloud mask. Taking HJ-1A/B CCD images as an example, this paper focused on the cloud and snow discrimination problem as follows: (1) paying special attention to the cloud and snow discrimination for HJ-1A/B CCD and HJ-1A/B CCD like SWIR band-lacking images; and (2) developing the practical cloud and snow discrimination method that can combine the spectral signature with spatio-temporal contextual information. The rest of this paper is organized as follows: Section 2 describes the study area, HJ-1A/B satellites, images pre-processing methods, and the data time series stack used in this paper. Section 3 presents the clouds and snow discrimination methodology. Section 4 shows the results and accuracy analysis. Section 5 discusses the advantages and limitations of the proposed methods. Section 6 presents the conclusions.

2. Study Area and Data

2.1. Study Area

In this study, Mt. Gongga region was chosen as the study area to evaluate the performance of the proposed method (Figure 1). This area was selected for the following reasons: (1) Mt. Gongga region is usually influenced by clouds and snow. This area is the highest peak in the eastern part of the Tibetan Plateau, and is one of the easternmost glacial areas in China [35]. It has persistent snow cover on the high mountains, especially in the upper snowline region where the altitude is about 4800 m. Besides, the annual precipitation of this region (3000 m, a.s.l) is about 1960 mm and most of the precipitation falls as rain from June to September, owing to the influence of Asian summer monsoons [36]. Because of the relatively concentrated precipitation during this period, the weather here is cloudy and the satellite images are usually contaminated by clouds. The frequent appearance of clouds and persistent snow make the area suitable for testing the algorithm's discrimination ability; (2) The land cover of the study area is complex and the seasonal variations are obvious. The topography over this area is an alpine terrain with huge vertical relief (altitude ranging from 890 m to 7556 m). The influence of terrain on the formation of vertical vegetation zonation is evident, which further makes the phenological changes of vegetation vary greatly in a very short distance. This kind of phenological changes might introduce uncertainty into the temporal contextual analysis. An area about 22,500 km^2 (150 km $\times$ 150 km) in the Mt. Gongga region was selected as the case region in this study.

2.2. HJ-1A/B Overview

The HJ-1A/B CCD images were chosen to develop and test the proposed method in this paper. Currently, the HJ-1 constellation consists of a pair of optical satellites (1A/B, launched in September 2008) and one microwave satellite (1C, launched in September 2013) [37]. The primary goal of the HJ-1A/B is to revisit any position in the world within two days for environmental monitoring and disaster mitigation. To achieve such a rapid global coverage, the two optical satellites, each of which has the four days revisit capability, are distributed in the same altitude and orbit plane, with a 180 degrees phase delay [38]. The HJ-1A/B are orbiting in a sun-synchronous circular orbit at 649.093 km altitude with a 10:30 a.m. $\pm$ 30 min descending node [39]. This satellite passing time is selected for the consideration of cloud cover and suitable sun illumination. It is also close to the Landsat local overpass time and matches Terra MODIS.

The payload of both HJ-1A/B includes two identical multi-spectral CCD image cameras that provide 30 meters spatial resolution and four bands observations from visible to near infrared: blue (Band 1, 0.43–0.52 µm), green (Band 2, 0.52–0.60 µm), red (Band 3, 0.63–0.69 µm) and near infrared (Band 4, 0.76–0.90 µm). The two CCD cameras are placed symmetrically with each other to the satellite nadir point. The placement makes the two CCD cameras equally divide the field of view and observe the earth side by side. The swath width of a single HJ1A/B CCD image is 360 km, while the combination of two CCD cameras obtains a swath width of 700 km. The CCD images can be freely obtained from the China Centre for Resource Satellite Data and Application (CRESDA). The relevant information is available at [40] with the interface in English.

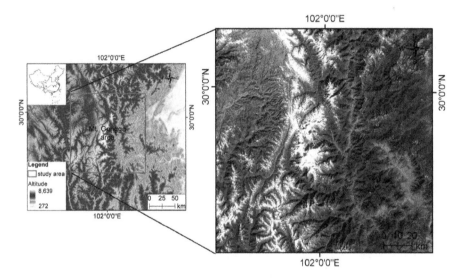

Figure 1. The geolocation and the digital elevation model of the study area. The right image is the subset of 19 November 2011 HJ-1B CCD1 image shown with 4, 3, and 2 bands in red, green and blue, respectively.

2.3. Image Pre-Processing

2.3.1. Geometric Correction

The HJ-1A/B images are distributed by the China Centre for Resource Satellite Data and Application (CRESDA), which are Level 2 products (after systematic geometric calibration). Precise registration of HJ CCD images was fulfilled by an automatic registration and topographic correction algorithm developed in [41]. The algorithm used the area-based image to image matching method to automatically select tie points between the geometrically corrected base images and HJ-1A/B CCD images. In this paper, the Global Land Survey (GLS) Landsat images from the year 2005 with geolocation error less than 30 m were used as base images [42]. The elevation data from the Shuttle Radar Topography Mission (SRTM) were used to correct the parallax errors caused by the local topographic relief [43].

2.3.2. Radiometric Calibration

The radiometric calibration is a fundamental process to eliminate the influence of the attenuation of the sensor photoelectric system [5]. It is especially important for images acquired from the constellation because of the significantly different attenuation of sensors in the different satellites [44]. The original digital number (DN) value recorded in HJ-1A/B multi-spectral CCD images were converted into Top Of Atmosphere (TOA) reflectance according to [45].

2.3.3. HJ-1A/B CCD Time Series Stack

In this paper, the HJ-1A/B CCD time series stack (HJTSS) refers to a sequence of HJ-1A/B CCD images acquired at a nominal temporal interval for the particular geographical region. Same with the temporal resolution of HJ-1A/B CCD images, the temporal interval of HJTSS is approximately two days. To get the same geographical region of different HJ-1A/B CCD images, tiles with fixed dimension (number of rows and columns) instead of original path/row were used to organize the HJTSS. The orbit configuration of HJ-1A/B constellation was the main factor considered when using the fixed tiles [46]. Data volume of a single file was considered to determine the tile dimension. The tile was finally determined as 5000 × 5000 pixels to ensure manageable file sizes. In this paper, in total, 181 scenes of images from the year 2011 with the paths 14–19 and row 80 were selected as the test images.

3. Methodology

To overcome the shortcomings caused by the limited HJ-1A/B spectral bands, the new clouds and snow discrimination method that combines spectral information with spatio-temporal context is described in this section. The proposed method consists of three major stages, as shown in the flowchart (Figure 2): the initial spectral test, the temporal context test, and the spatial context test. In the first stage, clouds and snow pixels are extracted together based on the whiteness and the Haze Optimized Transformation (HOT) spectral tests. Then, in the second stage, the spectral differences between HJTSS image and the cloud-free reference image are calculated to discriminate most cloud from snow pixels. To get the cloud-free reference images, the HJTSS in each month are firstly composed based on a simple modified maximum NDVI value method in this stage. Considering there might be residual clouds in the composites, the median value cloud screening method is used to detect the residual clouds pixels, and then the Saviziky–Golay (S-G) filter-based reconstruction method is used to reconstruct the signatures of these clouds pixels. The above two stages are conducted on the spectral signatures and temporal domain. In the third stage, the synthetic spatial texture information calculated from the distance of Regional Covariance Matrix (RCM) is performed to correct the commission and omission errors in the first two stages. In the result, the clouds and snow pixels are finally discriminated.

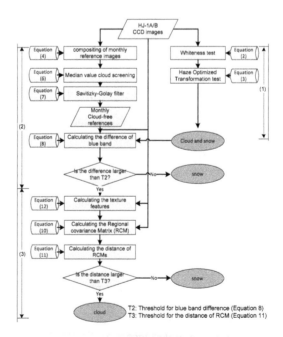

Figure 2. Flowchart of the algorithm development.

3.1. Initial Spectral Test for Cloud and Snow

Because of the similar spectral response of snow and cloud in the visible (VIS) and near-infrared (NIR) bands, two spectral tests were firstly used to extract both snow and cloud pixels for all of the HJTSS. The whiteness test (WT), originally proposed by Gomez-Chova *et al.* [47], was used as the fundamental test in this study. Since both thick cloud and snow are usually visually bright in the VIS and NIR band, the WT test is very effective for thick cloud and snow detection. It is based on following equations:

$$MeanVIS = (band1 + band2 + band3)/3 \qquad (1)$$

$$WT = \sum_{i=1}^{3} |(band\ i - meanVIS)/MeanVIS| < 0.3 \qquad (2)$$

where bands 1–3 are the blue (0.43–0.52 µm), green (0.52–0.60 µm) and red (0.63–0.69 µm) band of HJ-1A/B CCD images, respectively.

The WT test enhanced the brightness difference between the visible bands and the overall brightness. It works well in the ENVIronmental SATellite (ENVISAT) and Medium Resolution Imaging Spectrometer (MERIS) [1]. However, it has some drawbacks on thin clouds detection because the brightness of thin clouds is directly related to the land cover and is variable. If the underlying surfaces of the thin cloud have a high reflectance, the cloudy pixels are bright. Otherwise, the brightness of the thin cloudy pixels may be close to that of the bright and cloudiness pixels [13]. To fix this problem, the HOT test developed by Zhang *et al.* [48] was used here to detect haze or thin clouds. The basic assumption of HOT is that the spectral response to diverse surface cover classes under clear-sky conditions is highly correlated in visible bands, but the spectral response to haze and thin clouds is highly sensitive to both blue and red wavelengths. Therefore, this correlation change can be used to detect haze and thin clouds [1,21]. It can be expressed as:

$$HOTtest = band1 - 0.5 \times band3 - 0.08 > 0 \qquad (3)$$

where band 1 and 3 are the blue (0.43–0.52 µm) and red (0.63–0.69 µm) band of HJ-1A/B CCD images, respectively. It should be noted that due to the large reflectance of some bright pixels like barren rocks, turbid water or snow surface in the visible bands, the HOT might also include these pixels [1].

3.2. Separate Clouds from Snow Using the Temporal Context

As discussed earlier, because of the lack of SWIR band, separate clouds from snow using only visual to NIR spectral information is very difficult. However, from temporal aspect view, it is obvious that clouds cannot stay at the same place persistently due to its mobility characteristics. Therefore, it might be caused by the presence of clouds if surface reflectance has a big variation in time series observation [23]. The temporal context is the effective information and could be used to separate clouds from snow.

3.2.1. Compositing for the Monthly Cloud-Free Reference Images

A cloud-free reference image was firstly needed to detect the spectral variation of clouds in the HJTSS. Considering the cloud-free images are usually difficult to be acquired, reference images in this paper were composited from the partly cloudy images in each month. Many image compositing methods have been reported in the literature, which usually applied the minimum or maximum criteria [49–53]. Among these methods, the maximum NDVI value method is widely used because it can obtain the optimum vegetation observation and eliminate the cloud shadow to some extent. However, recent studies also noted that the compositing results from the maximum NDVI value method (Figure 3a) fails on the water surface because of the higher NDVI for cloud than water (As shown in area A) [53]. On the other hand, the composites produced from minimum blue band (Figure 3b) can chose the "clearest" pixel in a certain time window [52]. It performed better than the maximum NDVI over open water area, but it was still apparently affected by the contamination of cloud shadows (area B) because of the lower reflectance of cloud shadow than that of clear land. To take advantage of the maximum NDVI method on selecting vegetation pixels and the minimum blue method on choosing clear pixels, a simple combined criterion (Equation (4)) was used here for compositing of the references images.

$$\begin{cases} \max(NDVI_t) & \text{WaterTest} = 0 \\ \min(\rho_t^{\text{blue}}) & \text{WaterTest} = 1 \end{cases} \tag{4}$$

where

$$\text{Water Test} = (NDVI < 0.01 \text{ and Band } 4 < 0.11) \text{ or } (NDVI < 0.1 \text{ and Band } 4 < 0.05) \tag{5}$$

This combined criterion kept the open water information when clouds presented over the same geographic region with open water. It also preserved the optimal vegetation pixels in the compositing procedure. The water pixels were identified in the combined criterion firstly based on its physical characteristics: the lower signal observed in NIR wavelengths than visible wavelengths. The threshold of NDVI and Band 4 was inherited from the LEDAPS internal cloud masking algorithm where the number 0 represents none water pixel and 1 represents water pixel [21]. Figure 3c illustrated the image generated from the combined compositing criterion. Obviously, the composited results performed better than the above two methods and the quality of the composites was substantially improved.

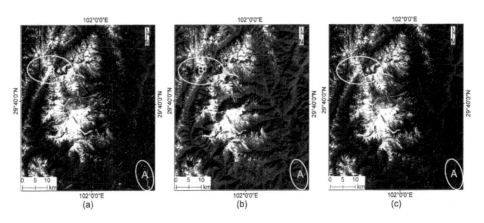

Figure 3. Cloud-free images generated by different composting methods: **(a)** Maximum NDVI; **(b)** Minimum blue; and **(c)** Combined compositing criterion.

3.2.2. Post-Processing for the Composites

Since some residual clouds may still exist in the above composites, the composites need to be further processed to get the cloud-free reference images. A median value cloud screening method was applied firstly to detect those residual clouds. This method is based on the assumption that the position of clouds will changed rapidly and clouds cannot stay at the same place persistently, and therefore the reflectance of clouds pixels will always be higher than the median values of the entire time series at the same location [9,54]. Because the short wavelengths are more sensitive to clouds, the blue band TOA reflectance was used in the median value cloud screening method. If the blue band TOA reflectance of a pixel in the composites is equal to or higher than the median value plus a constant, it is identified as a residual cloud pixel (Equation (6)).

$$\rho(\text{band } 1, x_j) \geq \text{median}(\rho(\text{band } 1, x_{\{1,2,3,\dots 12\}})) + T_1 \tag{6}$$

where, $\rho(\text{band1}, x_j)$ is the observed blue band TOA reflectance at month x for the jth pixel. T_1 is a constant to ensure that if the entire time series of a pixel is cloud free, pixels with blue band TOA reflectance higher than the median value will not be misidentified as clouds. Based on a test of all 181 images, the constant T_1 was determined as 0.04.

After successfully identified by the median value cloud screening method, the residual clouds pixels were then reconstructed by the S-G filter-based reconstruction method [55]. In the reconstruction method, the linear interpolation was firstly used to roughly predict the TOA reflectance of cloud pixels using the none-cloud TOA values from adjacent dates. Then, the S-G filter-based reconstruction method was used to estimate the TOA reflectance of those cloud pixels. The S-G filter is a simplified least-squares-fit convolution for smoothing and computing derivate of a set of consecutive values [56]. It uses a high-order polynomial instead of a constant to achieve the least-squares fitting within the sliding window to approximate the base function. Taking a fixed number of points in the vicinity point to fit a polynomial, it gives the smooth value of the vicinity point according to the polynomial during the fitting progress [57].

$$\hat{\rho}(b, x_j) = \sum_{i=-m}^{i=m} \frac{C_i \rho_{j+i}}{N} \tag{7}$$

where $\hat{\rho}(b, x_j)$ is the predicted TOA reflectance of the bth band at month j for the x cloud pixel, C_i is the coefficient given by the Savizky–Golay filter and N is the number of pixels in the smoothing window, which is equal to the smoothing window size $(2m + 1)$.

To illustrate the reconstruction effects for those residual clouds, Figure 4 shows the comparison before and after reconstruction of three typical areas. Obviously, the median method successfully captured the residual clouds in the composites, and after reconstruction, those cloudy areas transited smoothly with those clear grounds.

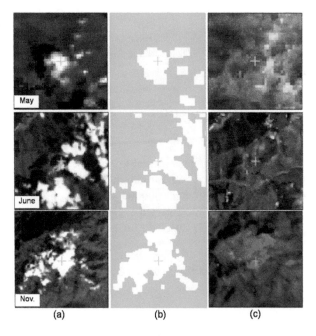

Figure 4. Comparison of residual clouds in the monthly composited images and their Savizky–Golay filter reconstruction results of May, June and November: (a) the residual clouds in the monthly composites; and (b) the residual clouds mask generated by the median value method, where the **yellow** color is clouds and **green** color is background. The identified clouds are a little larger in extent because of three pixel dilation of clouds to exclude some thin edges of clouds; (c) The reconstruction results by the Savizky–Golay filter.

3.2.3. Cloud and Snow Discrimination by the Reference Images

Based on Hagolle *et al.* [23], a pixel can be flagged as a cloud pixel if its blue band TOA reflectance satisfies the following multi-temporal criterion:

$$[\rho_{blue}(D) - \rho_{blue}(D_r)] > T_2 \times (1 + (D - D_r)/D_{T,n}) \tag{8}$$

where $\rho_{blue}(D)$ is the blue band TOA reflectance of a given pixel at date D, and $\rho_{blue}(D_r)$ is the corresponding blue band TOA reflectance of cloud-free reference images, which is the monthly composited reference images in this paper. $D - D_r$ is the number of days from the test date to reference date and expressed in days. It was calculated from the acquired date of a given image and the Day Of Year (DOY) data layer stored in the composited images. The blue band difference was used instead of other bands because of the relatively high reflectance for clouds and snow and low reflectance for most of the earth surface in this band [9].

When dates between the reference and cloud images are very close, the threshold T_2 for Equation (8) was set as 0.03 according to the reflectance difference analysis in [23], and the $D_{T,n}$ was set as a constant which was 30 days. However, due to the different set of spectral width, the optimal threshold for HJ CCD images needs to be adjusted. To find the optimal threshold for HJ CCD images, we analyzed the change of TOA reflectance for snow, cloud and clear land pixels at the two-day interval for all the HJ CCD images in 2011. In Figure 5, it is noted that a threshold of 0.05 can be better used to separate cloud from clear land. The $D_{T,n}$ was set according to the length of days in each month.

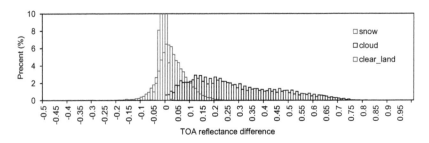

Figure 5. Histogram for Top Of Atmosphere (TOA) reflectance difference for the snow, clouds and clear ground pixels at two-day intervals.

Although the multi-temporal criterion is efficient to separate most of cloud from snow pixels, it still has some drawbacks. For example, some thin clouds might be regarded as clear ground since the spectral variation in multi-temporal images is quite slow. Besides, snow pixels with brighter reflectance than reference images may be mis-detected as clouds according to Equation (8). Therefore, the following spatial context criterion was used to further process the incorrectly detected and undetected clouds.

3.3. Separate Clouds from Snow Using the Synthesize Spatial Context

3.3.1. Theoretical Basis

To cope with the above problems, the spatial context of clouds and snow were further used in the third stage of the proposed method. The RCM that is widely used in the target detection and tracking [58,59] field was conducted to synthesize the multiple texture features in this stage. RCM is a fast region descriptor for object detection and classification [60]. Different from only using one texture feature, the RCM can synthesize several image statistics, including the spectral reflectance, gradient or filter responses as image features and then use the covariance of these features as the region descriptor. It is shown that large rotations and illumination changes can be absorbed by the RCM, and the noise corrupting individual samples can also be filtered out [61]. The similarity measurement of two RCM is the distance metric, which is defined on the positive definite symmetric matrices because the covariance matrices are not elements of the Euclidean space.

For the computing of the RCM, let I be the remote sensing images, and F be the $W \times H \times d$ dimensional features extracted from I:

$$F(x,y) = \phi(I,x,y) \tag{9}$$

where the function ϕ can be any mapping such as the gray values, color, gradients and filter response. x and y are the row and column in the image coordinate system. For a given rectangular $R \subset F$, let $\{z_k\}_{k=1,...,n}$ be the d-dimensional feature points inside R. The regional R can be represented with the $d \times d$ covariance matrix of the feature points:

$$C_R = \frac{1}{n-1} \sum_{k=1}^{n} (z_k - \mu)(z_k - \mu)^T \tag{10}$$

where μ is the mean of the points, n is the number of pixels.

Since the distance between RCM does not lie in Euclidean space, the similarity of two covariance matrices can be measured by Förstner et al. [62]:

$$d(C_0, C_T) = \sqrt{\sum_{i=1}^{n} \ln^2(\lambda_i(C_0, C_T))} \tag{11}$$

where $d(C_0, C_T)$ is the distance for RCM between reference C_0 and the detecting images C_T. $\{\lambda_i(C_0, C_T)\}_{i=1... n}$ is the generalized eigenvalues of C_0 and C_T.

3.3.2. RCM Implement for Cloud and Snow Discrimination

For cloud and snow discrimination, given a test and reference image, the aim of using RCM is to enhance the information of the cloudy area and weaken that of cloudless area. Since the presence of clouds makes both spectral signature and texture characters of cloud region change dramatically compared with reference images, the distance of RCM is sensitive to the cloudy area and insensitive to those clear grounds. When RCM of HJTSS and reference images shows a larger distance than the threshold, it may be caused by the clouds contamination.

Four bands TOA reflectance and the calculated NDVI were used for the development of RCM for both HJTSS and monthly reference composites. The responses from the Sobel operator and Laplace of Gaussian (LoG) operator, representing the first and second order derivatives, were also conducted as texture features and were included in RCM. Sobel operator is a non-linear edge enhancement detector filter that uses an approximation of the Sobel function to get the first order derivatives of a given image. It has a simple form and is widely used in the edge detection field. The LoG operator firstly uses the Gaussian convolution filter to reduce the image noise, and then adopts the Laplace operator to seek the zero crossing point of the second derivative of the image for edge detection, which improves the robustness to the noise and discrete points. The homogeneity feature, which is extracted from Gray Level Co-occurrence Matrix (GLCM) and measures the closeness of the distribution of elements in the GLCM to the GLCM diagonal [63], was also introduced into RCM. The eight-dimensional feature vector can be written in Equation (12) as defined in Equation (9):

$$F(x,y) = [b_{1-4}(x,y), NDVI(x,y), Sobel(x,y), LoG(x,y), Homo\_GLCM]^T \tag{12}$$

where, $b_{1-4}(x, y)$ are the four bands TOA reflectance of HJ-1A/B images; $NDVI(x, y)$ is calculated from band 4 and 3; $Sobel(x, y)$ is the blue band response from Sobel operator; $LoG(x, y)$ is the blue band response from Laplace of Gaussian operator; and Homo_GLCM is the homogeneity feature extracted from GLCM, (x, y) are the pixel locations.

The implementation steps were as follows. Firstly, a small region with 3×3 pixel matrix was placed inside the image, as showed in Figure 6. The RCMs of the small region were first calculated for the test (C_1) and reference image (C_2) through Equation (10). Then the distance between C_1 and C_2 was calculated according to Equation (11). A threshold of 6.50 was chosen for the optimal cloud and snow discrimination (Figure 6). This threshold was derived based on a test of all the 181 images. For most of clear and snow pixels, the distance between references and the HJTSS were always less than 6.50. While if there were clouds, even some thin clouds presented, the distance between RCMs increased dramatically and was usually larger than 6.50. Therefore, this threshold of 6.50 was helpful to discriminate the un-detected thin cloud and mis-detected snow pixels in the first two steps. After the RCM analysis, all cloudy pixels were dilated by three pixels in all eight connected directions to remove the surrounding pixels that may be partially influenced by clouds.

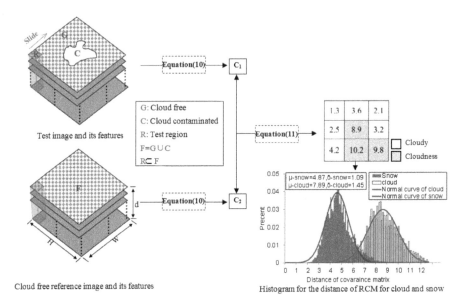

Figure 6. Image features construction and Regional Covariance Matrix (RCM) differences detection.

3.4. Accuracy Assessment

The following three different accuracies measures, which were the overall accuracy, precision and recall, were used to assess the accuracy of the algorithm results. Take the cloud as an example. Define the True Positives (TP) as the number of clouds correctly labeled as belonging to clouds in the algorithm, False Positives (FP) as the number of none-clouds incorrectly labeled as belonging to clouds, False Negatives (FN) as number of clouds incorrectly labeled as belonging to none-clouds, and True Negatives (TN) as number of none clouds which also labeled as belonging to none clouds. The accuracy, precision and recall are then defined as [64]:

$$Overall\,Accuracy = (TP + TN)/(TP + TN + FP + FN) \tag{13}$$

$$Precision = TP/(TP + FP) \tag{14}$$

$$Recall = TP/(TP + FN) \tag{15}$$

In the clouds case, precision denotes the proportion of truly cloud pixels in the cloud detection results, while recall is of all pixels that are actually clouds in the image, what fraction of them were detected as clouds. For precision and recall, they are better reflects the errors of omission and commission for the snow and cloud classes than overall accuracy. These three accuracies for cloud and snow were calculated separately.

4. Results and Analysis

4.1. Mask Results

Figure 7 depicts the cloud and snow detection results for HJ-1A/B images from four different dates. By visually comparing the results with the false color composites (RGB = 4, 3, 2), it is clear that the algorithm developed in this study can accurately separate the cloud and snow pixels. Figure 7a was a winter image acquired at 14 January 2011 with obvious clouds and snow over the whole image. The mask results showed its strong ability in excluding the snow pixels from cloud pixels. On the other hand, Figure 7b was a spring image acquired at 14 April 2011 with large variability in surface reflectance. The red quadrangles in the Figure 7b were covered by bare rocks (see red arrows) and snow (see yellow arrows), which were the bright earth surfaces in the current color scheme. The mask results worked well in terms of identifying clouds in areas with very bright land surface. Figure 7c was an image acquired at 12 June 2011, with many thick and thin cirrus clouds (see the yellow arrows). It is noted that the thin clouds were also accurately identified through the proposed method. Figure 7d was acquired at 1 September 2011 with bare rocks (see yellow arrow) and thick aerosols in the image (see the red arrow). In the mask result, the aerosols and bare bright rocks were successfully excluded and the cloudy pixels were also accurately identified.

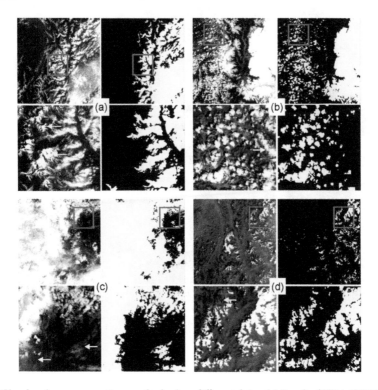

Figure 7. Cloud and snow separation results for four different dates. (**a**) Result of HJ1A-CCD1-20110114 scene; (**b**) Result of HJ1A-CCD1-20110414 scene; (**c**) Result of 12 June HJ1A-CCD2-20110612 scene; and (**d**) Result of HJ1B-CCD1-20110901 scene. For each date, (**Upper left**) and (**Upper right**) show false color composited HJ images and the corresponding cloud mask; and (**Lower left**) and (**Low right**) images are enlargements of (**Upper left**) and (**Upper right**) images, respectively with a size of 1000 pixels × 1000 pixels.

Figure 8 provides an illustration of the algorithm performance for the four middle month of every season of the year. The standard seasonal definition for the Northern Hemisphere adopted by the climate modeling community is used where spring is defined by the months March to May. Each column contained 14 images. Time intervals for most of these images were two days because of the two days revisiting period for the HJ-1A/B constellation. The mobility of clouds and the temporal contextual information provided by different periods over the same geographical regions, clearly visible in these images, was critical to the success of the developed algorithm. Generally, the algorithm developed in this paper worked well for most of the images for every season. The performance of the algorithm was robust regardless of the snow brightness. In general, it achieved the best performance when the surface condition was stable, for example in July and October with durable snow cover. However, the algorithm tended to overestimate cloudiness during sudden snow fall or snowmelt period when surface changes were rapid and the composites reference had less snow than these days. For instance, the first column, which represented the winter season, showed overestimating of some clouds. The qualitative evaluation was an important aspect of the development of the algorithm. To more rigorously assess its accuracy, reference data were used.

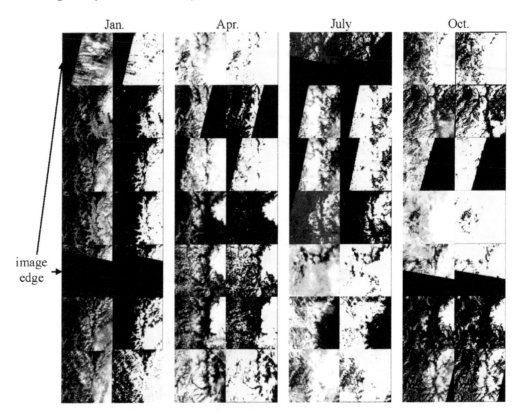

Figure 8. *Cont.*

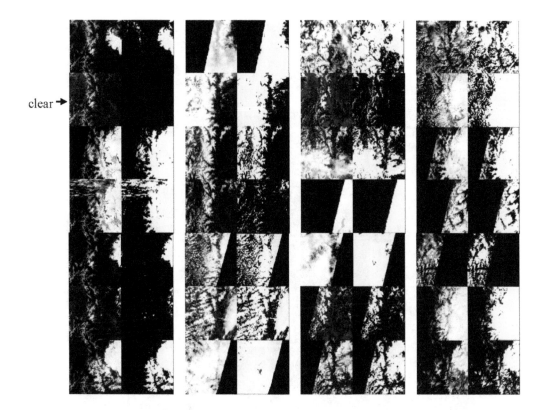

Figure 8. Examples of performance of the proposed algorithm for 150 km × 150 km Mt. Gongga area for the middle month of each season of the year 2011. The standard seasonal definition for the Northern Hemisphere adopted by the climate modeling community is used, where spring is defined by the months March to May. False color composited images with bands 4, 3, and 2 in **red**, **green** and **blue**, respectively, are shown on the left, and cloud masks are shown on the right with yellow color.

4.2. Performance of Each Stage for the Cloud and Snow Discrimination

To illustrate the effectiveness of each stage for cloud and snow discrimination in the proposed method, Figure 9 shows the HJ-1A/B images acquired in four different dates, the corresponding monthly reference composites, initial mask results from spectral test, mask results after temporal contextual test, final mask results after spatial contextual test, and the distance of RCM between references images and HJ-1A/B images, respectively.

In general, the improvement for the cloud and snow discrimination for each stage can be evidently elucidated from Figure 9. For the initial spectral test (Figure 9, row iii), it is apparent that the snow and clouds were accurately separated from clear grounds. However, snow and clouds pixels can only be extracted together in this stage due to their spectral similarity. On the other hand, after the temporal contextual test, many snow pixels were dramatically eliminated from clouds. According to statistics for the four case areas in Figure 9, approximately 79.98% misclassified clouds pixels from initial spectral test were successfully separated by temporal contextual test. However, it is also noted that some snow pixels still existed in the temporal test results (see red arrow). Particularly, after using of RCM the snow pixels are efficiently excluded from clouds (in the red ellipse). The RCM test further eliminated 20.02% of the misclassified cloud pixels from the temporal contextual test.

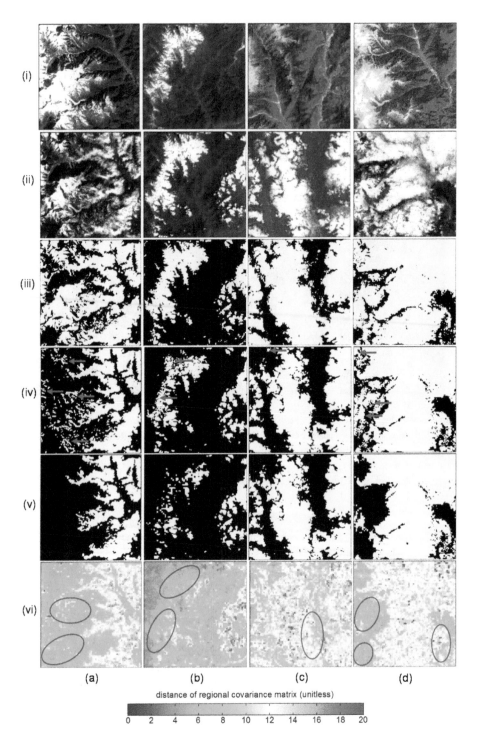

Figure 9. The comparison before and after using the regional covariance matrix (RCM). Column (**a–d**) are the HJ images acquired in 6 January, 30 April, 29 June and 10 October and its corresponding cloud and snow discriminating results in each stage, respectively. The first and second row are the corresponding monthly composited images and the original images shown in false color with bands 4, 3, and 2 in **red**, **green** and **blue**, respectively. The third, fourth and fifth rows are the cloud masks for spectral test, temporal test, and texture test, respectively. The sixth row is the distance of RCM with its color bar under the bottom of this figure.

4.3. Pixel Accuracy Assessment

Based on the visual assessment, clouds and snow pixels in HJTSS from different dates can be effectively discriminated. To further quantitatively assess the accuracy of the results, a reference cloud mask for each image was derived and the pixel-level accuracy was evaluated. The reference cloud masks were derived by a supervised classification method and manual editing method as follows. First, the object-based classification algorithm was used to segment the images into objects. Then, sample objects of both clouds and snow were chosen as the training datasets to construct a decision tree to classify the segmented objects into clouds, snow and clear grounds. Last, the classification results were edited manually to eliminate misclassification errors, leading to the production of the final reference masks.

To reduce the manual editing effort, a pseudo-random number function was used to randomly select a tile with 500 pixels × 500 pixels dimension within each image as the reference cloud mask. The HJTSS were firstly divided into 10 × 10 tiles, which were numbered consecutively from 1 to 100 from left to right and top to bottom (Figure 10a). Then, a random positive number, representing the selected tile, was selected from 1 to 100 using the psedo-random number function. The final selected reference tiles for all the HJTSS are shown in Figure 10b. Each number in the grid represents how many times this tile was chose.

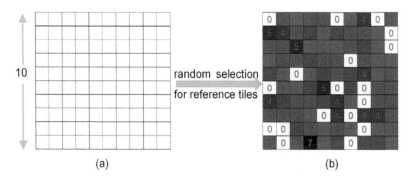

random selection for reference tiles

(a)   (b)

Figure 10. (a) is the tiles grid for the reference masks and **(b)** is the selection results for each tile.

Figure 11 displays the scatter plots of cloud cover percentage between all the reference cloud masks and the algorithm cloud masks. Overall, estimates of percent cloud cover from our proposed method were very accurate (Figure 11), with an R^2 of more than 0.9. The slope of the regression line was 0.94, with a very small interception (3.29%), and relatively small Root Mean Square Error (RMSE) (8.89%). However, from the above comparison of all the reference masks, it is also noted that the agreement for some of the days were relatively lower than other days. These disagreements were mainly caused by the sudden snowfall event after examination, which will be further discussed in Section 5.3.

Figure 12 illustrates the histograms of the overall accuracy, precision and recall for clouds and snow, respectively. At the pixel scale, for clouds, the average overall accuracy was 91.32% with a small standard deviation of 6.5% (Figure 12a). The average precision was 85.33% (Figure 12b) with a standard deviation of 14.17%. Moreover, the average recall was 81.82% (Figure 12c) with a standard deviation of 12.31%. For snow, the average overall accuracy was 92.80% with a small standard deviation of 5.3% (Figure 12d). The average precision was 82.18% (Figure 12e) with a standard deviation of 15.28%. Moreover, the average recall was 82.81% (Figure 12f) with a standard deviation of 14.78%. It is noted that the overall accuracy and precision of clouds were slightly lower than that of snow. The reasons were mainly attributed to the three pixels dilated (in all eight connected directions) for all cloud pixels. The lower precision of clouds was reasonable because un-detected clouds will more greatly influence future image applications than a little lost data by buffered cloud edges. Because the spectral variation of clouds was larger than snow pixels, especially for those thin clouds, clouds will be more easily detected as clear ground. Therefore, the recall of clouds was slightly lower than that of snow.

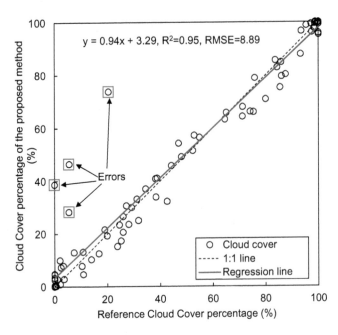

Figure 11. Visual cloud cover *vs.* detected cloud cover for the all the reference masks in the year 2011.

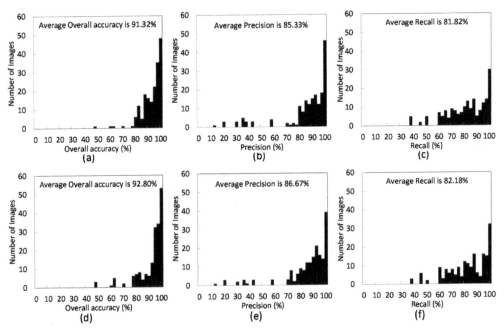

Figure 12. Histogram of the cloud overall accuracy, precision and recall: (**a–c**) accuracy, precision and recall for clouds, respectively; and (**d–f**) for snow.

5. Discussions

5.1. The Effectiveness of the Temporal Contextual Information for Cloud and Snow Discrimination

Temporal characteristics of satellite images offer important textural information for the discrimination of clouds and snow. With the added temporal information, only three optical and one NIR band were used in the proposed approach, making up for the SWIR band-lacking shortcomings for HJ-1A/B CCD images. The temporal information can be used in the following two ways for the separation of cloud and snow: (1) using the multi-temporal images to compose a cloud-free reference image and then make the spectral difference analysis [23,65]; and (2) making the time series analysis for all the multi-temporal images and then detect the sudden changes caused by clouds [8,9,24]. In this paper, the above two methods were applied to generate the monthly time series of cloud-free reference images. Those reference images were successfully used for clouds and snow discrimination.

For the compositing of the cloud-free reference images, the setting of time interval is an important factor that will influence final detection accuracy [13]. If the time interval is too short, the available images in the time window will be very limited, and the composites might still have many residual clouds. Two kinds of cloud underestimation condition will be caused when the reference images still have clouds. One is that snow in the HJTSS appears at the same location of clouds. At this condition, the difference between HJTSS and reference images would be the difference between snow and clouds (close to 0) and therefore omission errors will be introduced. The other underestimation condition is that clouds in the HJTSS appear in the same clouds region. This condition will also cause omission errors because the composites still have residual clouds. On the contrary, if the time interval is too long such as the seasonal compositing, although increased observation frequency could further eliminate the clouds influence for composites, the regular phenological changes of geographical features might be another important uncertainty source. The following two aspects were taken into consideration when the time interval was set as monthly composites in this paper. Firstly, for the revisiting period of HJ-1A/B is two days, there will be enough images to composite a nearly clear-sky reference image in one month [46]. According to the spectral changes analysis and the temporal filter, the totally cloud-free references images can be acquired for every month. Secondly, considering the spectral variation for most of vegetation and land surface are slightly in a month, monthly composites is representative and suitable for the cloud detection. Despite this, a clear-sky sub-monthly composite reference image such as the semi-month would be better than the monthly composites due to the shorter temporal difference between the test and reference images.

5.2. The Usefulness of Spatial Contextual Information for Cloud and Snow Discrimination

Texture refers to the repeated local patterns and their regular arrangement of the ground objects in the images. The texture features can better reflect the macroscopic properties and the detail structures of the ground objects than a single pixel. Correctly understanding the texture difference of clouds and snow can provide the important basis for their discrimination. To our knowledge, the texture of clouds is a type of random texture. Even though the texture element of clouds is variable and unpredictable, it is greatly different from the texture feature of ground object and snow area. For instance, the edge of the cloud contains gray level jump characteristics, and the part of the cloud is similarity to the whole. The cloud cluster has a certain fractals similarity [30]. On the other hand, the texture of snow in remote sensing images usually is bumpy because the influence from terrain relief, vegetation or man-made features. The gray gradient of snow usually is larger than that of clouds.

The frequently-used texture features of cloud are the average gradient, fractal dimension and GLCM [63]. In this paper, texture features derives from gray gradient (the first and second order derivatives) and the spatial correlation properties of gray (that is the GLCM) were used for clouds and snow discrimination. These texture features has been demonstrated effectively for cloud or snow detection. However, some other texture features such as GLCM [63], pixel shape index [66], morphological profiles [67], and wavelet-based texture [68] can also evidently improve the accuracy of snow or cloud cover extraction. However, increasing the number of texture features does not consequently lead to higher accuracy but will result in a large computing cost [29,32]. Therefore, more efforts in the future can be put into find the representative and less computing-consuming features to improve the efficiency and accuracy.

The RCM was used to synthesize all the spectral and texture features into a distance index between HJTSS and reference images to express their similarity at the regional scales. The RCM enhanced the cloud and snow region difference at the regional scale, which is helpful for the discriminating snow from clouds for the following two cases. The first case is that snow areas in HJTSS are larger than that in reference images. In this condition, since the reference images were composited from the HJTSS in each month, the smaller snow areas in reference image than HJTSS will cause the overestimate of clouds and consequently underestimate snow pixels. Another case is that the snow spectral signature in HJTSS varied higher than the cloud spectral threshold (T_1 in Equation (6)). In this condition, due to the melt effects of snow or differences in observation conditions, the reflectance difference between snow in HJTSS and reference image might also introduce overestimate of clouds. Fortunately, since the RCM was calculated from a series of regional texture information, it has a strong capability of filtering out the noise corrupting individual samples. Therefore, both of the above cases can be eliminated at the regional scales by RCM.

5.3. Error Sources of the Proposed Method

Overall, the cloud and snow detection results indicated that the proposed method can achieve a good performance for the discrimination of cloud and snow in HJ-1A/B images with only four bands within a whole year. However, several error sources which might influence the algorithm accuracy should also be pointed out. The first error source came from the way which the cloud boundary was treated. In order to remove the cloud surrounding pixels that may be partially contaminated, the proposed method dilated all cloud pixels by three pixels in all eight connected directions, which might overestimate clouds areas. Therefore, the overall accuracy and precision of clouds were slightly lower than that of snow (As shown in Figure 12). Another error source might be the undetected cirrus clouds. To our knowledge, the detection of cirrus clouds is very challenging since regions covered by cirrus clouds not only contain signature and texture information from clouds but also from the ground features [13,69]. For example, Kovalskyy and Roy [12] recently found that about 7% cirrus contaminated pixels in historical conterminous United States Landsat archive were undetected. These thin and cirrus clouds might also be omitted because of the lack of cirrus cloud sensitive bands (e.g., Landsat-8 cirrus band). Finally, while sudden snowfall happens (As shown in Figure 11), some snow pixels might be incorrectly labeled as clouds. An auxiliary cloud and snow mask with the same temporal resolution such as from MODIS may be helpful for the discrimination of clouds and sudden snowfalls.

5.4. Applicability of the Developed Methods in the Future

The proposed method is highly automatic and efficient when processing a tremendously large volume of imagery in near-real time. It can be easily implemented on a parallel processor. Since the HJ-1A/B has been in orbit for over seven years and these satellites still run well, the proposed methods can be used for cloud and snow discrimination for all the historical archive images, and increasing more areas in the foreseeable future. Besides, it should be also noted that although the proposed method was designed for use with HJ-1A/B images, it can be adapted for images acquired by the similar satellite instruments such as Sentinel-2A/B [70], which have similar spectral bands and temporal resolutions. The method in this paper is general and efforts in the future will be put into the test for other regions with different environments.

6. Conclusions

Accurate detection of clouds for satellite images is the fundamental pre-processing step for a variety of remote sensing applications. This paper presented a new practical method for cloud and snow discrimination by combining ideas from many past approaches and integrating the spectral signatures with spatio-temporal contextual information. The methodology included three closely related major stages, initial spectral test, temporal context test, and spectral context test.

Visual assessment revealed that the method developed in this paper can accurately separate the cloud and snow pixels. The pixel-level accuracy assessment was performed by comparing the detection results with the reference cloud masks generated by a random-tile validation scheme. Good agreements were found between detection results and reference cloud masks, with the average overall accuracy, precision and recall for clouds being 91.32%, 85.33% and 81.82%, respectively. The temporal contextual test can exclude approximately 79.98% misclassified clouds pixels from initial spectral test, while the spatial contextual test can further exclude the 20.02% residual misclassified clouds pixels. Generally, the proposed method exhibited high accuracy for clouds and snow discrimination of SWIR-lacking HJ-1A/B CCD images and was an improvement over the traditional spectral-based algorithms. It can provide an accurate cloud mask for the on-going HJ-1A/B images and the similar satellites with the same temporal and spectral settings. The calculating efficiency and accuracy of the result can be improved by comparing the effectiveness of different texture features in the future.

Acknowledgments: This research was funded jointly by the National Natural Science Foundation project of China (41271433, 41571373), the International Cooperation Key Project of CAS (Grant No. GJHZ201320), the International Cooperation Partner Program of Innovative Team, CAS (Grant No. KZZD-EW-TZ-06), the "Hundred Talents" Project of Chinese Academy of Sciences (CAS), and the Hundred Young Talents Program of the Institute of Mountain Hazards and Environment. We are grateful to all the contractors, image providers and the anonymous reviewers for their valuable comments and suggestions.

Author Contributions: All authors have made major and unique contributions. Jinhu Bian developed the algorithm and drafted the preliminary version of this paper. Ainong Li designed the framework of this research, and finished the final version of this paper. Qiannan Liu processed the major data sources and assisted in the validation work and Chengquan Huang assisted in the manuscript revision.

Conflicts of Interest: The authors declare no conflict of interest.

References

1. Zhu, Z.; Woodcock, C.E. Object-based cloud and cloud shadow detection in Landsat imagery. *Remote Sens. Environ.* **2012**, *118*, 83–94. [CrossRef]
2. Li, A.; Jiang, J.; Bian, J.; Deng, W. Combining the matter element model with the associated function of probability transformation for multi-source remote sensing data classification in mountainous regions. *ISPRS J. Photogramm.* **2012**, *67*, 80–92. [CrossRef]
3. Huang, C.Q.; Coward, S.N.; Masek, J.G.; Thomas, N.; Zhu, Z.L.; Vogelmann, J.E. An automated approach for reconstructing recent forest disturbance history using dense Landsat time series stacks. *Remote Sens. Environ.* **2010**, *114*, 183–198. [CrossRef]
4. White, J.C.; Wulder, M.A.; Hobart, G.W.; Luther, J.E.; Hermosilla, T.; Griffiths, P.; Coops, N.C.; Hall, R.J.; Hostert, P.; Dyk, A.; *et al.* Pixel-based image compositing for large-area dense time series applications and science. *Can. J. Remote Sens.* **2014**, *40*, 192–212. [CrossRef]
5. Roy, D.P.; Ju, J.C.; Kline, K.; Scaramuzza, P.L.; Kovalskyy, V.; Hansen, M.; Loveland, T.R.; Vermote, E.; Zhang, C.S. Web-Enabled Landsat Data (WELD): Landsat ETM plus composited mosaics of the conterminous United States. *Remote Sens. Environ.* **2010**, *114*, 35–49. [CrossRef]
6. Baret, F.; Hagolle, O.; Geiger, B.; Bicheron, P.; Miras, B.; Huc, M.; Berthelot, B.; Niño, F.; Weiss, M.; Samain, O. Lai, fapar and fcover cyclopes global products derived from vegetation: Part 1: Principles of the algorithm. *Remote Sens. Environ.* **2007**, *110*, 275–286. [CrossRef]
7. Li, A.; Huang, C.; Sun, G.; Shi, H.; Toney, C.; Zhu, Z.L.; Rollins, M.; Goward, S.; Masek, J. Modeling the height of young forests regenerating from recent disturbances in Mississippi using Landsat and ICEsat data. *Remote Sens. Environ.* **2011**, *115*, 1837–1849. [CrossRef]

8.   Lyapustin, A.; Wang, Y.; Frey, R. An automatic cloud mask algorithm based on time series of MODIS measurements. *J. Geophys. Res.-Atmos.* **2008**. [CrossRef]

9.   Zhu, Z.; Woodcock, C.E. Automated cloud, cloud shadow, and snow detection in multitemporal Landsat data: An algorithm designed specifically for monitoring land cover change. *Remote Sens. Environ.* **2014**, *152*, 217–234. [CrossRef]

10. Zhu, Z.; Wang, S.; Woodcock, C.E. Improvement and expansion of the fmask algorithm: Cloud, cloud shadow, and snow detection for Landsats 4–7, 8, and Sentinel 2 images. *Remote Sens. Environ.* **2015**, *159*, 269–277. [CrossRef]

11. Platnick, S.; King, M.D.; Ackerman, S.A.; Menzel, W.P.; Baum, B.A.; Riedi, J.C.; Frey, R.A. The MODIS cloud products: Algorithms and examples from terra. *IEEE Trans. Geosci. Remote Sens.* **2003**, *41*, 459–473. [CrossRef]

12. Kovalskyy, V.; Roy, D. A one year Landsat 8 conterminous United States study of cirrus and non-cirrus clouds. *Remote Sens.* **2015**, *7*, 564–578. [CrossRef]

13. Shen, H.F.; Li, H.F.; Qian, Y.; Zhang, L.P.; Yuan, Q.Q. An effective thin cloud removal procedure for visible remote sensing images. *ISPRS J. Photogramm.* **2014**, *96*, 224–235. [CrossRef]

14. Amato, U.; Antomadis, A.; Cuomo, V.; Cutillo, L.; Franzese, M.; Murino, L.; Serio, C. Statistical cloud detection from SEVIRI multispectral images. *Remote Sens. Environ.* **2008**, *112*, 750–766. [CrossRef]

15. Hansen, M.C.; Roy, D.P.; Lindquist, E.; Adusei, B.; Justice, C.O.; Altstatt, A. A method for integrating MODIS and Landsat data for systematic monitoring of forest cover and change in the Congo basin. *Remote Sens. Environ.* **2008**, *112*, 2495–2513. [CrossRef]

16. Huang, C.Q.; Thomas, N.; Goward, S.N.; Masek, J.G.; Zhu, Z.L.; Townshend, J.R.G.; Vogelmann, J.E. Automated masking of cloud and cloud shadow for forest change analysis using Landsat images. *Int. J. Remote Sens.* **2010**, *31*, 5449–5464. [CrossRef]

17. Zhu, Z.; Woodcock, C.E.; Olofsson, P. Continuous monitoring of forest disturbance using all available Landsat imagery. *Remote Sens. Environ.* **2012**, *122*, 75–91. [CrossRef]

18. Dozier, J. Spectral signature of alpine snow cover from the Landsat Thematic Mapper. *Remote Sens. Environ.* **1989**, *28*, 9–12. [CrossRef]

19. Hall, D.K.; Riggs, G.A.; Salomonson, V.V. Development of methods for mapping global snow cover using moderate resolution imaging spectroradiometer data. *Remote Sens. Environ.* **1995**, *54*, 127–140. [CrossRef]

20. Choi, H.; Bindschadler, R. Cloud detection in Landsat imagery of ice sheets using shadow matching technique and automatic normalized difference snow index threshold value decision. *Remote Sens. Environ.* **2004**, *91*, 237–242. [CrossRef]

21. Vermote, E.; Saleous, N. Ledaps Surface Reflectance Product Description. 2007. Avaiable online: https://dwrgis.water.ca.gov/documents/269784/4654504/LEDAPS+Surface+Reflectance+Product+Description.pdf (accessed on 27 September 2015).

22. Masek, J.G.; Vermote, E.F.; Saleous, N.E.; Wolfe, R.; Hall, F.G.; Huemmrich, K.F.; Gao, F.; Kutler, J.; Lim, T.K. A land surface reflectance dataset for north America, 1990–2000. *IEEE Geosci. Remote Sens. Lett.* **2006**, *3*, 68–72. [CrossRef]

23. Hagolle, O.; Huc, M.; Pascual, D.V.; Dedieu, G. A multi-temporal method for cloud detection, applied to Formosat-2, Venmus, Landsat and Sentinel-2 images. *Remote Sens. Environ.* **2010**, *114*, 1747–1755. [CrossRef]

24. Goodwin, N.R.; Collett, L.J.; Denham, R.J.; Flood, N.; Tindall, D. Cloud and cloud shadow screening across queensland, australia: An automated method for Landsat TM/ETM Plus time series. *Remote Sens. Environ.* **2013**, *134*, 50–65. [CrossRef]

25. Tseng, D.C.; Tseng, H.T.; Chien, C.L. Automatic cloud removal from multitemporal SPOT images. *Appl. Math. Comput.* **2008**, *205*, 584–600. [CrossRef]

26. Tso, B.; Olsen, R.C. A contextual classification scheme based on MRF model with improved parameter estimation and multiscale fuzzy line process. *Remote Sens. Environ.* **2005**, *97*, 127–136. [CrossRef]

27. Liu, D.S.; Kelly, M.; Gong, P. A spatial-temporal approach to monitoring forest disease spread using multi-temporal high spatial resolution imagery. *Remote Sens. Environ.* **2006**, *101*, 167–180. [CrossRef]

28. Li, C.-H.; Kuo, B.-C.; Lin, C.-T.; Huang, C.-S. A spatial–contextual support vector machine for remotely sensed image classification. *IEEE Trans. Geosci. Remote Sens.* **2012**, *50*, 784–799. [CrossRef]

29. Hughes, G.P. On the mean accuracy of statistical pattern recognizers. *IEEE Trans. Inform. Theory* **1968**, *14*, 55–63. [CrossRef]

30. Li, P.; Dong, L.; Xiao, H.; Xu, M. A cloud image detection method based on SVM vector machine. *Neurocomputing* **2015**, *169*, 34–42. [CrossRef]

31. Vasquez, R.E.; Manian, V.B. Texture-based cloud detection in MODIS images. *Proc. SPIE* **2003**. [CrossRef]

32. Zhu, L.; Xiao, P.; Feng, X.; Zhang, X.; Wang, Z.; Jiang, L. Support vector machine-based decision tree for snow cover extraction in mountain areas using high spatial resolution remote sensing image. *J. Appl. Remote Sens.* **2014**, *8*, 084698. [CrossRef]

33. Racoviteanu, A.; Williams, M.W. Decision tree and texture analysis for mapping debris-covered glaciers in the Kangchenjunga area, eastern Himalaya. *Remote Sens.* **2012**, *4*, 3078–3109. [CrossRef]

34. Townshend, J.R.G.; Justice, C.O. Selecting the spatial-resolution of satellite sensors required for global monitoring of land transformations. *Int. J. Remote Sens.* **1988**, *9*, 187–236. [CrossRef]

35. Luo, J.; Chen, Y.; Wu, Y.; Shi, P.; She, J.; Zhou, P. Temporal-spatial variation and controls of soil respiration in different primary succession stages on glacier forehead in Gongga Mountain, China. *PLoS ONE* **2012**, *7*, e42354. [CrossRef] [PubMed]

36. Li, Z.; He, Y.; Pang, H.; Yang, X.; Jia, W.; Zhang, N.; Wang, X.; Ning, B.; Yuan, L.; Song, B. Source of major anions and cations of snowpacks in Hailuogou No. 1 glacier, Mt. Gongga and Baishui No. 1 glacier, Mt. Yulong. *J. Geogr. Sci.* **2008**, *18*, 115–125. [CrossRef]

37. Wang, Q.; Wu, C.; Li, Q.; Li, J. Chinese HJ-1A/B satellites and data characteristics. *Sci. China Earth Sci.* **2010**, *53*, 51–57. [CrossRef]

38. Chen, J.; Cui, T.W.; Qiu, Z.F.; Lin, C.S. A three-band semi-analytical model for deriving total suspended sediment concentration from HJ-1A/CCD data in turbid coastal waters. *ISPRS J. Photogramm.* **2014**, *93*, 1–13. [CrossRef]

39. Lu, S.L.; Wu, B.F.; Yan, N.N.; Wang, H. Water body mapping method with HJ-1A/B satellite imagery. *Int. J. Appl. Earth Obs.* **2011**, *13*, 428–434. [CrossRef]

40. China Centre for Resources Satellite Data and Application. Avaialble online: Http://www.Cresda.Com/site2/satellite/7117.Shtml (accessed on 27 September 2015).

41. Bian, J.; Li, A.; Jin, H.; Lei, G.; Huang, C.; Li, M. Auto-registration and orthorecification algorithm for the time series HJ-1A/B CCD images. *J. Mt. Sci.-Engl.* **2013**, *10*, 754–767. [CrossRef]

42. Gutman, G.; Huang, C.Q.; Chander, G.; Noojipady, P.; Masek, J.G. Assessment of the NASA-USGS Global Land Survey (GLS) datasets. *Remote Sens. Environ.* **2013**, *134*, 249–265. [CrossRef]

43. Wolfe, R.E.; Nishihama, M.; Fleig, A.J.; Kuyper, J.A.; Roy, D.P.; Storey, J.C.; Patt, F.S. Achieving sub-pixel geolocation accuracy in support of MODIS land science. *Remote Sens. Environ.* **2002**, *83*, 31–49. [CrossRef]

44. Li, J.; Chen, X.; Tian, L.; Feng, L. Tracking radiometric responsivity of optical sensors without on-board calibration systems-case of the Chinese HJ-1A/B CCD sensors. *Opt. Express* **2015**, *23*, 1829–1847. [CrossRef] [PubMed]

45. Zhao, Z.; Li, A.; Bian, J.; Huang, C. An improved DDV method to retrieve AOT for HJ CCD image in typical mountainous areas. *Spectrosc. Spect. Anal.* **2015**, *35*, 1479–1487. (In Chinese).

46. Bian, J.; Li, A.; Wang, Q.; Huang, C. Development of dense time series 30-m image products from the Chinese HJ-1A/B constellation: A case study in zoige plateau, china. *Remote Sens.* **2015**, *7*, 16647–16671. [CrossRef]

47. Gomez-Chova, L.; Camps-Valls, G.; Calpe-Maravilla, J.; Guanter, L.; Moreno, J. Cloud-screening algorithm for ENVISAT/MERIS multispectral images. *IEEE Trans. Geosci. Remote Sens.* **2007**, *45*, 4105–4118. [CrossRef]

48. Zhang, Y.; Guindon, B.; Cihlar, J. An image transform to characterize and compensate for spatial variations in thin cloud contamination of Landsat images. *Remote Sens. Environ.* **2002**, *82*, 173–187. [CrossRef]

49. Goward, S.N.; Turner, S.; Dye, D.G.; Liang, S. The University-of-Maryland improved global vegetation index product. *Int. J. Remote Sens.* **1994**, *15*, 3365–3395. [CrossRef]

50. Holben, B.N. Characteristics of maximum-value composite images from temporal AVHRR data. *Int. J. Remote Sens.* **1986**, *7*, 1417–1434. [CrossRef]

51. Roy, D.P. The impact of misregistration upon composited wide field of view satellite data and implications for change detection. *IEEE Trans. Geosci. Remote* **2000**, *38*, 2017–2032. [CrossRef]

52. Liang, S.; Zhong, B.; Fang, H. Improved estimation of aerosol optical depth from MODIS imagery over land surfaces. *Remote Sens. Environ.* **2006**, *104*, 416–425. [CrossRef]

53. Luo, Y.; Trishchenko, A.P.; Khlopenkov, K.V. Developing clear-sky, cloud and cloud shadow mask for producing clear-sky composites at 250-meter spatial resolution for the seven MODIS land bands over canada and north America. *Remote Sens. Environ.* **2008**, *112*, 4167–4185. [CrossRef]

54. Flood, N. Seasonal composite Landsat TM/ETM+ images using the Medoid (a multi-dimensional median). *Remote Sens.* **2013**, *5*, 6481–6500. [CrossRef]

55. Bian, J.H.; Li, A.N.; Song, M.Q.; Ma, L.Q.; Jiang, J.G. Reconstructing ndvi time-series data set of modis based on the Savitzky-Golay filter. *J. Remote Sens.* **2010**, *14*, 725–741.

56. Chen, J.; Jonsson, P.; Tamura, M.; Gu, Z.H.; Matsushita, B.; Eklundh, L. A simple method for reconstructing a high-quality NDVI time-series data set based on the Savitzky-Golay filter. *Remote Sens. Environ.* **2004**, *91*, 332–344. [CrossRef]

57. Savitzky, A.; Golay, M.J.E. Smoothing and differentiation of data by simplified least squares procedure. *Anal. Chem.* **1964**, *36*, 1627–1639. [CrossRef]

58. Zhou, H.Y.; Yuan, Y.; Zhang, Y.; Shi, C.M. Non-rigid object tracking in complex scenes. *Pattern Recogn. Lett.* **2009**, *30*, 98–102. [CrossRef]

59. Zhang, X.G.; Liu, H.H.; Li, X.L. Target tracking for mobile robot platforms via object matching and background anti-matching. *Robot. Auton. Syst.* **2010**, *58*, 1197–1206. [CrossRef]

60. Tuzel, O.; Porikli, F.; Meer, P. Region covariance: A fast descriptor for detection and classification. *Lect. Notes Comput. Sci.* **2006**, *3952*, 589–600.

61. Wang, Y.H.; Liu, H.W. A hierarchical ship detection scheme for high-resolution SAR images. *IEEE Trans. Geosci. Remote Sens.* **2012**, *50*, 4173–4184. [CrossRef]

62. Förstner, W.; Moonen, B. A metric for covariance matrices. In *Geodesy—The Challenge of the 3rd Millennium*; Grafarend, W.E., Krumm, F.W., Schwarze, V.S., Eds.; Springer-Verlag: Berlin, Germany; Heidelberg, Germany, 2003; pp. 299–309.

63. Haralick, R.M.; Shanmuga, K.; Dinstein, I. Textural features for image classification. *IEEE Trans. Syst. Man Cybern.* **1973**, *SMC-3*, 610–621. [CrossRef]

64. Powers, D.M. Evaluation: From precision, recall and f-measure to ROC, informedness, markedness and correlation. *J. Mach. Learn. Technol.* **2011**, *2*, 37–63.

65. Derrien, M.; le Gléau, H. Improvement of cloud detection near sunrise and sunset by temporal-differencing and region-growing techniques with real-time SEVIRI. *Int. J. Remote Sens.* **2010**, *31*, 1765–1780. [CrossRef]

66. Zhang, L.P.; Huang, X.; Huang, B.; Li, P.X. A pixel shape index coupled with spectral information for classification of high spatial resolution remotely sensed imagery. *IEEE Trans. Geosci. Remote Sens.* **2006**, *44*, 2950–2961. [CrossRef]

67. Fauvel, M.; Benediktsson, J.A.; Chanussot, J.; Sveinsson, J.R. Spectral and spatial classification of hyperspectral data using SVMS and morphological profiles. *IEEE Trans. Geosci. Remote Sens.* **2008**, *46*, 3804–3814. [CrossRef]

68. Huang, X.; Zhang, L.; Li, P. A multiscale feature fusion approach for classification of very high resolution satellite imagery based on wavelet transform. *Int. J. Remote Sens.* **2008**, *29*, 5923–5941. [CrossRef]

69. Li, H.F.; Zhang, L.P.; Shen, H.F.; Li, P.X. A variational gradient-based fusion method for visible and SWIR imagery. *Photogramm. Eng. Remote Sens.* **2012**, *78*, 947–958. [CrossRef]

70. Drusch, M.; del Bello, U.; Carlier, S.; Colin, O.; Fernandez, V.; Gascon, F.; Hoersch, B.; Isola, C.; Laberinti, P.; Martimort, P.; *et al.* Sentinel-2: ESA's optical high-resolution mission for GMES operational services. *Remote Sens. Environ.* **2012**, *120*, 25–36. [CrossRef]

Permissions

All chapters in this book were first published in Remote Sensing, by MDPI; hereby published with permission under the Creative Commons Attribution License or equivalent. Every chapter published in this book has been scrutinized by our experts. Their significance has been extensively debated. The topics covered herein carry significant findings which will fuel the growth of the discipline. They may even be implemented as practical applications or may be referred to as a beginning point for another development.

The contributors of this book come from diverse backgrounds, making this book a truly international effort. This book will bring forth new frontiers with its revolutionizing research information and detailed analysis of the nascent developments around the world.

We would like to thank all the contributing authors for lending their expertise to make the book truly unique. They have played a crucial role in the development of this book. Without their invaluable contributions this book wouldn't have been possible. They have made vital efforts to compile up to date information on the varied aspects of this subject to make this book a valuable addition to the collection of many professionals and students.

This book was conceptualized with the vision of imparting up-to-date information and advanced data in this field. To ensure the same, a matchless editorial board was set up. Every individual on the board went through rigorous rounds of assessment to prove their worth. After which they invested a large part of their time researching and compiling the most relevant data for our readers.

The editorial board has been involved in producing this book since its inception. They have spent rigorous hours researching and exploring the diverse topics which have resulted in the successful publishing of this book. They have passed on their knowledge of decades through this book. To expedite this challenging task, the publisher supported the team at every step. A small team of assistant editors was also appointed to further simplify the editing procedure and attain best results for the readers.

Apart from the editorial board, the designing team has also invested a significant amount of their time in understanding the subject and creating the most relevant covers. They scrutinized every image to scout for the most suitable representation of the subject and create an appropriate cover for the book.

The publishing team has been an ardent support to the editorial, designing and production team. Their endless efforts to recruit the best for this project, has resulted in the accomplishment of this book. They are a veteran in the field of academics and their pool of knowledge is as vast as their experience in printing. Their expertise and guidance has proved useful at every step. Their uncompromising quality standards have made this book an exceptional effort. Their encouragement from time to time has been an inspiration for everyone.

The publisher and the editorial board hope that this book will prove to be a valuable piece of knowledge for researchers, students, practitioners and scholars across the globe.

List of Contributors

Cheng Wang, Xiaohuan Xi, Dong Li, Shaobo Xia and Pinghua Wang
Key Laboratory of Digital Earth Science, Institute of Remote Sensing and Digital Earth, Chinese Academy of Sciences, Beijing 100094, China

Shezhou Luo
Key Laboratory of Digital Earth Science, Institute of Remote Sensing and Digital Earth, Chinese Academy of Sciences, Beijing 100094, China
Department of Geography and Program in Planning, University of Toronto, 100St. George St., Room 5047, Toronto, ON M5S 3G3, Canada

Hongcheng Zeng
Faculty of Forestry, University of Toronto, 33 Willcocks Street, Toronto, ON M5S 3B3, Canada

Caixia Liu
State Key Laboratory of Remote Sensing Science, Institute of Remote Sensing and Digital Earth, Chinese Academy of Sciences (CAS), Beijing 100101, China

Xiaoyi Wang and Huabing Huang
State Key Laboratory of Remote Sensing Science, Institute of Remote Sensing and Digital Earth, Chinese Academy of Sciences (CAS), Beijing 100101, China
Department of Environmental Sciences, Policy & Management, University of California, Berkeley, CA 94720, USA

Gregory S. Biging and Yanlei Chen
Department of Environmental Sciences, Policy & Management, University of California, Berkeley, CA 94720, USA

Jun Yang
Ministry of Education Key Laboratory for Earth System Modeling, Center for Earth System Science, Tsinghua University, Beijing 100084, China

Peng Gong
State Key Laboratory of Remote Sensing Science, Institute of Remote Sensing and Digital Earth, Chinese Academy of Sciences (CAS), Beijing 100101, China
Ministry of Education Key Laboratory for Earth System Modeling, Center for Earth System Science, Tsinghua University, Beijing 100084, China
Joint Center for Global Change Studies, Beijing 100875, China

Qinchuan Xin
School of Geography and Planning, Sun Yat-sen University, Guangzhou 510275, China

Xiaoli Yin and Guang Liu
Key Laboratory of Digital Earth Science, Institute of Remote Sensing and Digital Earth, Chinese Academyof Sciences, No. 9 Dengzhuang South Road, Beijing 100094, China

Li Zhang
Key Laboratory of Digital Earth Science, Institute of Remote Sensing and Digital Earth, Chinese Academyof Sciences, No. 9 Dengzhuang South Road, Beijing 100094, China
Hainan Key Laboratory of Earth Observation, Hainan 572029, China

Binghua Zhang
Key Laboratory of Digital Earth Science, Institute of Remote Sensing and Digital Earth, Chinese Academyof Sciences, No. 9 Dengzhuang South Road, Beijing 100094, China
Hainan Key Laboratory of Earth Observation, Hainan 572029, China
College of Resources and Environment, University of Chinese Academy of Sciences, No. 19A Yuquan Road, Beijing 100049, China

Dong Xie
Key Laboratory of Digital Earth Science, Institute of Remote Sensing and Digital Earth, Chinese Academy of Sciences, No. 9 Dengzhuang South Road, Beijing 100094, China
Department of Mathematics, The George Washington University, 2115 G St. NW, Washington, DC 20052, USA

Chunjing Liu
Key Laboratory of Digital Earth Science, Institute of Remote Sensing and Digital Earth, Chinese Academy of Sciences, No. 9 Dengzhuang South Road, Beijing 100094, China
College of Information Science and Engineering, Shandong Agricultural University, No. 61 Daizong Road, Taian 271018, China

Dan J. Krofcheck and Marcy E. Litvak
Department of Biology, University of New Mexico, Albuquerque, NM 87131, USA

Jan U. H. Eitel and Lee A. Vierling
Geospatial Laboratory for Environmental Dynamics, University of Idaho, Moscow, ID 83844, USA
McCall Outdoor Science School, University of Idaho, McCall, ID 83638, USA

Christopher D. Lippitt
Department of Geography and Environmental Studies, University of New Mexico, Albuquerque, NM 87131, USA

Urs Schulthess
CIMMYT-Bangladesh, House 10/B, Road 53, Gulshan-2, Dhaka 1213, Bangladesh

Alberto Rodriguez-Ramirez
Global Change Institute, The University of Queensland, St Lucia, QLD 4072, Australia

Tadzio Holtrop
Global Change Institute, The University of Queensland, St Lucia, QLD 4072, Australia
Institute for Biodiversity and Ecosystem Dynamics (IBED), University of Amsterdam, P.O. Box 94248, Amsterdam 1090 GE, The Netherlands

Yeray González-Marrero
Global Change Institute, The University of Queensland, St Lucia, QLD 4072, Australia
Department of Biology, Ghent University, Ghent 9000, Belgium

Manuel González-Rivero
Global Change Institute, The University of Queensland, St Lucia, QLD 4072, Australia
Australian Research Council Centre of Excellence for Coral Reef Studies, St Lucia, QLD 4072, Australia

Oscar Beijbom
Global Change Institute, The University of Queensland, St Lucia, QLD 4072, Australia
Department of Electrical Engineering & Computer Sciences, University of California, Berkeley, CA 94709, USA

Ove Hoegh-Guldberg and Anjani Ganase
Global Change Institute, The University of Queensland, St Lucia, QLD 4072, Australia
Australian Research Council Centre of Excellence for Coral Reef Studies, St Lucia, QLD 4072, Australia
School of Biological Sciences, The University of Queensland, St Lucia, QLD 4072, Australia

Chris Roelfsema and Stuart Phinn
Remote Sensing Research Centre, School of Geography, Planning and Environmental Management, The University of Queensland, St Lucia, QLD 4072, Australia

Patricio Xavier Molina
Gestión de Investigación y Desarrollo, Instituto Geográfico Militar, Seniergues E4-676 y Gral, Telmo Paz y Miño, El Dorado 170403, Quito, Ecuador
Technical University of Madrid (UPM), C/ Ramiro de Maeztu, 7, Madrid 28040, Spain

Mercedes Farjas Abadía, Juan Carlos Ojeda Manrique and Luis Alberto Sánchez Diez
Technical University of Madrid (UPM), C/Ramiro de Maeztu, 7, Madrid 28040, Spain

Gregory P. Asner
Department of Global Ecology, Carnegie
Institution for Science, 260 Panama Street,
Stanford, CA 94305, USA

Renato Valencia
Laboratorio de Ecología de Plantas, Escuela
de Ciencias Biológicas, Pontificia Universidad
Católica del Ecuador, Apartado 17-01-2184,
Quito, Ecuador

**Susana Del Pozo, Jesús Herrero-Pascual,
Pablo Rodríguez-Gonzálvez and Diego
González-Aguilera**
Department of Cartographic and Land
Engineering, University of Salamanca,
Hornos Caleros, 05003 Ávila, Spain

**Beatriz Felipe-García and David Hernández-
López**
Institute for Regional Development (IDR),
Albacete, University of Castilla La Mancha,
02071 Albacete, Spain

Qi Chen
Key Laboratory of Carbon Cycling in Forest
Ecosystems and Carbon Sequestration of
Zhejiang Province, School of Environmental &
Resource Sciences, Zhejiang A&F University,
Lin An 311300, China
Department of Geography, University of
Hawaii at Manoa, Honolulu, HI 96822, USA

Yunyun Feng
Key Laboratory of Carbon Cycling in Forest
Ecosystems and Carbon Sequestration of
Zhejiang Province, School of Environmental &
Resource Sciences, Zhejiang A&F University,
Lin An 311300, China

Changwei Wang
Department of Geography, University of
Hawaii at Manoa, Honolulu, HI 96822, USA

Dengsheng Lu
Key Laboratory of Carbon Cycling in Forest
Ecosystems and Carbon Sequestration of
Zhejiang Province, School of Environmental &
Resource Sciences, Zhejiang A&F University,
Lin An 311300, China

Center for Global Change and Earth
Observations, Michigan State University, East
Lansing, MI 48823, USA

**Maiza Nara dos-Santos and Edson Luis
Bolfe**
Brazilian Agricultural Research
Corporation — Embrapa, Campinas, SP
13070-115, Brazil

Michael Keller
Brazilian Agricultural Research
Corporation — Embrapa, Campinas, SP
13070-115, Brazil
USDA Forest Service, International Institute
of Tropical Forestry, San Juan, PR 00926,
USA

Don Hillger and Dan Lindsey
NOAA/NESDIS Center for Satellite
Applications and Research (StAR), Fort
Collins, CO 80523, USA

Tom Kopp
The Aerospace Corporation, El Segundo, CA
90245, USA

**Curtis Seaman, Steven Miller and Jeremy
Solbrig**
CIRA, Colorado State University, Fort Collins,
CO 80523, USA

Eric Stevens
Geographic Information Network of Alaska
(GINA), Fairbanks, AK 99775, USA

William Straka III
CIMSS, University of Wisconsin, Madison,
WI 53706, USA

Melissa Kreller
NWS, Fairbanks, AK 99775, USA

Arunas Kuciauskas
NRL, Marine Meteorology Division,
Monterey, CA 93943, USA

Amanda Terborg
Aviation Weather Center, NWS, Kansas, MO
64153, USA

Lilong Zhao and Zhihong Jiang
Key Laboratory of Meteorological Disaster of Ministry of Education, Collaborative Innovation Center on Forecast and Evaluation of Meteorological Disasters, Nanjing University of Information Science and Technology, Najing 210044, China

Jianjun Xu
Global Environment and Natural Resources Institute (GENRI), College of Science, George Mason University, Fairfax, WV 22030, USA

Alfred M. Powell
NOAA/NESDIS/STAR, College Park, ML 20740, USA

Donghai Wang
China State Key Laboratory of Severe Weather Chinese Academy of Meteorological Sciences, Beijing 100081, China

Bo Xu, Wanshou Jiang and Jing Zhang
State Key Laboratory of Information Engineering in Surveying, Mapping and Remote Sensing, Wuhan University, Wuhan 430072, China

Jie Shan
Lyles School of Civil Engineering, Purdue University, West Lafayette, IN 47907, USA

Lelin Li
National-Local Joint Engineering Laboratory of Geo-Spatial Information Technology, Hunan University of Science and Technology, Xiangtan 411201, China

Ainong Li
Institute of Mountain Hazards and Environment, Chinese Academy of Sciences, Chengdu 610041, China

Jinhu Bian and Qiannan Liu
Institute of Mountain Hazards and Environment, Chinese Academy of Sciences, Chengdu 610041, China
University of Chinese Academy of Sciences, Beijing 100049, China

Chengquan Huang
Department of Geography, University of Maryland, College Park, MD 20742, USA

Cheng Wang, Xiaohuan Xi, Dong Li, Shaobo Xia and Pinghua Wang
Key Laboratory of Digital Earth Science, Institute of Remote Sensing and Digital Earth, Chinese Academy of Sciences, Beijing 100094, China

Shezhou Luo
Key Laboratory of Digital Earth Science, Institute of Remote Sensing and Digital Earth, Chinese Academy of Sciences, Beijing 100094, China
Department of Geography and Program in Planning, University of Toronto, 100St. George St., Room 5047, Toronto, ON M5S 3G3, Canada

Hongcheng Zeng
Faculty of Forestry, University of Toronto, 33 Willcocks Street, Toronto, ON M5S 3B3, Canada

Caixia Liu
State Key Laboratory of Remote Sensing Science, Institute of Remote Sensing and Digital Earth, Chinese Academy of Sciences (CAS), Beijing 100101, China

Xiaoyi Wang and Huabing Huang
State Key Laboratory of Remote Sensing Science, Institute of Remote Sensing and Digital Earth, Chinese Academy of Sciences (CAS), Beijing 100101, China
Department of Environmental Sciences, Policy & Management, University of California, Berkeley, CA 94720, USA

Gregory S. Biging and Yanlei Chen
Department of Environmental Sciences, Policy & Management, University of California, Berkeley, CA 94720, USA

Jun Yang
Ministry of Education Key Laboratory for Earth System Modeling, Center for Earth System Science, Tsinghua University, Beijing 100084, China

Peng Gong
State Key Laboratory of Remote Sensing Science, Institute of Remote Sensing and Digital Earth, Chinese Academy of Sciences (CAS), Beijing 100101, China
Ministry of Education Key Laboratory for Earth System Modeling, Center for Earth System Science, Tsinghua University, Beijing 100084, China
Joint Center for Global Change Studies, Beijing 100875, China

Qinchuan Xin
School of Geography and Planning, Sun Yat-sen University, Guangzhou 510275, China

Xiaoli Yin and Guang Liu
Key Laboratory of Digital Earth Science, Institute of Remote Sensing and Digital Earth, Chinese Academy of Sciences, No. 9 Dengzhuang South Road, Beijing 100094, China

Li Zhang
Key Laboratory of Digital Earth Science, Institute of Remote Sensing and Digital Earth, Chinese Academy of Sciences, No. 9 Dengzhuang South Road, Beijing 100094, China
Hainan Key Laboratory of Earth Observation, Hainan 572029, China

Binghua Zhang
Key Laboratory of Digital Earth Science, Institute of Remote Sensing and Digital Earth, Chinese Academy of Sciences, No. 9 Dengzhuang South Road, Beijing 100094, China
Hainan Key Laboratory of Earth Observation, Hainan 572029, China
College of Resources and Environment, University of Chinese Academy of Sciences, No. 19A Yuquan Road, Beijing 100049, China

Dong Xie
Key Laboratory of Digital Earth Science, Institute of Remote Sensing and Digital Earth, Chinese Academy of Sciences, No. 9 Dengzhuang South Road, Beijing 100094, China

Department of Mathematics, The George Washington University, 2115 G St. NW, Washington, DC 20052, USA

Chunjing Liu
Key Laboratory of Digital Earth Science, Institute of Remote Sensing and Digital Earth, Chinese Academy of Sciences, No. 9 Dengzhuang South Road, Beijing 100094, China
College of Information Science and Engineering, Shandong Agricultural University, No. 61 Daizong Road, Taian 271018, China

Dan J. Krofcheck and Marcy E. Litvak
Department of Biology, University of New Mexico, Albuquerque, NM 87131, USA

Jan U. H. Eitel and Lee A. Vierling
Geospatial Laboratory for Environmental Dynamics, University of Idaho, Moscow, ID 83844, USA
McCall Outdoor Science School, University of Idaho, McCall, ID 83638, USA

Christopher D. Lippitt
Department of Geography and Environmental Studies, University of New Mexico, Albuquerque, NM 87131, USA

Urs Schulthess
CIMMYT-Bangladesh, House 10/B, Road 53, Gulshan-2, Dhaka 1213, Bangladesh

Alberto Rodriguez-Ramirez
Global Change Institute, The University of Queensland, St Lucia, QLD 4072, Australia

Tadzio Holtrop
Global Change Institute, The University of Queensland, St Lucia, QLD 4072, Australia
Institute for Biodiversity and Ecosystem Dynamics (IBED), University of Amsterdam, P.O. Box 94248, Amsterdam 1090 GE, The Netherlands

Yeray González-Marrero
Global Change Institute, The University of Queensland, St Lucia, QLD 4072, Australia
Department of Biology, Ghent University, Ghent 9000, Belgium

Manuel González-Rivero
Global Change Institute, The University of Queensland, St Lucia, QLD 4072, Australia
Australian Research Council Centre of Excellence for Coral Reef Studies, St Lucia, QLD 4072, Australia

Oscar Beijbom
Global Change Institute, The University of Queensland, St Lucia, QLD 4072, Australia
Department of Electrical Engineering & Computer Sciences, University of California, Berkeley, CA 94709, USA

Ove Hoegh-Guldberg and Anjani Ganase
Global Change Institute, The University of Queensland, St Lucia, QLD 4072, Australia
Australian Research Council Centre of Excellence for Coral Reef Studies, St Lucia, QLD 4072, Australia
School of Biological Sciences, The University of Queensland, St Lucia, QLD 4072, Australia

Chris Roelfsema and Stuart Phinn
Remote Sensing Research Centre, School of Geography, Planning and Environmental Management, The University of Queensland, St Lucia, QLD 4072, Australia

Patricio Xavier Molina
Gestión de Investigación y Desarrollo, Instituto Geográfico Militar, Seniergues E4-676 y Gral, Telmo Paz y Miño, El Dorado 170403, Quito, Ecuador
Technical University of Madrid (UPM), C/ Ramiro de Maeztu, 7, Madrid 28040, Spain

Mercedes Farjas Abadía, Juan Carlos Ojeda Manrique and Luis Alberto Sánchez Diez
Technical University of Madrid (UPM), C/Ramiro de Maeztu, 7, Madrid 28040, Spain

Gregory P. Asner
Department of Global Ecology, Carnegie Institution for Science, 260 Panama Street, Stanford, CA 94305, USA

Renato Valencia
Laboratorio de Ecología de Plantas, Escuela de Ciencias Biológicas, Pontificia Universidad Católica del Ecuador, Apartado 17-01-2184, Quito, Ecuador

Susana Del Pozo, Jesús Herrero-Pascual, Pablo Rodríguez-Gonzálvez and Diego González-Aguilera
Department of Cartographic and Land Engineering, University of Salamanca, Hornos Caleros, 05003 Ávila, Spain

Beatriz Felipe-García and David Hernández-López
Institute for Regional Development (IDR), Albacete, University of Castilla La Mancha, 02071 Albacete, Spain

Qi Chen
Key Laboratory of Carbon Cycling in Forest Ecosystems and Carbon Sequestration of Zhejiang Province, School of Environmental & Resource Sciences, Zhejiang A&F University, Lin An 311300, China
Department of Geography, University of Hawaii at Manoa, Honolulu, HI 96822, USA

Yunyun Feng
Key Laboratory of Carbon Cycling in Forest Ecosystems and Carbon Sequestration of Zhejiang Province, School of Environmental & Resource Sciences, Zhejiang A&F University, Lin An 311300, China

Changwei Wang
Department of Geography, University of Hawaii at Manoa, Honolulu, HI 96822, USA

Dengsheng Lu
Key Laboratory of Carbon Cycling in Forest Ecosystems and Carbon Sequestration of Zhejiang Province, School of Environmental & Resource Sciences, Zhejiang A&F University, Lin An 311300, China
Center for Global Change and Earth Observations, Michigan State University, East Lansing, MI 48823, USA

Maiza Nara dos-Santos and Edson Luis Bolfe
Brazilian Agricultural Research Corporation –
Embrapa, Campinas, SP 13070-115, Brazil

Michael Keller
Brazilian Agricultural Research Corporation –
Embrapa, Campinas, SP 13070-115, Brazil
USDA Forest Service, International Institute
of Tropical Forestry, San Juan, PR 00926, USA

Don Hillger and Dan Lindsey
NOAA/NESDIS Center for Satellite
Applications and Research (StAR), Fort
Collins, CO 80523, USA

Tom Kopp
The Aerospace Corporation, El Segundo, CA
90245, USA

**Curtis Seaman, Steven Miller and Jeremy
Solbrig**
CIRA, Colorado State University, Fort Collins,
CO 80523, USA

Eric Stevens
Geographic Information Network of Alaska
(GINA), Fairbanks, AK 99775, USA

William Straka III
CIMSS, University of Wisconsin, Madison,
WI 53706, USA

Melissa Kreller
NWS, Fairbanks, AK 99775, USA

Arunas Kuciauskas
NRL, Marine Meteorology Division,
Monterey, CA 93943, USA

Amanda Terborg
Aviation Weather Center, NWS, Kansas, MO
64153, USA

Lilong Zhao and Zhihong Jiang
Key Laboratory of Meteorological Disaster
of Ministry of Education, Collaborative
Innovation Center on Forecast and
Evaluation of Meteorological Disasters,
Nanjing University of Information Science
and Technology, Najing 210044, China

Jianjun Xu
Global Environment and Natural Resources
Institute (GENRI), College of Science, George
Mason University, Fairfax, WV 22030, USA

Alfred M. Powell
NOAA/NESDIS/STAR, College Park, ML
20740, USA

Donghai Wang
China State Key Laboratory of Severe Weather
Chinese Academy of Meteorological Sciences,
Beijing 100081, China

Bo Xu, Wanshou Jiang and Jing Zhang
State Key Laboratory of Information
Engineering in Surveying, Mapping and
Remote Sensing, Wuhan University, Wuhan
430072, China

Jie Shan
Lyles School of Civil Engineering, Purdue
University, West Lafayette, IN 47907, USA

Lelin Li
National-Local Joint Engineering Laboratory
of Geo-Spatial Information Technology,
Hunan University of Science and Technology,
Xiangtan 411201, China

Ainong Li
Institute of Mountain Hazards and
Environment, Chinese Academy of Sciences,
Chengdu 610041, China

Jinhu Bian and Qiannan Liu
Institute of Mountain Hazards and
Environment, Chinese Academy of Sciences,
Chengdu 610041, China
University of Chinese Academy of Sciences,
Beijing 100049, China

Chengquan Huang
Department of Geography, University of
Maryland, College Park, MD 20742, USA

Index

A

Aboveground Biomass (agb), 36, 38, 41, 124-125
Aboveground Carbon Density (acd), 93
Advanced Microwave Sounding Unit-a (amsu), 165
Advanced Very High Resolution Radiometer, 142
Aerosol Concentration, 204
Afforestation, 20-21, 32
Agroforestry, 124-126, 129-130, 135-139
Airborne Discrete, 1, 18
Airborne Lidar, 2, 15-18, 20-21, 27, 33-34, 55, 93-94, 97, 104-106, 124-125, 128-130, 136, 138-139, 181, 202-203
Allometry, 94, 96, 101, 104, 124, 135, 137, 139

B

Biomass Estimation, 16, 26, 36, 38, 41, 52-55, 94, 96, 105, 128, 138-140

C

Calibration and Validation (cal/val), 141
Canopy Height Model (chm), 129
Carbon Density Estimation Model, 93
Casi (compact Airborne Spectrographic Imager), 1
Close Range Photogrammetry, 108
Cloud Detection Methods, 205
Coupled Model Intercomparison Project (cmip5), 165
Crop Yield Estimation, 2
Cultural Heritage, 108, 122
Cumulative Effect, 74

D

Day/night Band (dnb), 142
Dynamic Pasture Growth, 37

E

Ecological Measurements, 73
Environmental Data Record (edr), 141, 163

F

Feature Clustering, 181
Fixed-effects Regression Models, 124-125

Forest Census Data, 96
Forest Inventory, 20, 94, 101, 104, 138-139
Forest Monitoring, 20
Frequent Disturbance, 57-58

G

Geo-referenced Field Data, 73
Global Time-series, 20
Grassland Biomass Estimation, 36
Gross Primary Production (gpp), 57
Ground Track Mercator (gtm), 142

H

Humid Mesothermal Climate, 125
Hyperspectral Data, 1, 4, 15-18, 139, 226

J

Joint Polar Satellite System (jpss), 141

K

Key Performance Parameter (kpp), 141-143

L

Land Cover Classification, 1-4, 6, 9-12, 15-18, 33, 54, 140, 204
Landsat Imagery, 16, 20-21, 23, 30-32, 54, 223-224
Landtrendr Algorithm, 20, 26
Laser Scanning, 2, 16-18, 35, 104, 106, 108, 121-123, 138, 202-203
Lidar (light Detection and Ranging), 1, 125
Lidar Point Clouds, 4, 181, 202-203
Low Flat Plateaus, 125
Lower Stratosphere, 165-166, 174

M

Marine Settings, 73
Maximum Likelihood (mlc) Classifiers, 1
Maximum Likelihood Sac (mlesac), 182
Microwave Sounding Units (msu), 165-166
Minimal Annual Precipitation, 57
Mixed-effects Models, 124-125, 129, 132-133, 135-137

Model Fitting, 100, 181-182

Moderate-resolution Imaging Spectroradiometer (modis), 36-37

Multi-decadal Change, 20

Multispectral Camera, 108, 110, 112, 116, 119, 121-122

Multispectral Radiometric Analysis, 108

N

Near Constant Contrast (ncc), 142

Normalized Difference Red Edge (ndre), 57

Normalized Difference Snow Index (ndsi), 60, 205

Normalized Difference Vegetation Index (ndvi), 36-37, 40

Normalized Difference Wetness Index (ndwi), 57, 59

Nutrient Run-off, 74

O

Optical Remote Sensing, 2, 71, 74, 89

P

Passive Remote Sensing, 54, 108

Pathologies, 108, 111, 114, 116-117, 121-122

Piñon-juniper Woodlands, 57-59

Planted Forest Cover, 20

Plot-aggregate Allometric Approach, 94, 99, 103

Poor Segmentation, 181

Principal Components Analysis (pca), 1

R

Radioactive Gases, 165

Radiometric Calibration, 21, 40, 108-109, 111-112, 121-123

Random Sample Consensus (ransac), 181-182

Rapid Carbon Stock Accumulation, 58

Region Growing (rg), 181

Regional Covariance Matrix, 204, 208, 214

Regression Coefficients, 124-125

Remote Sensing, 1-2, 9-10, 15-20, 33-34, 36-37, 53-55, 57-60, 68-74, 89, 94, 104-105, 108, 123, 125, 137-138, 163, 181, 204, 213, 221, 223-225

Roof Plane Segmentation, 181-182, 200

Root-mean-square Error (rmse), 36

S

Satellite Imagery, 16, 54-55, 89, 136, 141, 153, 157, 225-226

Satellite Observations, 165-166, 169, 179

Sedimentation, 74

Semi-arid Grasslands Management, 36

Semi-automated Field Image Collection, 73

Simple Models, 57

Small-footprint Lidar, 93

Snow Discrimination, 204-206, 208, 212-213, 221-223

Spatial And Temporal Adaptive Reflectance Fusion Model (starfm), 36-37

Spatial Heterogeneity, 73

Spatially-explicit Testing, 93

Spatio-temporal Context, 204-206, 208

Spectral Signature, 61, 204-206, 213, 222, 224

Stratospheric Sounding Units (ssu), 165-166

Support Vector Machine (svm), 1, 9-10, 26, 36, 38, 78

T

Terrestrial Forest Ecosystems, 20

Terrestrial Laser Scanner, 108, 110, 122, 202

Top-of-canopy Height (tch), 94

Topographic Features, 93, 103

Tropical Rainforest, 93, 106

U

Upper Atmospheric Temperature, 165, 179

V

Vegetation Indices (vis), 57-58

Vegetation Types, 15, 17, 103, 125, 129, 136-137, 139

Visible/infrared Imaging Radiometer Suite (viirs), 141

W

Wood Density, 94, 96, 99, 102-103, 124, 127-129, 131-132, 136, 139

World Climate Research Program (wcrp), 165